International Conference on Automotive Innovation & Green Energy Vehicle (AIGEV 2020)

IOP Conference Series: Materials Science and Engineering
Volume 1068

Pekan, Malaysia
10 - 11 November 2020

Editors:

Mohammad Razali Hanipah
Devarajan Ramasamy
Sudhakar Kumarasamy

ISBN: 978-1-7138-3343-7
ISSN: 1757-8981

Printed from e-media with permission by:

Curran Associates, Inc.
57 Morehouse Lane
Red Hook, NY 12571

Some format issues inherent in the e-media version may also appear in this print version.

Institute of Physics
Dirac House, Temple Back
Bristol BS1 6BE UK

Phone: 44 1 17 929 7481
Fax: 44 1 17 920 0979

techtracking@iop.org

Additional copies of this publication are available from:

Curran Associates, Inc.
57 Morehouse Lane
Red Hook, NY 12571 USA
Phone: 845-758-0400
Fax: 845-758-2633
Email: curran@proceedings.com
Web: www.proceedings.com

TABLE OF CONTENTS

The 4th International Conference on Automotive Innovation Green Energy Vehicle - (AiGEV 2020)

10-11, November, Pekan Campus, Universiti Malaysia Pahang, Malaysia

Selected, peer-reviewed papers from the 4th International Conference on Automotive Innovation Green Energy Vehicle -(AiGEV 2020)

Edited by:

Ts. Dr. Mohd Razali Hanipah (Chairman)

Director, Automotive Engineering Centre,Universiti Malaysia Pahang,Malaysia

Assoc. Prof. Dr. Devarajan Ramasamy (Editor in Chief)

College of Engineering, Universiti Malaysia Pahang,Malaysia

Dr. Sudhakar Kumarasamy (Editor)

Faculty of Mechanical and Automotive Engineering Technology, Universiti Malaysia Pahang,Malaysia

IOP Conf. Series: Materials Science and Engineering 1068 (2021) 011001 doi:10.1088/1757-899X/1068/1/011001

Preface

The **4th International Conference on Automotive Innovation Green Energy Vehicle (AiGEV 2020)** was organised by Automotive Engineering Centre (AEC),Universiti Malaysia Pahang (UMP) in collaboration with University College TATI at Pekan Campus, UMP Malaysia during November 10-11, 2020. This year due to Coronavirus (COVID-19) pandemic, the 4th International Conference on Automotive Innovation and Green Energy Vehicle (AiGEV) 2020, was held virtually. All the Keynote sessions and technical session of AiGEV 2020 were successfully conducted through hybrid platform (Online and offline).

The objective of AiGEV 2020 Conference is to provide a venue for researchers and practitioners in related fields from academia, industry and government to meet in a forum where the latest research results are presented and prospect for future developments on automotive engineering". The papers submitted to AiGEV 2020 Conference were single-blind peer-reviewed, at least by two independent referees mainly consists of researchers and passed through several stages of editing procedure.

A brief overview on the AiGEV 2020 Conference reveals the following: 36 presented papers in the conference, 2 invited keynote speeches, 1 plenary presentation, 6 conference sections in which the papers were distributed, **122** participants from 6 countries (Malaysia, Germany, UK, India, Indonesia and Thailand). The invited keynote speakers have been professors, outstanding specialists in the field of Automotive engineering from UTP and UTM, Malaysia. The technical sessions presentation was allocated 15 minutes for presentation and 5 minutes for comments, questions and answers, and for each paper the total amount of time was 20 minutes (both presentation and Q&A). Based on the technical, quality and presentation, best papers were selected and awarded during each technical session.

We express our gratitude to the Organising Committee for their valuable contribution to this successful edition of AiGEV 2020.We express our sincere gratitude to all people who contributed to the review process and did all their best to enhance the scientific merit and quality of the Proceedings. We are also grateful to the authors for their contributions, to the speakers and participants of the AiGEV 2020. We hope to welcome them along with new participants to the next event AiGEV 2022.

We also address thanks to all those involved in organising this edition of the conference, as well as to our sponsor Taat Bistari Sdn Bhd and to IOP Conference Series: Materials Science and Engineering

IOP Conf. Series: Materials Science and Engineering 1068 (2021) 011001 doi:10.1088/1757-899X/1068/1/011001

Ts. Dr. Mohd Razali Hanipah,

Chairman of AiGEV 2020,

Director, Automotive Engineering Centre,

Universiti Malaysia Pahang 26600 Pekan, Pahang,

MALAYSIA

The 4th International Conference on Automotive Innovation Green Energy Vehicle - (AiGEV 2020)

MAIN COMMITTEE

Honorary Chairman:

Professor Ir. Dr. Wan Azhar Wan Yusoff

Advisor:

Assoc. Prof. Dr. Abdul Adam Abdullah

Chairman:

Ts. Dr. Mohd Razali Hanipah

Event Co-ordinator:

Dr. Izwan Ismail

Project Leader:

Rosidah Mohd. Norsat

Secretary & Bursary:

Baharudin Basiron

Secretariat & Sponsorship:

Rosidah Mohd. Norsat (Head)

Nawal Nazihah Nasaruddin

Hamizaliza Hamizan

Promotion & Publicity:

Muhamad Imran Mohmad Sairaji (Head)

Mohd Aminuddin Ayob

Mahdhir Mohd Yusof

Technical & Logistics:

Daniel Mohamed (Head)

IOP Conf. Series: Materials Science and Engineering 1068 (2021) 011001 doi:10.1088/1757-899X/1068/1/011001

Mohd Idzwanrosli Mohd Ramli

Khairul Anuar Rani

Muhammad Hilmi Che Kamarudin

Zulkifly Zakaria

Mohd Al'Afis Ismail

Proceeding and Journals Publication:

Assoc. Prof. Dr. Devarajan Ramasamy (Editor in Chief)

Dr. Sudhakar Kumarasamy (Editor)

Assoc. Prof. Dr. Saiful Anwar Che Ghani

Assoc. Prof. Dr. Wan Azmi Wan Hamzah

Prof. Ir. Dr. Ahmad Razlan Yusof

Assoc. Prof. Dr. Hj, Mohd Shahrir Mohd Sani

Assoc. Prof. Dr. Sharifah Nur Aqida Syed Ahmad

Dr. Azri Alias

Dr. Farzad Jaliliantabar

REVIEWERS

Madderla.sandhya	Sreenath sukumaran	Rajan.kumaresan
Dr Mohd Adnin	L.O. Afolabi	A.A. Alias
Azri Alias	N. Asim	Dr Nurizzatul Atikha
Dr Nor Atiqah	Dr Mohd Azri	Aklilu.T Baheta
P. Baredar	F. Basrawi	Saiful Anwar Che Ghani
F. Dzaharudin	Mr Che Ku Eddy	Dr Ashok Gandhi
Mebrahitom Asmelash Gebremariam	Ftwi Yohaness Hagos	Razali Hanipah
D. Ibrahim	M.Y. Ismail	Farzad Jaliliantabar
Kumaran Kadirgama	Dr Ram Kiran	A. Mahdavian
M.R.A. Mansor	Mr Idris Mat Sahat	Mohd Shahrir Mohd Sani
Ahmed Nurye Oumer	A. Pandey	M.H. Peeie
Mm Quazi	M.T.A. Rahman	Md Mustafizur Rahman
Devarajan Ramasamy	N. Razak	M. Samykano
Dr Akash Shukla	K Sudhakar	M.K. Sued
Sharifah Nur Aqida Syed Ahmed	Dr Jayaseelan V	Wan Azmi Wan Hamzah
Wan Sharuzi Wan Harun	Ahmed Razlan Yusoff	A.F. Yusop

IOP Conf. Series: Materials Science and Engineering 1068 (2021) 011001 doi:10.1088/1757-899X/1068/1/011001

INVITED KEYNOTE SPEAKERS

Professor Dr. Abdul Rashid Abdul Aziz, Director, Institute of Transport Infrastructure, Universiti Teknologi Petronas, Malaysia

Topic: COVID-19 – Transportation at the Cross Road

Professor Ts. Ir. Dr. Saiful Amri Mazlan , Professor, Malaysia-Japan International Institute Of Technology, Jabatan Kejuruteraan Mekanikal Kejituan, Universiti Teknologi Malaysia Kuala Lumpur, Malaysia

Topic : Recent Progress in Magnetorheological Materials Towards Automotive Applications

CHAIR PERSON KEYNOTE SESSION

Dr. Sudhakar Kumarasamy

Assoc Prof. Dr. Syarifah Nur Aqida Syed Ahmad

CHAIRPERSON TECHNICAL SESSION

Ir. Dr. Addie Irawan Hashim

Dr. Ahmad Najmuddin Ibrahim

Assoc. Prof. Dr. Wan Azmi Wan Hamzah

Assoc. Prof. Dr. Saiful Anwar Che Ghani

IOP Conf. Series: Materials Science and Engineering 1068 (2021) 011002 doi:10.1088/1757-899X/1068/1/011002

Peer review declaration

All papers published in this volume of *IOP Conference Series: Materials Science and Engineering* have been peer reviewed through processes administered by the Editors. Reviews were conducted by expert referees to the professional and scientific standards expected of a proceedings journal published by IOP Publishing.

- **Type of peer review: Single-blind**
- **Conference submission management system: Easychair.org**
- **Number of submissions received: 58**
- **Number of submissions sent for review: 58**
- **Number of submissions accepted: 26**
- **Acceptance Rate (Number of Submissions Accepted / Number of Submissions Received X 100): 45%**
- **Average number of reviews per paper: 2**
- **Total number of reviewers involved: 47**
- **Any additional info on review process: The papers were evaluated based on the technical criteria, quality criteria and presentation criteria. The referees argued detailed their reviews and provided an overall evaluation (strong accept, accept, weak accept, borderline paper, weak reject, reject, strong reject), specifying their confidence (expert, high, medium, low, none) regarding qualifications to review the allocated paper. For each paper, the similarity index was determined by the organisers using Turnitin software at Universiti Malaysia Pahang.**
- **Contact person for queries: Associate Prof. Dr Devarajan Ramasamy; deva@ump.edu.my**

Road Peculiarities Detection using Deep Learning for Vehicle Vision System

M.R. Rani[1]*, M.Z.C. Mustafar[1], N.H.F. Ismail[1], M.S.F. Mansor[1], Z. Zainuddin[1]
[1]Safety & Intelligent Drive, PROTON, Shah Alam, Malaysia

*E-mail: rahairirani@proton.com

Abstract. Recent development of Advance Driver Assistance System (ADAS) has seen various advancement in object detection for vehicle vision system, particularly on the detection of other vehicles, pedestrians, road lane and signage. While these detections can provide assistant to avoid road accidents, they still lack to include road condition factors that also contributed to road accidents in Malaysia. This paper proposes a detection of the road peculiarities such as pothole and road bumps to act as additional safety feature in ADAS. With the breakthrough of deep learning in solving image recognition problems, this work takes advantage of Single Shot Detector (SSD)-MobileNetV2 as the detection algorithm, implemented on the real-time. The training images for potholes and road bumps taken from the Malaysia roads are fed into the detection model, and then the pre-trained weights are fine-tuned over the training process. The results show that the detection algorithm can predicts the potholes and road bumps, while exhibit the detection accuracy and confidence limitation due to the variety of shape and pattern of potholes and road bumps. Testing the detection algorithm with NVIDIA Jetson Nano yielded about 20 frames per second (fps), suitable for real-time applications.

1. Introduction

Over the last decade, safety has been a noteworthy concern in late vehicle innovation and tremendous researches and works has been conducted guided towards the course. Investigations regarding the safety of vehicle occupant have established two standards [1]: passive and active safety systems. Formerly, the passive safety system was developed to reduce any injury to the vehicle occupant through passive systems such as seat belts and air bags subsequent the unfortunate incidences. They have become standard vehicle safety equipment and have been successful in reducing injuries or fatalities in accidents. However, the air bags are deployed after the accident has already occurred and it would be great if the accident can be fully avoided. There has been an increasing interest in active safety and have seen a growing trend among automotive industries since then. The main purpose of active systems is to assist the driver in avoiding the accidents before they occur. These systems are also known as Advance Driver Assistant Systems (ADAS) and design to help the driver by taking the pressure off the driver in standard situations. A notable example of ADAS are Forward Collision Warning (FCW), Autonomous Emergency Braking (AEB), Lane Keeping System (LKS) and Lane Change Assist (LCA). In recent years, these ADAS has been integrated to enable full autonomous driving. To develop such system, robust and accurate object detection is the important area to be covered.

The role of object detection in understanding and analysing the driving scene is of great importance in order to build more intelligent driver assistance systems. This feature give ability to the road signs and speed limits in road signage detection. Object detection also can be incorporating with other sensors such as radar and light detection and ranging (LIDAR) to form the vehicle perception.

Road peculiarities such as potholes and road bumps will change the vehicle dynamic motion. Potholes will cause the vehicle to accelerate suddenly and further decreases the friction between the tyres and the road. These situations also lead to a dangerous vehicle deviation or out of control. In addition, potholes in the road can cause damage to the wheel, suspension and chassis components. Pothole size and vehicle speed were the main factors for the damage. The larger pothole can cause severe damage or alignment issues. Road bump acts for slowing traffic in low speed limit areas such as car parks, residential areas, office complexes and industrial areas. Detection of road peculiarities such as potholes and road bumper are crucial for vehicle driving dynamics and road safety. The objective of this paper tries to extend the capability of vehicle object detection system with the detection of road peculiarities. According to [2], the road condition factors correspondingly about 0.4 times less compared to the absence of such factors to the road accidents. This detection expected to increase the driving safety especially in the developing country roads such ASEAN.

2. Methodology

Object detection is the computer vision technique for locating the instance of objects in image or video. Object detection algorithms typically leverage machine learning or deep learning to produce meaningful results. When humans look at images or video, they can recognize and locate objects of interest within a matter of moments. The goal of object detection is to replicate this kind of ability using a computer.

Figure 1. Object detection for an image with its bounding boxes

Method for implementing object detection generally fall into either machine learning (ML) based approaches or deep learning (DL) based approaches. In ML, it is necessary to first define features and then using method such as Support Vector Machine (SVM) to do the classification. This feature selection technique is best for classification of clear feature different in classes, however, is less likely to work in distinguish between a car and a truck.

Contrary to ML, DL able to do end-to-end object detection without specifically defining the features as show in Figure 2. This means the algorithm only take the images and their bounding boxes as input in the learning process. DL in object detection is typically based on convolution neural network (CNN). There are a number of popular object detection algorithm based on deep learning such as:

i. Region-based Convolutional Neural Network (R-CNN) [3]
ii. Single Shot Detector (SSD) [4]
iii. You Only Look Once (YOLO) [5]

iv. Retina-Net [6]

Due to its accuracy and performance in real-time, this paper uses SSD as the object detector in the algorithm and MobileNetV2 as the feature extractor.

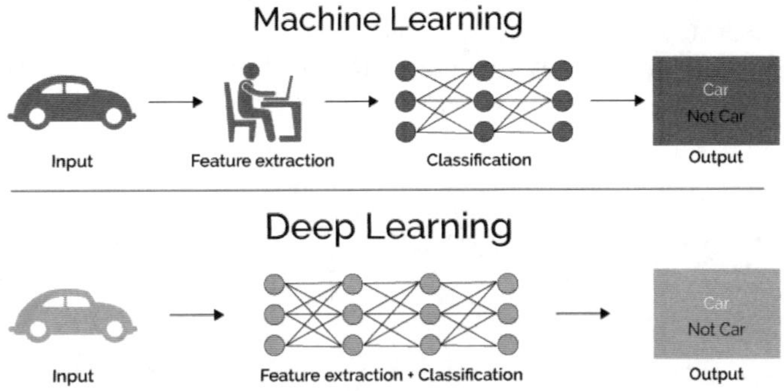

Figure 2. The difference between machine learning and deep leaning lies on the feature extraction technique

2.1 Single Shot Detector (SSD) MobileNetV2

SSD MobileNetV2 consists of two major part in its algorithm, SSD as the object detector and MobineNetV2 as the base network or feature extractor. Feature extractor works to extract the significant information from the image before being passed to object detector in order to determine the object classes and bounding boxes in the image.

2.1.1 MobileNetV2 - MobileNetV2 is the successor of MobileNetV1 [7] that intends to design an efficient neural network based on the mobile devices. This refinement from the V1 that makes it even more efficient and powerful [8].

The idea behind MobileNetV1 is that convolutional layers, which are essential to computer vision tasks but are quite expensive to compute, can be replaced by so-called depth wise separable convolutions.

The job of the convolution layer is split into two subtasks: first there is a depth wise convolution layer that filters the input with rectified linear unit (ReLU), followed by a 1×1 (or pointwise) convolution layer that combines these filtered values to create new features.

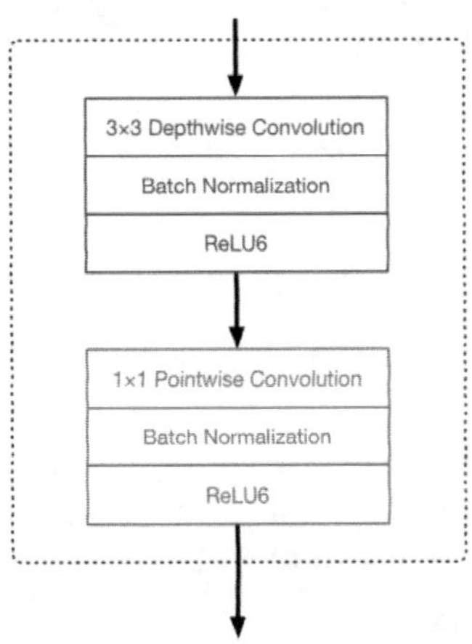

Figure 3. Depthwise separable convolution block in MobileNet

MobileNetV2 still uses depth wise separable convolutions, but there are major changes to its main building block as show in Figure 4.

There are three convolutional layers in the block. The last two are: a depth wise convolution that filters the inputs, followed by a 1×1 pointwise convolution layer. However, this 1×1 layer now has a different job.

In V1 the pointwise convolution either kept the number of channels the same or doubled them. In V2 it does the opposite: it makes the number of channels smaller. This is why this layer is now known as the projection layer, it projects data with a high number of dimensions (channels) into a tensor with a much lower number of dimensions.

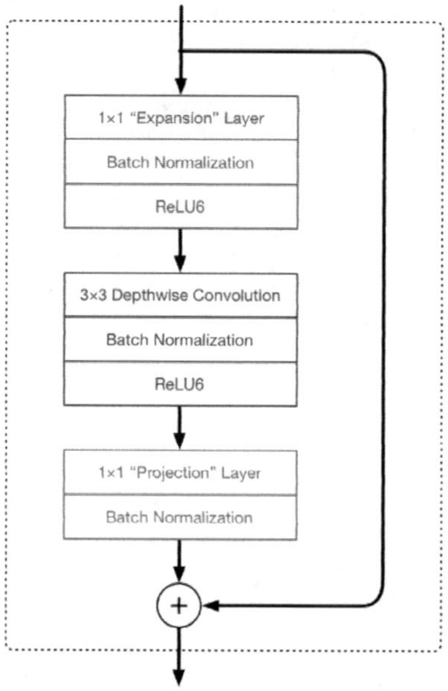

Figure 4. MobileNetV2 main block

The second new thing in MobileNetV2's building block is the residual connection. This works just like in ResNet [9] and exists to help with the flow of gradients through the network.

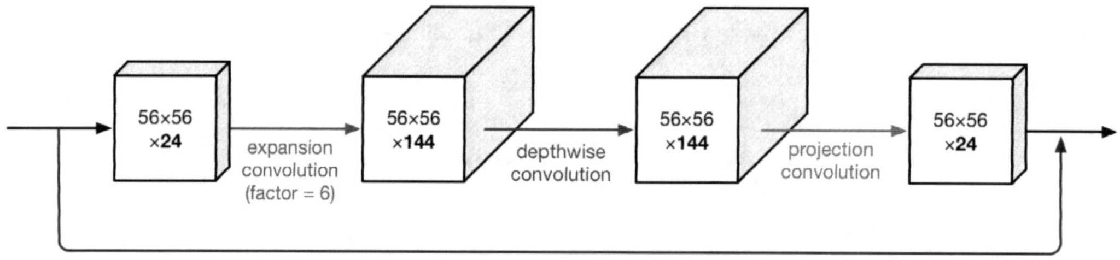

Figure 5. Residual connection in MobileNetV2

The idea behind V1 was to replace expensive convolutions with cheaper ones, even if it meant using more layers. That was a great success. The main changes in the V2 architecture are the residual connections and the expansion/projection layers.

2.1.2 Single Shot Detector (SSD) - SSD is designed to be independent of the base network (feature extractor), and so it can run on top of pretty much anything, including Mobile Net V2 or V1. The complete network can be shown as in Figure 6.

The network not only takes the output of the last base network layer but also the outputs of several previous layers, and these outputs are fed into the SSD layers. The job of the Mobile Net layers are to convert the pixels from the input image into features that describe the contents of the image, and pass these along to the other layers. Hence, Mobile Net is used here as a feature extractor for a second neural network.

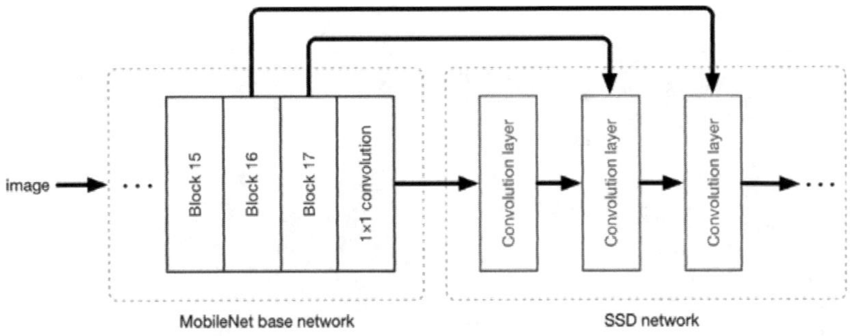

Figure 6. Complete network of SSD-Mobile Net

In the case of object detection with SSD, not just these high-level features are processed but also lower-level ones, which is why the previous layers are also forwarded. Since object detection is more complicated than classification, SSD adds many additional convolutional layers on top of the base network. Thus, it is crucial to have a base network that is fast, and that exactly what MobileNetV2 is.

2.2 Training the Dataset
In deep learning, there are two main processes involved, which are training and inference. In training, the network is guided to converge and archive a desired accuracy based on the labelled dataset. This requires running a thousand or perhaps millions of experiments by adjusting the model weights in the network. The training process is compute-extensive and often done in data centre or a powerful computer.

Deep learning inference is a process of using a trained neural network model to make prediction against previously untrained data. The inference process typically running on a smaller device and required further optimization of the trained model.

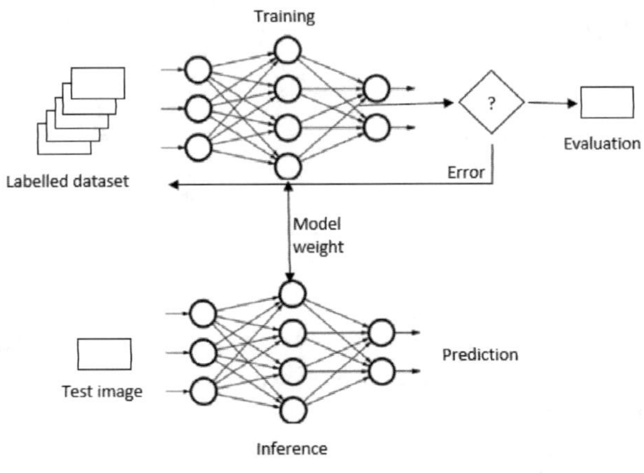

Figure 7. The training and inference process in deep learning

2.2.1 Creating the Labelled Dataset - In order to perform deep learning training, a dataset of road peculiarity must first be established. The dataset consists of 500 images of road bumps and potholes for each class that taken along the Malaysia roads with camera was mounted at vehicle windscreen. The images have varied in shape and structure of the road bumps and potholes to make it a robust dataset. These images went through a manual labelled process in order to specify the bounding box of the classes in the images according to its class.

(a) Bounding box for pothole

(b) Bounding box for road bump

Figure 8. Bounding boxes are manually created for the dataset for (a) pothole and (b) road bump

3. Results

Training a new dataset in deep learning from scratch is competitively expensive. One way to speed up the training is to employ the transfer learning technique, which uses the weights that has been trained for other classes. For this purpose, a SSD MobileNetV2 model that has been trained for COCO dataset [10] will be used as the initial weights. The training configuration for the transfer learning as Table 1.

The objective of the learning process is minimizing the total loss, which is the combination of the classification loss and the localization loss. By readjusting the neural network parameters in the next step, the algorithm tries to reduce the total loss over time.

Table 1. Training configuration for SSD MobileNetV2 Transfer Learning

Parameters	Setting
No. of steps	200 000
Batch size	4
No. of classes	2
Initial learning rate	0.004
Learning rate decay steps	800720
Deep learning library	Tensor flow v1.5
Central Processing Unit (CPU)	Intel i5 2400 – 12 GB
Graphic Processing Unit (GPU)	NVIDIA GTX 1060 6GB
Initial weights	SSD MobileNetV2-COCO – 2018-03-29

Classification loss measures the predicted class probability towards the ground truth. For the classification loss function, these probabilities are measures towards the true probability distribution with cross entropy, which quantifies the difference between two probability distributions. Equation 1 presents cross entropy, where y^ is the predicted value, y is the ground truth value and M is the number of classes.

$$classification\ loss = \sum_{i}^{M} y_i \log(\hat{y_i}) \quad (1)$$

The transfer learning of the classification loss versus the number of steps is shown as in Figure 9. The faded signal is the actual loss while the lighten signal is the smoothed loss.

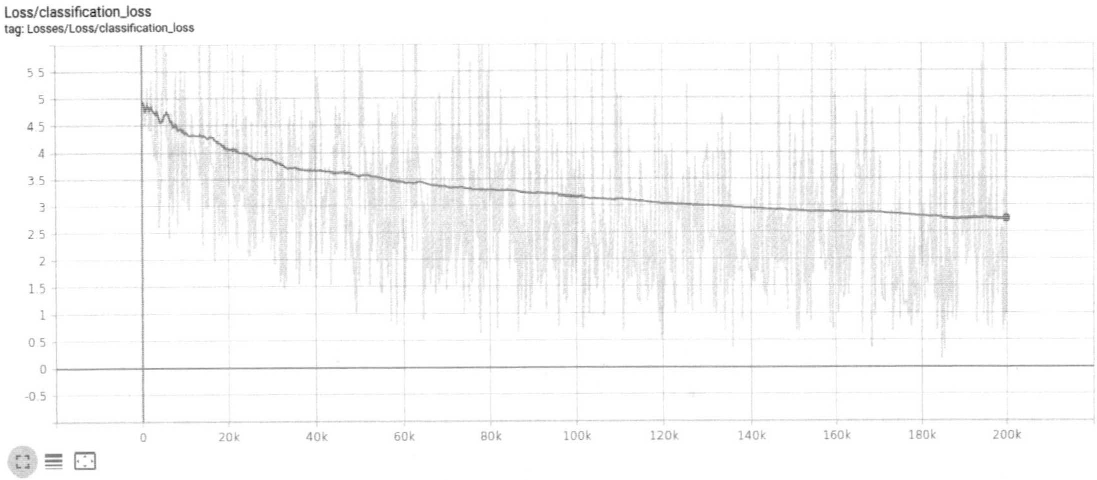

Figure 9. Classification loss of SSD-MobileNetV2 transfer learning

Localization is the ability of the network to locate an object in an image. The localization loss is the computed error between the ground truth and the predicted bounding boxes. Smooth L1 localization loss is defined by Equation 2 and as a penalty term to the loss function with the purpose to avoid over-fitting [6].

$$localization\ loss(x) = \begin{cases} 0.5x^2 & if\ |x| < 1 \\ |x| - 0.5 & if\ otherwise \end{cases} \quad (2)$$

The localization loss for the transfer learning is shown as in Figure 10. Like the classification loss, the localization loss also made decrease as the step increases.

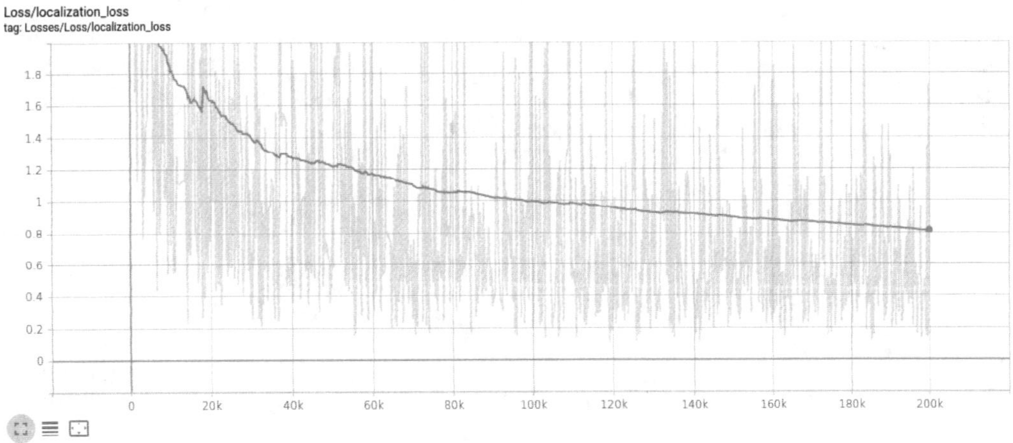

Figure 10. The localization loss of SSD-MobileNetV2 transfer learning

The total loss is summation of the classification and localization loss. Equation 3 defines the total loss and the total loss of the transfer learning is shown as in Figure 11.

$$Loss_{Total} = \frac{1}{N}(Loss_{classification} + \alpha Loss_{localization}) \quad (3)$$

where N is the number of matched bounding boxes and α balances the weights between two losses, picked by cross validation.

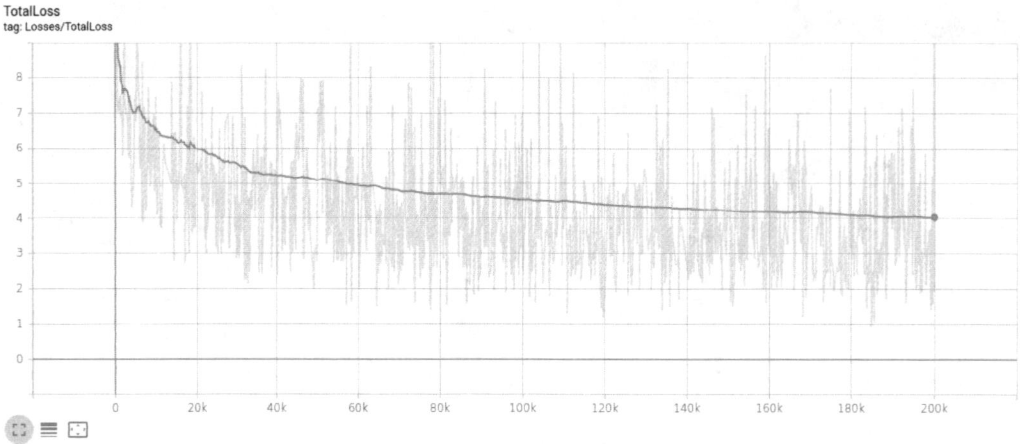

Figure 11. Total loss of SSD-MobileNetV2 transfer learning

At the end of the transfer learning, the training results can be summarized by Table 2.

Table 2. Training results for of SSD-MobileNetV2 transfer learning

	Results
Classification loss	0.9812
Localization loss	0.3604
Total loss	1.932
Total training duration	8 days 8 hour 44 minutes

4. Discussion

The trained model produced in the training process is infeasible to run in the vehicle. Typically, the vision controller in the vehicle uses small form computer, exactly what NVIDIA Jetson Nano is. In order to optimize the trained model, the batch size is reduced to single batch and the model is convert to low precision model using NVIDIA TensorRT software.

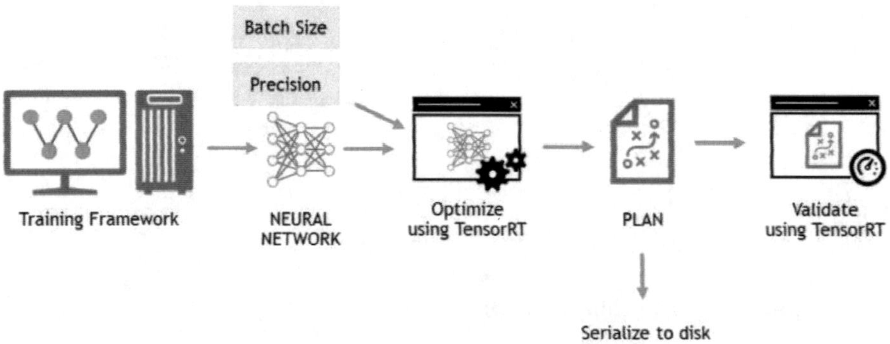

Figure 12. The optimization process from the trained model to inference model in NVIDIA TensorRT

4.1 Testing the model

In order to test the accuracy and robustness of the optimized inference model, the test images that different in the dataset are used as input. The example of the results of the testing can be shown as in Figure 13.

(a) Potholes detection (b) Road bump detection

Figure 13. Road peculiarities detection with SSD MobileNetV2 for (a) pothole and (b) road bump

The summary of the detection result from 10 test images for each class can be illustrated in the Table 3.

Table 3. Summary of road peculiarity detection using SSDMobileNetV2

Road Peculiarity	Accuracy Percentage	Average Confidence
Potholes	60%	0.65
Road bumps	70%	0.43

In potholes detection, the model is able to detect the potholes even the holes are filled with water. This is due to the training dataset also contains the wet potholes. However not all the potholes are detected, only 60% accuracy in the test images since large and distant potholes could not be recognized. This may be due to the dataset does not contains enough data for that situations.

Road bumps detection is a quite challenging task due to the variety of their patterns and the bigger size of bounding box. Bigger the size of bounding box causes more unwanted noise present in the dataset. Although a clear road bump is using as the test image, the detection only yields 0.43 of average confident level.

Running the optimized SSD MobileNetV2 with NVIDIA Jetson Nano yields ~20 frame per seconds (fps), good enough to be considered as real time applications. The detection speed is crucial in this work since the detection system is built for vehicle system. The vehicle able to react quickly when the detection speed is increasing.

5. Conclusion

SSD MobileNetV2 has been implemented to detect road peculiarities such as potholes and road bumps in this paper. The results show the detection algorithm able to recognize the road peculiarities but with the accuracy limitations in potholes and detection confidence limitation in road bumps. From our observation, the dataset size plays an important role in order to make the detection more robust and accurate. The detection model could adapt more patterns and situations when the dataset size is increased.

In future, we would like to increase the detection performance by increasing the training dataset size (over 5000 images for each class) in order to make detection algorithm more robust.

Acknowledgment

The authors would like to thank to Perusahaan Otomobil Nasional Berhad (PROTON) for funding this work.

References

[1] Asean N. ASEAN NCAP Roadmap 2018 *Malaysian Institute of Road Safety Research* (MIROS), Kajang; 2021-2025.

[2] Musa, M. F., Hassan, S. A., & Mashros, N. 2020 The impact of roadway conditions towards accident severity on federal roads in Malaysia. *PLoS one*, **15(7),** e0235564.

[3] Girshick, R., Donahue, J., Darrell, T., & Malik, J. 2014 Rich feature hierarchies for accurate object detection and semantic segmentation. *In Proceedings of the IEEE conference on computer vision and pattern recognition* pp. 580-587.

[4] Liu, W., Anguelov, D., Erhan, D., Szegedy, C., Reed, S., Fu, C. Y., & Berg, A. C. 2016 Ssd: Single shot multibox detector. In European conference on computer vision pp. 21-37. *Springer, Cham.*

[5] Redmon, J., Divvala, S., Girshick, R., & Farhadi, A. 2016. You only look once: Unified, real-time object detection. *In Proceedings of the IEEE conference on computer vision and pattern recognition* pp. 779-788.

[6] Goyal, P. and Kaiming, H., 2018. Focal loss for dense object detection. *IEEE transactions on pattern analysis and machine intelligence*, **39**, pp.2999-3007.

[7] Howard, A.G., Zhu, M., Chen, B., Kalenichenko, D., Wang, W., Weyand, T., Andreetto, M. and Adam, H., 2017. Mobilenets: Efficient convolutional neural networks for mobile vision applications. *arXiv preprint arXiv*:**1704**.04861.

[8] Sandler, M., Howard, A., Zhu, M., Zhmoginov, A. and Chen, L.C., 2018. Mobilenetv2: Inverted residuals and linear bottlenecks. *In Proceedings of the IEEE conference on computer vision and pattern recognition* pp. 4510-4520.

[9] He K, Zhang X, Ren S, Sun J. Deep residual learning for image recognition. 2016. p. 770-8.

[10] Lin, T.Y., Maire, M., Belongie, S., Hays, J., Perona, P., Ramanan, D., Dollár, P. and Zitnick, C.L. 2014 *Microsoft coco: Common objects in context. In European conference on computer vision* pp. 740-755

Degradation behaviour of plasticized PLA/CNTs nanocomposites prepared by the different technique of blending

H. Norazlina[1*], A. Suhaila[1], A. Nabihah[1], M.M Rabiatul[1], I. Zulhelmie[2], Y. Yusoh[2]
[1]Faculty of Engineering Technology, University College TATI, 24000 Kemaman Terengganu Darul Iman, Malaysia.
[2]College of Engineering, Universiti Malaysia Pahang, 26600 Pekan, Pahang Darul Makmur, Malaysia.

*E-mail: norazlinatatiuc@gmail.com

Abstract. The present research goals to investigate how the preparation technique became the factor to develop materials with a good combination of properties and optimum degradation ability. PLA/CNTs nanocomposites were prepared via melt blending and solution blending that were involved of unmodified carbon nanotubes (CNTs) and modified CNTs (mCNTs) at 1.5 wt.% loading. The surface morphology of nanocomposites was viewed by Field Emission Scanning Electron Microscopy (FESEM). The effect of 5 wt.% poly (ethylene glycol) (PEG) as plasticizer on nanocomposites were determined. The weight loss in soil degradation study was run for 6 months. The morphology study by FESEM confirmed the finding through the existence of a smooth fracture surface especially when PEG was loaded. In soil degradation analysis, neat PLA exhibited a low weight loss rate after 6 months. The maximum weight loss for both techniques was shown by PLA/PEG/CNTs from melt blending technique and PLA/PEG/mCNTs from solution blending, believed from the pore occurred bring to poor properties.

1. Introduction

Non-biodegradability plastics synthesized from petroleum-based products grip main consideration due to an environmental problem which is causes the waste disposal problem. Besides, these petroleum-based polymers are also limited sources and fluctuations in price [1]. Consequently, the awareness in the development of an alternate and biodegradable polymer from renewable resources can be seen lately. Concerning the processing properties, it is worth noticing that the tallying of plasticizer into highly filled poly (lactic acid) (PLA) can achieve lubrication in improving filler dispersion and assisting in viscosity control. An article reported the preparation of plasticized PLA by using poly (ethylene glycol) (PEG) by melt blending [2]. The dielectric properties and shielding effectiveness were improved by loading of 4.0 wt.% MWCNTs. Da Silva and coworkers used PEG and dibutyl sebacate (DBS) as plasticizer at 20 wt.% during melt blending of PLA/Gr [3].

Another study that is quite similar to the current study was reported by Chieng in 2014 [4]. They used graphene nanoplatelets at 0.3 wt.% and lower molecular weight of PEG, which was 200 g/mol. More particularly, plasticizer aids in the processing and extruding of the polymer at lower temperatures. Thus, it can increase flexibility and reduce cracking tendencies of finished products. In this context, the ideal plasticizer for PLA can be any biodegradable product necessarily non-volatile that has a relatively low molecular weight to produce a substantial lessening of glass transition (T_g) and increase in ductility. Besides, the PEG also gives benefit to enhancing the biodegradation rate since the neat PLA has a low tendency to degrade.

In the packaging of foods, microbial degradation of PLA should be considered especially microorganisms including fungi and lactic acid bacteria for their feasible capabilities of PLA degradation [5]. Numerous studies have been stated on soil degradation of PLA composites such as blending with natural filler and carbon-based materials. The polymer degradation rates could be controlled by blending with plasticizers, additives and often inorganic fillers. Indeed, several research works have reported on the biodegradation or hydrolytic degradation of different aliphatic polyesters, due to the hydrophilicity of the nanoparticles [6]. However, other work stated the nanofillers can be able to impede the degradation of the polyesters chain, assigning this effect to the enhanced barrier properties of nanocomposites [7].

In current study, poly (lactic acid) (PLA) as a matrix has been selected to be blended with carbon-based materials. The blending of these nanofillers with various loading was made by melt blending and solution blending to determine the effect of processing methods on properties and degradation behaviour of nanocomposites. The addition of PEG during blending was also performed. The nanocomposites were then analyzed for degradation, which covered soil degradation behaviour and hydrolysis study by solution immersion. The advantage of PEG was highlighted since the plasticizer showed good assistance in nanofiller dispersion in PLA and could increase the biodegradation rate of nanocomposites compared with neat PLA.

2. Materials and Methods

2.1 Chemicals and raw materials

Multi-wall carbon nanotubes (CNTs) were supplied by Global Science Resources with >90wt% purity and ash <1.5wt%. It has outer diameter close to 20–40 nm and length between 10–30 μm. PLA was supplied from Nature Works. Poly (ethylene glycol) PEG with average molecular mass 950–1050 g mol^{-1}, HNO_3 65% and H_2SO_4 95-97% were provided by Merck KGaA.

2.2 Preparation of samples

About 1.0 g of CNTs was added into 20 % of HNO_3 solutions and stirred in a few minutes. Then, the solution was sonicated for 2 hours. After that, the CNTs were washed with deionized water (DI) and isolated by filtering the product until no acidic. Then, the product was put in an oven at 40 °C for drying. The modification was pursued by heating the CNTs into 1:3 mixtures of concentrated HNO_3 and H_2SO_4 at 100 °C for 100 minutes. Then, isolated mCNTs was filtered and neutralized by washing with deionized water (DI) was dried in an oven. Since the current study was purposely to determine the effect of preparation techniques and plasticizer on the rate of degradation with a different type of nanofillers, the loading of nanofiller was set at 1.5 wt.%.

In the melt blending method, the material was blended via an internal mixer and the twin-screw rotator functioned at 180 °C for front and back screws. The speed was fixed between 3 rpm–5 rpm. PLA was fed into an internal mixer and molten followed by the addition of nanofillers till homogenous blending was gained. For 5 wt.% PEG loading, it was added after the PLA was in the molten state.

In the solution blending method, the amount of 1.0 g PLA was completely dissolved in 20 g of chloroform at 45 °C. The 1.5 wt.% of CNTs and mCNTs were dispersed in chloroform with the amount of 0.2 mg/mL. The PLA solution was consequently mixed with dispersed nanofiller in Teflon bottle and stirred for 4 h at 40 °C to obtain homogenous PLA nanocomposites solution. After sonication for another 10 min to continue the stability, the mixed solution was transferred into a petri dish and the solvent was evaporated in a fume hood to gain the film. For the nanocomposites with

PEG, the 5 wt.% PEG was added in a well-dispersed PLA solution. After that, the addition of dispersed nanofiller was done.

2.3 Characterization of samples

2.3.1 Field emission scanning electron microscopy (FESEM)

FESEM analysis topological and morphology analysis was performed at a magnification of 10–300,000×. The elemental analysis could be done at low voltage, low landing energy and large depth of focus with resolution of 0.8 nm at 15 kV and 1.2 nm at 1 kV. Before testing, the CNTs nanofiller was finely ground to powder form. As for the nanocomposite, the cross-sectional area of the fractured surface was used to evaluate the dispersion of nanofiller inside PLA nanocomposite.

2.3.2 Soil burial of biodegradable PLA nanocomposites

The biodegradability of the nanocomposite was studied by calculating weight loss of blends over time in a soil environment. A series of square-shaped samples, measuring around 10 mm x 10 mm were weighed and then buried in loamy sand-type soil. The samples were buried at a depth of 15 cm. After every 30 day of incubation till reached 180 days, samples were cleaned by washing with distilled water and dried to a constant weight in the oven. The degradation of samples was measured by weight loss of samples during the experiments. The calculation as in equation (1) where ΔW_{loss} is weight loss, W_f is final weight and W_i is the initial weight.

$$\Delta W_{loss} = [(W_i - W_f)/ W_i] \times 100 \quad (1)$$

3. Results and Discussions

3.1 Morphology study by FESEM

The morphological images of PLA/CNTs shown in Figure 1 are for (a) unfilled PEG and (b) filled PEG nanocomposites. From Figure 1(a), it was observed that the undispersed CNTs nanofiller disappeared in Figure 3.2(b). This can be explained by the aid of the plasticizing effect of PEG in the nanofiller dispersion of the PLA matrix. The rough surface morphology in PLA/mCNTs as shown in Figure 1(c) was compared with the smooth surface fracture when PEG was loaded as viewed in Figure 1(d). The good dispersion of CNTs nanofillers in the PLA matrix was believed to be responsible for the enhancement of properties. In common, surface modification of CNTs enhances adhesion between CNTs and the PLA matrix due to functional groups. Due to high specific surface area, large aspect ratio and strong interfacial interaction of –COOH and –OH on mCNTs with PLA matrix, the PLA molecules can be easily bounded on the surface of CNTs, which can form an adequate number of nucleation centre and play the roles of crystal nucleus. A similar finding was reported by Yousefzade et al. [8].

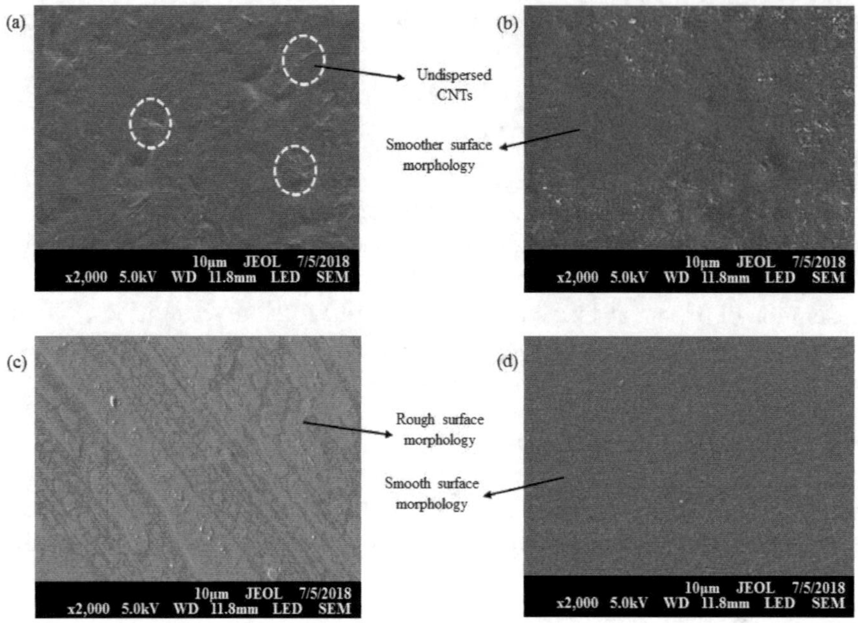

Figure 1. FESEM micrograph of (a) PLA/CNTs; (b) PLA/PEG/CNTs; (c) PLA/mCNTs and (d) PLA/PEG/mCNTs prepared by melt blending technique.

Even without nanofiller modification, the dispersion of pristine carbon nanotube bundles could enhance the deagglomeration by covalent bonding created between the ester group in PLA and CNTs as shown in Figure 2. The loading of 1.5 wt.% mCNTs in PLA provided the smooth surface of thin-film where even a few pores can be spotted on the cross-section. One interesting finding to be highlighted in the current study is even high loading of mCNTs was inserted (1.5 wt.%), the smooth surfaces of nanocomposites can be viewed. An opponent with a study by Luo et al. in 2018 [9], they obtained the rough surface morphology of PLA thin film with 0.4 wt.% mCNTs. They prepared the film by solution casting using dichloromethane (DCM) solvent.

Furthermore, the good distribution and dispersion affect the degree of crystallinity of nanocomposites. The homogenous structure with smooth fracture means excellence interaction between filler and matrix are recognized bringing to high crystallinity, thus enhanced the properties of nanocomposites. Hence, the barrier property result of nanofiller that acted as a nucleating agent may hinder the movement of plasticizer and contributed to the constancy of the mixture, preventing the plasticizer transmission to the surface. The stability of plasticizer is important to confirm it is well-functioned in nanocomposites.

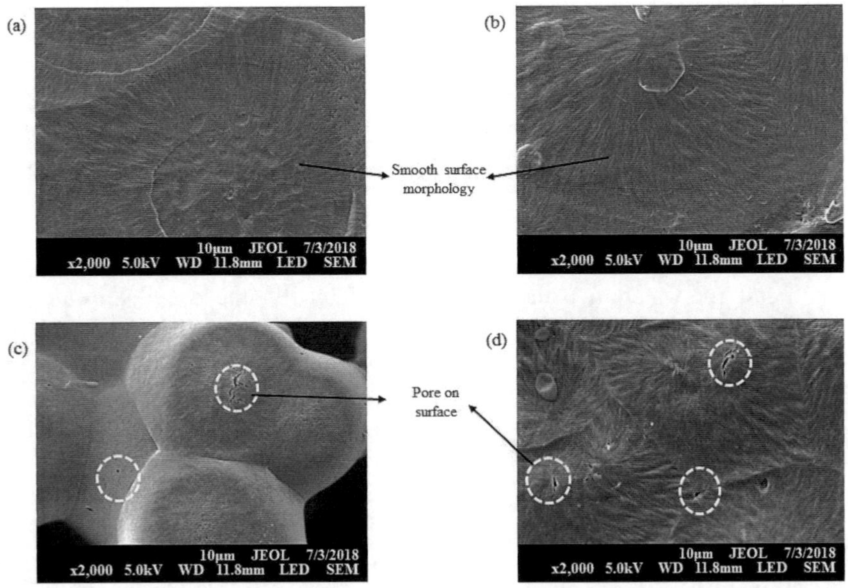

Figure 2. FESEM micrograph of (a) PLA/CNTs; (b) PLA/PEG/CNTs; (c) PLA/mCNTs and (d) PLA/PEG/mCNTs prepared by solution blending technique.

3.2 Biodegradation analysis by soil burial study

Usually, the biological attack on polymer composite materials is influenced by fungal development or soil burial. Fungal growth is potential only in humid conditions and their mechanical properties are affected after long-term disclosure to microorganisms. The degradation of polymer composites due to biological factors is reliant on the chemistry of the composites, salinity and acidity of the soils [10]. From a physical viewpoint, two types of polymer degradation are heterogeneous and homogeneous, which is likewise known as surface and intramolecular degradation of the polymer.

Figure 3 shows the trend of the degradation of melt blending samples. It was found that PLA degradation occurred mainly through ester bonds' scission. The degradation of PLA under compost is a process containing four main phenomena. They are namely (1) water absorption; (2) ester cleavage and formation of oligomer fragments; (3) solubilization of oligomer fragments, and finally; (4) diffusion of soluble oligomers by bacteria, respectively. Hence, among the factor, which enlarged the hydrolysis tendency of neat PLA, eventually controls the degradation of PLA. The neat PLA results exhibited the slightest degradation which was only 0.04 % weight loss after 6 months as compared to PLA nanocomposites. It is worth highlighting that even though PLA degrades at a slow rate, it has the tendency to have complete biodegradability after a certain time and not polluting the environment after biodegradation [11]. Pantani and Sorrentino in 2013 examined the consequence of crystallinity on the biodegradation rate of injection-molded PLLA parts [12]. Their results confirmed that the degradation rate was affected by initial morphology. They resolved that the crystals shaped during degradation have less connectivity and are thus less effective at inhibiting whole swelling of the polymer. In difference way, they suggested that crystals formed before degradation present higher connectivity that promises a denser structure that is more resistant to enzymatic attack and oligomer diffusion.

One to be highlight in this research was the significant enhancement of biodegradability in certain PLA nanocomposites. PLA/PEG/CNTs had the highest weight loss which reached 12.00 % after six months. This can be relating to the performance of this nanocomposite in mechanical testing. The hardness of PLA/PEG/CNTs is low compared to other nanocomposites, which can be concluded the less compatibility of filler in the matrix. Hence, the weak interfacial bonding contributes to ease of the breakage of ester bonds in the matrix. It can be observed the GO and mCNTs have a similarity in % weight loss with the presence of PEG which was revealed to range from 5.48–5.54 %.

Interestingly, after the addition of hydrophilic PEG, the degradation rate enhanced higher than nanocomposites without 5 wt.% PEG loading for all nanocomposites. The environmental degradation rate rose simultaneously with the loading of PEG fillings. The occurrence of carboxyl groups in modified nanofiller increased the hydrophobicity of PLA. It was leading to improving water resistivity which later led to slow biodegradation. This discovered that the modified nanofiller improve the stability of PLA by resisting the microbial attack during composting. Since the hydrophobic nature of PLA, it was difficult for water to penetrate the materials. But the excellent water solubility and hydrophilic characters of PEG in blends could aid absorb and keep the water. In the soil burial analysis, both PLA and nanofiller would carry out hydrolysis reaction therefore the linear chains would break down at the ester bond. This caused the loss of molecular weight and mass of nanocomposites.

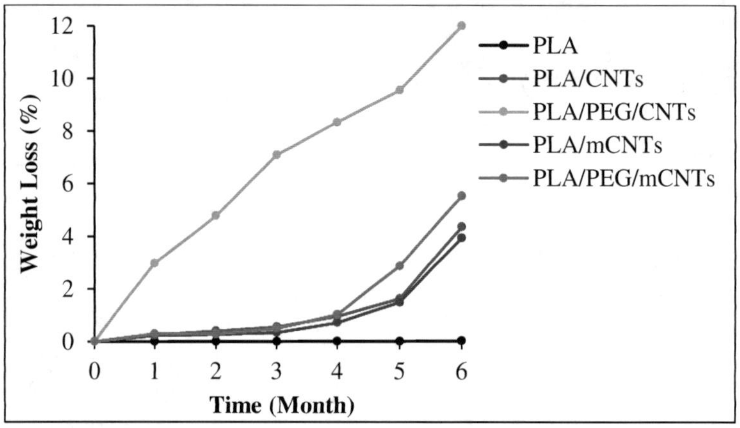

Figure 3. Weight loss of neat PLA and PLA nanocomposites with 1.5 wt.% nanofillers prepared by melt blending technique.

Figure 4 exhibits the % of weight loss for thin film samples prepared by solution blending. The percentage of weight loss increased with rise in the period. The PLA exhibited a small rate of weight loss after the sixth month which was 0.06 % but slightly higher than PLA prepared from the melt blending method. The presence of nanofillers enhanced the degradation rate compared with neat PLA. The trend showed a positive result for Gr and mCNTs where they exhibited a high weight loss after the third month of analysis. Therefore, the merger of carbon-based materials into a decomposable polymer matrix represented a promising approach not only for enhancing the polymer performance but also for growing its biodegradation rate in composting conditions. Nevertheless it is well known that the biodegradation mechanism of PLA comprises chemical hydrolysis, the role of microorganisms and how they are affected by the existence of nanoparticles is still not well understood [12].

Also, instead of improving the dispersion capability in the PLA matrix and decreases the brittleness of PLA nanocomposites, the presence of PEG is useful to resolve the plastic-based disposal problem especially. This is verified through the trend performed by nanocomposites added with a plasticizer. Their results showed high weight loss compares with unfilled PEG. The maximum weight loss was obtained from PLA/PEG/mCNTs where it recorded an 8.94 % of weight loss. About the melt blending sample, this value increased up to 2.40 % confirmed the pores observed on a surface fracture from FESEM analysis affected the rate of biodegradation.

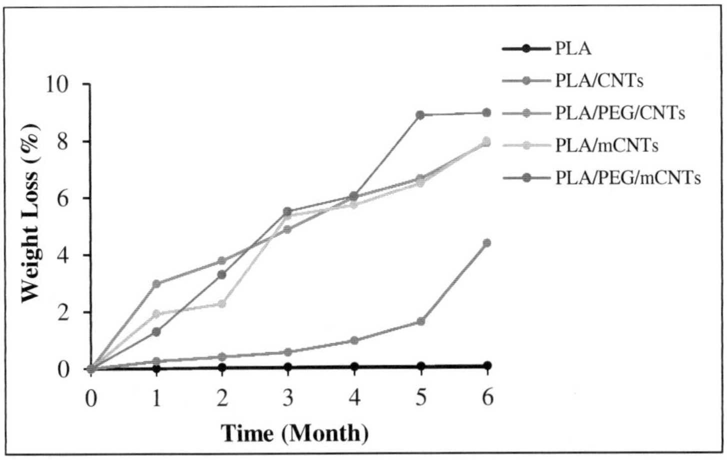

Figure 4. Weight loss of neat PLA and PLA nanocomposites with 1.5 wt.% nanofillers prepared by solution blending technique.

The degradation of PLA and nanocomposites can occur in three different ways: (1) scission of the main chains, (2) scission of the side chains and (3) scission of the intersectional chains. Besides, the degradation ability is induced by a series of factors such as photodegradation, oxidation, hydrolysis, thermolysis, enzymolysis and biodegradation [13]. As shown in Figure 5, biodegradation of polymers generally takes place in two main steps: primary degradation, in which fragmentation of the polymer chain ensues due to hydrolysis or another oxidative reaction, and ultimate biodegradation, where the microorganisms assimilate the low PLA M_w chains are formed.

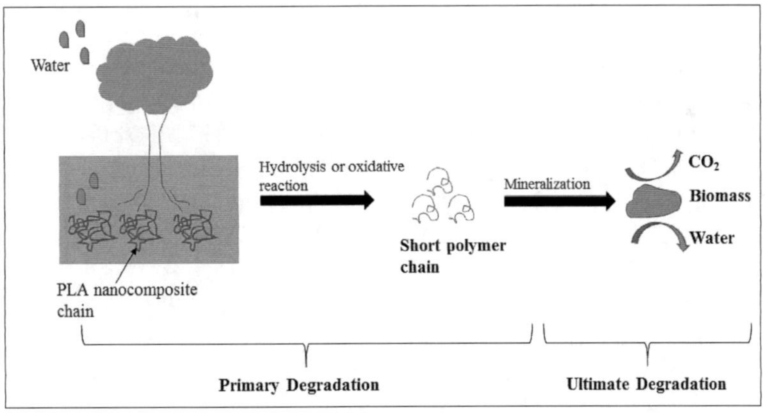

Figure 5. Schematic of PLA biodegradation mechanism.

PLA degradation upon disposal in the environment is more challenging because PLA is largely unaffected to attack by microorganisms in soil or sewage below ambient conditions. Based on a research [14], the polymer must first be hydrolyzed at elevated temperatures at about 60 °C to decrease the M_w before biodegradation can commence. Under surroundings of high temperature and high humidity, as inactive compost, for example, the PLA degrades rapidly and disintegrates within weeks to months. The primary mechanism of degradation occurs in a two-step process which also starts with hydrolysis and is followed by the bacterial attack on the fragmented residues. During the initial phases of degradation, the high M_w polyester chains hydrolyzed to lower M_w oligomers. Microorganisms presented in the soil then began to digest the lower M_w lactic acid oligomers, producing carbon dioxide and water. The rate of hydrolysis was accelerated by acids or bases and was reliant on temperature and moisture content.

4. Conclusion

The nanocomposites prepared by melt blending and solution blending was successfully done. FESEM analysis showed the PLA/PEG/mCNTs from melt blending had smooth morphology. Nanocomposites from solution blending exhibited pore on the surface. The morphology study is parallel with the degradation ability of samples. The sample with the smooth surface has low degradation which expected has better properties. In biodegradability analysis, neat PLA from both techniques showed from 0.04–0.06 % of weight loss. Even the degradation was at a low rate, this proves that this biopolymer has a future in reducing the environmental problem. PLA/PEG/CNTs from melt blending had the highest weight loss that reached 12.00 % after six months. The maximum weight loss from solution blending obtained from PLA/PEG/mCNTs where it recorded an 8.94 % of weight loss.

Acknowledgement

The authors appreciatively acknowledge University College TATI for providing laboratory equipment. Also, we are thankful to Universiti Malaysia Pahang for facilities and giving supports in this study.

References

[1] Zhong Y, Godwin P, Jin Y and Xiao H 2020 *Advanced Industrial and Engineering Polymer Research* **3** 27
[2] Ahmad F A et al 2020 *Polymers* **12** 427
[3] da Silva J A, Dalmolin C, Pachekoski W M and Becker D 2018 Journal of Polymer Applied Science **46745** 1
[4] Chieng B W, Ibrahim N A, Yunus W M Z W and Hussein M Z 2014 *Polymers (Basel)* **6** 93
[5] Pattanasuttichonlakul W, Sombatsompop N and Prapagdee B 2018 *International Biodeterioration Biodegradation* **132** 74
[6] Gai S, Xu J, Zhang H, Yin R and Zhang W 2020 *Tissue Engineering Part B: Reviews* **26** 484
[7] Luo Y, Lin Z and Guo G 2019 *Nanoscale Research. Letter* **14** 56
[8] Yousefzade O, Franco L, Nami M, Puiggali J and Garmabi H 2019 *Journal of Composite Material* **53** 2131
[9] Luo J, Wang H, Zuo D, Ji A and Liu Y 2018 *Micromachines* **9** 635
[10] Haider T P, Völker C, Kramm J, Landfester K and Wurm F R 2019 *Angewandte Chemie International Edition* **58** 50
[11] Qi X, Ren Y and Wang X 2017 *International Biodeterioration Biodegradation* **117** 215
[12] Papageorgiu G Z 2018 *Polymers* 202
[13] Castro-Aguirre E, Auras R, Selke S, Rubino M and Marsh T 2018 *Polymers (Basel)* **10** 202
[14] Lunt J 1998 *Polymer Degradation Stability* **59** 145

Comparative study between furnace brazing and laser brazing

Y. Erny A. M. Y.[1], Fadhil A. R.[1], M. H. Aiman[1*], M. Ishak[1], M. M. Quazi[1]

[1]Joining, Welding and Laser Processing Lab (JWL), Faculty of Mechanical and Automotive Engineering Technology, Universiti Malaysia Pahang, 26600 Pekan, Malaysia

*E-mail: aimanmh@ump.edu.my

Abstract. Nowadays, brazing has been widely used in many industries, especially in the automotive application. The brazing process is introduced as it can be used to join the different metals together without melting the parent material. In this present study, furnace brazing and laser brazing of Ti-6Al- 4V titanium alloy and 316 L stainless steel (SS) with silver-based, BAg 8-1.5Ti filler metal were studied. There are significantly different between furnace brazing, also known as conventional brazing method and laser brazing in terms of joining strength and microstructure reaction. Furnace brazing was performed at 870°C and 880°C with 30 minutes of heating duration. Meanwhile, laser brazing was performed using a 200Watt continuous wave laser with varying laser power. Both of brazing method was conducted with a vacuum pressure of 3×10^{-3} Pa. Besides, to maintain the accuracy of the temperature measurement of laser brazing, an infrared thermometer is used. The tensile test was conducted to analyse the mechanical properties. The cross-sections of the brazed joints have been examined using an optical microscope. The brazed joints of the furnace brazing show an average tensile strength of 55.89 kPa for 880°C and 43.16 kPa for 870°C. Nonetheless, the maximum tensile strength of laser brazed joints was 27.95 kPa, which is lower than furnace brazed joints.

1. Introduction

Brazing is one of the most traditional methods for joining dissimilar materials due to its simple operation, great repeatability, and relatively small thermal impact on a substrate [1]. The brazing process is applied to join two metals (similar or dissimilar) using heat and filler metals which have a melting temperature of above (450◦C) [2, 3]. The joining of two metals made by heating them to the melting point of the filler metal to ensure that molten filler flows through the capillary action between the two mating surface [2]. Brazing makes a metallurgical bond between the filler metal and the surfaces of the two metals being joined [4]. Nevertheless, brazing includes insignificant microstructural changes and thermal distortion of the base metal, which makes it desirable over welding in many manufacturing circumstances [5]. Brazing method has been typically used when high joint strength is required. However, it is challenging to achieve a good joint [6].

Some of the prominent types of brazing techniques are furnace brazing and laser brazing. Generally, furnace brazing is a semi-automated process which allows joining of genuinely complex multi-joint assemblies. It offers the adaptability to join a wide range of metals. Vacuum furnace brazing additionally called as conventional furnace brazing, is the right choice for bonding materials challenging to join by a traditional welding process. In contrast, laser brazing is a combination of laser heating and brazing with filler metal, which has the basic characteristics of brazing [7]. Some of the advantages of laser processes are that they allow not only localized fusion with a constraint of 'heat affected zones', but also the possibility of automation and ideal control of the quality of the joints [8]. Consequently, laser brazing is becoming a well-established laser joining technique that is completely incorporated into industrial production lines [9]. It offers efficient and attractive solutions to the automotive sector due to its high strength and smooth surfaces. Moreover, laser brazing, as well as furnace brazing, is used for a diverse array of applications, including aerospace components, automotive components, and biomedical assemblies [10].

Comparing between furnace brazing and laser brazing, it has a few comparisons, including brazing processing time, energy efficiency, heat distribution, and adoption cost. Focusing on the brazing process, conventional furnace brazing is a prolonged process with the unnecessary heating of the whole job at high temperature. Meanwhile, laser brazing possesses a short and fast process because only the selected part should be heated, and the amount of heat input is small. Also, laser brazing has distinct advantages that pay off the long term, legitimating the higher costs for the technology. Most significant benefits are high processing speeds and low heat input [11]. Nevertheless, conventional furnace brazing is suitable for mass production of small parts and does not require flux or post-cleaning. It can allow inert or vacuum atmospheres protecting from oxidation. Of little difference, lasers eliminate the need to heat the whole part and far more energy-efficient. Laser- braze parts are available for immediate onward processing with minimal cool- downtime.

Depending on the brazing temperature and time, an increase in the content of the active element causes a decrease in the joint strength. These issues emerge because conventional furnace brazing requires long treatment times in order to ensure the heating and cooling of the whole segment [12]. These issues are suppressed by laser brazing due to the fast heating and cooling time of a few moments. In the brazing process, brazing temperature is one of the parameters that influencing the joining strength. The effect of the brazing temperature on the microstructure of brazed joints is not so unambiguous. Microstructures may be different due to the rapid cooling and heating during the process as well as the strength of the joint.

On the other hand, the higher the brazing temperature is, the more intensive reaction occurs between the filler metal and base metal. Consequently, the thicker intermetallic layer might be formed at the interface [13]. Tight process (time/ temperature) control and the chance of a clean metallurgical joint at the beginning of brazing are attractive attributes. Above all, brazing is a relatively quick process that offers adequate strength in the joint of dissimilar materials [14]. However, when joining Ti to steel, it is restricted by their minimal solid solubility (0.1 at. %) at ambient conditions and common solid solubility in liquid conditions [15]. The research focused on joining of dissimilar material (316L SS and titanium alloy, Ti6Al4V) by using silver-based (BAg 8-1.5Ti) filler metal alloy to investigate the comparison between the effect of laser power and furnace brazing temperature on joining strength of furnace brazing and laser brazing. Also, the microstructure interaction between the base metal and filler metals are comprehensively studied.

2. Method and materials

Dissimilar metals of a 316L stainless steel plate and a Ti-6Al-4V titanium rod were employed in the brazing process. The samples were divided into rectangular plates with 50 mm × 10 mm × 2 mm dimension for stainless steel. And 6 mm × 6 mm dimension for Ti-6Al-4V. These specimens were then, grind using grinding papers 800 grit and subsequently were cleaned by acetone before brazing. The physical and mechanical properties of the materials and their chemical composition are recorded in table 1 and table 2, separately. The filler metal used was BAg- 8 1.5Ti with thickness was around 100 μm. The filler metal was cut into 5 mm × 5 mm and cleaned in acetone prior to brazing. Afterward, it is sandwiched between the overlapped areas of the base metals, as shown in figure 1.

Table 1. Physical and Mechanical properties of alloys

Alloy	Melting point (^{0}C)	Thermal conductivity (W/ (m.K))	Tensile strength (MPa)
Ti-6Al-4V	1604	6.7	950

316L SS	1375	16.2	515
BAg 8- 1.5 Ti	790	-	-

Table 2. Chemical composition of the alloys used

Material	Chemical composition (wt.%)								
Ti-6Al-4V	V	Al	C	O	N	Fe	H	Ti	
	3.83	6.01	0.008	0.088	0.003	0.083	0.002	Bal.	
316L SS	C	Mn	Si	P	S	Cr	Ni	Mo	Fe
	0.03	2.00	0.75	0.045	0.03	16.0-18.0	10.0-14.0	2.00-3.00	Bal.
BAg 8- 1.5 Ti	Ag	Cu	Ti						
	71.25	27.25	1.5						

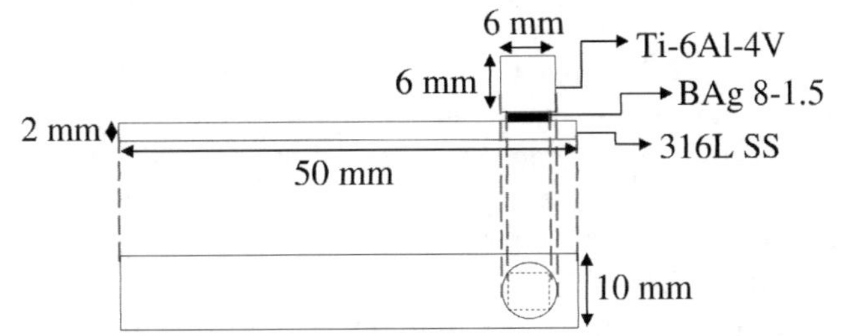

Figure 1. Schematic of material configuration (T-joint)

Subsequently, titanium rod will undergo laser surface modification (LSM) to modify the surface roughness which will contribute to the high absorption of heat during the laser brazing process using laser surface marking machine, operated in pulse mode. The laser marked the titanium rod surface within 7 mm in diameter. After the LSM process, the materials will be set up, as shown in figure 1. Since the surface area of 316L SS plate is big, it was arranged to be at the bottom of the Ti rod to reduce the energy disperse towards the material.

The furnace brazing was performed with pressure vacuum of 3×10^{-3} Pa. For this reason, the parts were heated in a vacuum furnace up to 880°C for 30 minutes (heating rate: 29.33°C/ min) as illustrated in figure 3(a). The samples were easily and quickly cooled at the room temperature in a vacuum furnace. Three samples each were furnace brazed for assessing the process. Meanwhile, laser brazing, a continuous wave of the fiber laser machine with a maximum laser power of 200 Watt, is used. The experiment was carried out in a vacuum chamber to avoid oxidation during the brazing process. The laser beam was focused on the surface of the titanium rod and irradiated at an angle of 18°. The parameters used in this study are shown in table 3.

Table 3. Furnace brazing and laser brazing parameter

Parameter	Furnace brazing	Laser brazing
Brazing temperature/ Laser power	870°C, 880°C	140 W, 150 W
Heating duration	30 mins; 5 mins HT	60 s
Vacuum pressure	3×10^{-3} Pa	3×10^{-3} Pa
Cooling rate	Natural cooling	Natural cooling

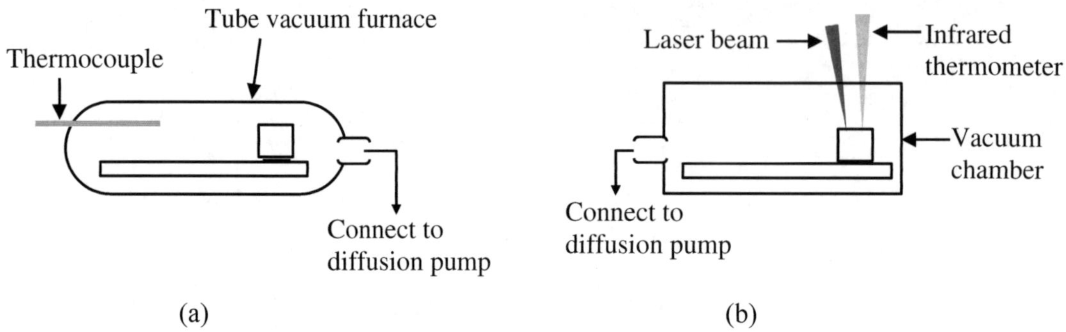

Figure 2. Set up of (a) Conventional furnace brazing; (b) Laser brazing

Besides, to maintain the accuracy of the temperature measurement of laser brazing, an infrared thermometer CTLM2HBF300- C3 Micro-Epsilon with temperature limitation of 385°C- 1600°C was used due to non-contact measurement which measured the surface temperature based on the emitted energy of the specimens. Brazing temperature was influencing the brazing joints as they can improve joining strength because of it is exceptionally relying upon the energy absorbance. Concerning furnace brazing, it was not a big deal since the temperature set will heat the whole sample; thus, the temperature will be the same. The temperature for laser brazing was measured at one point on the titanium rod where the laser beam is irradiated. The laser temperature sensor is set nearby the laser beam on the titanium rod surface, as shown in figure 2(b). As the laser energy is transmitted to the metal, the metal heats up, resulting in an increase of temperature as seen in figure 3(b). During the laser brazing, regardless of the workpiece fixture, the instantaneous melting of the material by laser and the subsequent cooling process behind the beam trigger thermal cycles.

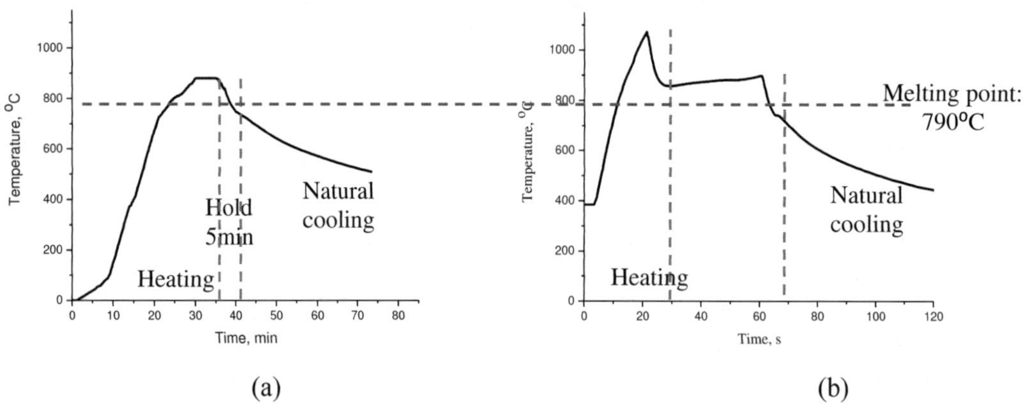

Figure 3. Temperature profile for (a) Furnace brazing; (b) Laser brazing

The tensile test was conducted using Universal Testing Machine to assess the joint strength of the samples. Figure 4 illustrates the schematic configuration of the shear test. The chosen samples were cut, mounted, polished and etched for microscopic evaluation. The cross-section of the brazed joints was etched with the chemical composition of the Killers' Etchant for titanium rod and Adler's Etchant for stainless steel 316L for about approximately 20 seconds. Meanwhile, for AgCuTi filler metal microstructure, a nitric acid solution is used and immersed for 10 seconds. The samples were checked using a 3D measuring laser microscope to observe and identify the pattern for both methods of the brazing process.

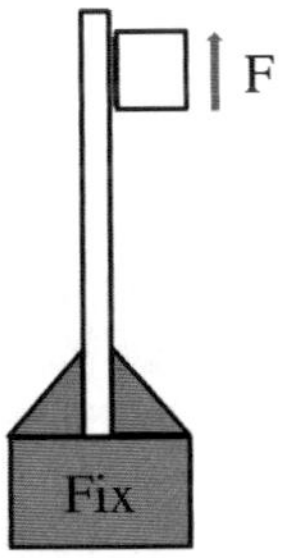

Figure 4. Schematic of the shear test

3. Results and discussions

3.1. The microstructure of furnace brazing joints

The microstructure of the brazed joints at 870°C and 880°C for 30 minutes are shown in figure 5. The microstructure of the brazed joint gradually changed with an increase in brazing temperature. There are also reaction layers between the brazed alloy and the growth of the interfacial reaction layer, which resulted in the loss of elements in the braze alloy. The microstructure of the brazed joints changed significantly with diminishing brazing temperatures to 870°C. The microstructure examined at a relatively high brazing temperature of 880°C is shown in figure 5(c) and (d). The titanium microstructure is an equiaxed pattern. Each parameter has almost the same titanium microstructure pattern. The different pattern of this microstructure will indeed show the differences in the joining strength. Besides, it should be acknowledged that erosion may occur in the furnace brazing process because of a long process time than the laser brazing process time.

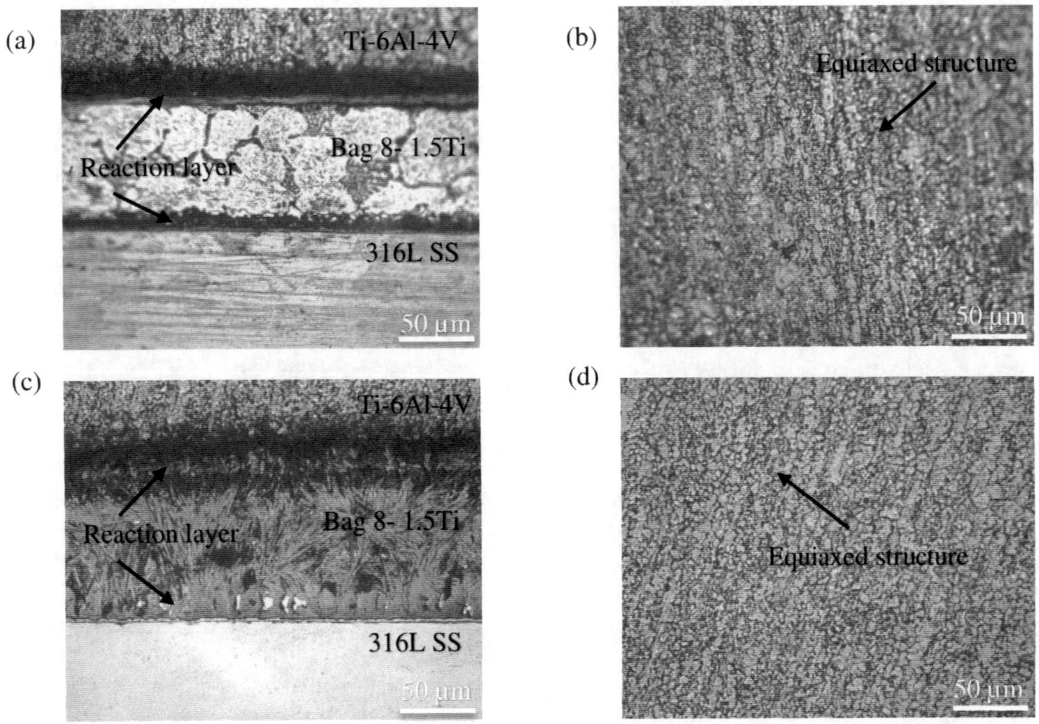

Figure 5. (a) Interface microstructure at 870°C (50X); (b) Titanium microstructure at 870°C (50X); (c) Interface microstructure at 880°C (50X); (d) Titanium microstructure at 880°C (50X)

3.2. The microstructure of the laser brazed joints

Figure 6 portrays the microstructure of the cross-section of the laser brazed samples performed with different laser power. In this figure, it can be seen the titanium microstructure look like Widmanstatten pattern, also known as Thomson structures which indicate that the molten metal cooled very slow. Apart from that, the Widmanstatten pattern form because of the development of the new phases within the grain boundaries of the parent metal, which lead to the increasing of the hardness and brittleness of the metal. Subsequently, the pattern for each parameter looks alike. As for filler metal, the microstructures are different where it is a bit rough when power used is 140 Watt compared to the power of 150 Watt. However, with the growth of reaction between Ti and SS, as with the increase in the thickness of IMC layer, vacancies for voids presumably become to an ever- increasing extent, likely form in IMC layer which may decrease the joint strength.

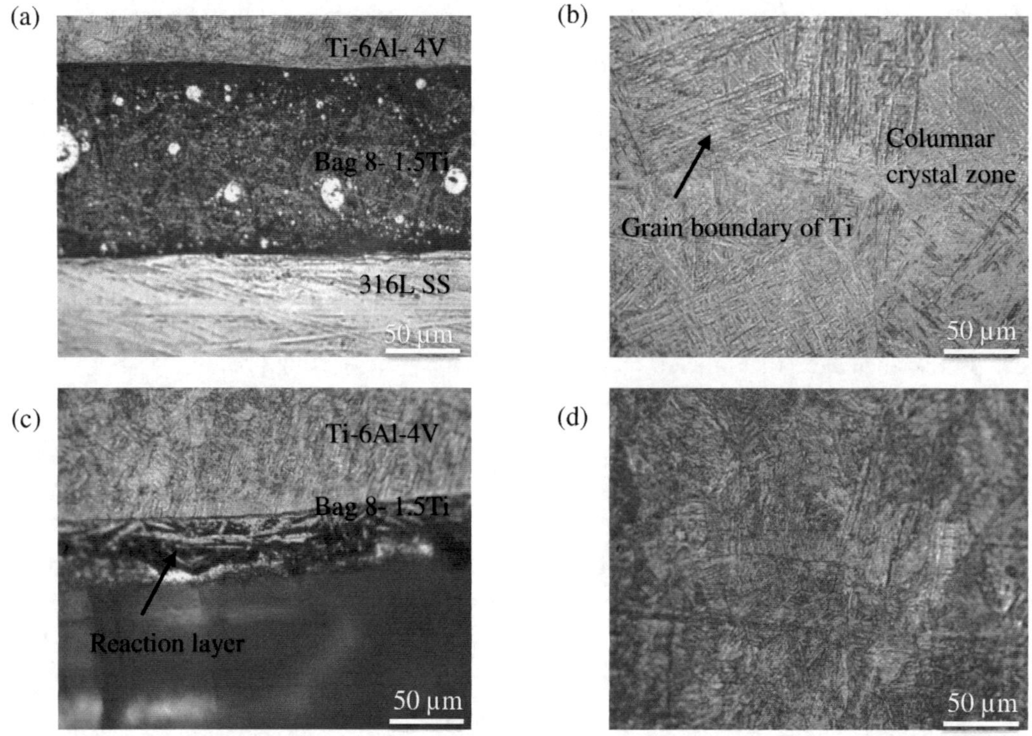

Figure 6. (a) Interface microstructure at 140 W (50X); (b) Titanium microstructure at 140 W (50X); (c) Interface microstructure at 150 W (50X); (d) Titanium microstructure at 150 W (50X)

The observed microstructure was a difference due to the different thermal behaviour of each parameter used. The reason is the used of the low energy density, a low laser power, lower peak temperature and shorter thermal cycle. Thus, the quenching effect that was realized by thermal transmission through the substrate was strong. The thickness of interfacial reaction layers, as shown in figure 5 and 6, was found to depend on the temperature and time. The microstructure also was affected by the heating and cooling rate. However, the significantly enhanced cooling rate and solidification rate refined microstructure and will promote the phase transformation as well. Therefore, the microstructure of laser brazed samples is different from the furnace brazed ones.

3.3. Tensile shear strength

With an increase in brazing temperature and laser power, the shear strength of joints showed a general ability to increase. The average shear results obtained are displayed in figure 7. The tensile strength of brazed furnace samples is shown in table 4. The highest tensile strength of the brazed sample of the furnace was 55.89 KPa. The tensile strength of the ideal laser brazed sample is therefore approximately

41.77% lower than the tensile strength for furnace brazed samples. However, based on figure 7, the graph's pattern indicates that the tensile strength increases when laser power and furnace temperature increase to 150 Watt and 880°C, respectively.

The microstructure affects the tensile strength values. As the filler metal spreads completely over the brazed region, the thickness of the reaction layer and alloying the elements are the most critical parameters for the joint strength. Subsequently, the filler metal was not spread entirely, and there is very minimal reaction layer form in laser brazing with laser power of 140 Watt. Hence, the tensile strength is low among the others. The low joint strength, however, is also attributed to the thermally induced stresses created due to stress concentration generating as a result of a variation in the phase present and thermal expansion coefficient. For the laser brazed sample, the maximum tensile strength, 32.54 kPa, is obtained at a laser power of 150 Watt and a duration of 60 s.

Table 4. Tensile shear strength of furnace brazed samples

Furnace temperature, °C	870	880
Tensile strength (KPa)	43.16	55.89

Table 5. Tensile shear strength of laser brazing samples

Laser power, Watt	140	150
Tensile strength (KPa)	27.95	32.54

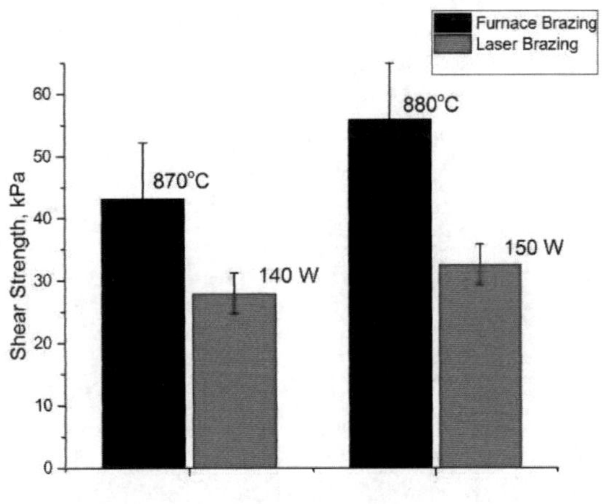

Figure 7. Shear strength of furnace brazing and laser brazing

Comparing the overall tensile strength of the brazed joint between furnace brazing and laser brazing, it is clearly shown that the furnace brazing has a higher joint strength. It might be due to the difference in heating rate and the cooling rate, which also plays a crucial role other than the microstructure in determining the joint strength. After all, laser brazing can be said having a faster heating rate and cooling rate, which would degrade the mechanical properties compared to the furnace brazing.

From a microstructural perspective, the difference in joint strength with the brazing parameters was, to a great extent, depending on the joint microstructure. The thickness and type of reaction layer between the brazed alloy are basics factors in determining the brazed joint strength. Clearly, the thickness of the layers increased with an increase in brazing temperature and time. It is suggested that the shear strength of the joint was affected by the formation of the required reaction layer in order to achieve good strength for the joint and the growth of the interaction layer, which degrades the strength.

4. Conclusions

The current study assessed the microstructure and effect of furnace brazing temperature and laser power on the joining strength of the brazed joints between the titanium alloy and stainless steel with BAg 8-1.5Ti filler metal. There is an effect of increasing the brazing temperature either using laser or furnace as a heating method.

The difference in joint strength with the brazing parameters was largely based on the thickness of the reaction layer. The maximum shear strength test can be achieved at 55.89 kPa when using furnace brazing at temperature 880°C meanwhile the lowest shear strength test can be observed at 27.95 kPa when using laser power of 140 Watt of laser brazing.

Acknowledgments
The authors would like to thank Ministry of Education Malaysia and University Malaysia Pahang for financial support given through research grant RDU192608 (RACER/1/2019/TK03/UMP//3), RDU1903118, and PGRS200306. Moreover, we are very much thankful to Kotaro Matsu from Tokyo Braze Co. Ltd for providing the brazing filler materials. Also, laboratory facilities support such as 3D measuring laser microscope from Faculty of Manufacturing and Mechatronic Engineering Technology of University Malaysia Pahang, are gratefully acknowledge.

References
[1] Han G H, Wang Y F, Zhao H Y, Song X G, Cao J and Feng J C 2017 Vacuum brazing of TZM alloy to ZrC particle reinforced W composite using Ti-28Ni eutectic brazing alloy *Int. J. Refract. Met. Hard Mater.* **69** 240–6
[2] Khathem S, Mohammed M and Hashim F 2019 The effect of brazing temperature on shear strength of nickel based superalloy (Inconel 738LC) joints *Acta Phys. Pol. A* **135** 919–21
[3] Sulaiman A J, Aiman M H, Quazi M M, Ishak M and Ariga T 2020 Enhancement of mechanical properties of Copper Brazed by laser surface modification *IOP Conf. Ser. Mater. Sci. Eng.* **788**
[4] Singh T, Sagar D and Singh G 2015 Feasibility of brazing as " Joining Process " **8354** 569–75
[5] Bridges D, Ma C, Palmer Z, Wang S, Feng Z and Hu A 2017 Laser brazing of Inconel® 718 using Ag and Cu-Ag nanopastes as brazing materials *J. Mater. Process. Technol.* **249** 313–24
[6] Xia Y, Dong H, Hao X, Li P and Li S 2019 Vacuum brazing of Ti6Al4V alloy to 316L stainless steel using a Ti-Cu-based amorphous filler metal *J. Mater. Process. Technol.* **269** 35–44
[7] Lu F, Lu B, Tang X and Yao S 2008 Study of influencing factors and joint performance of laser brazing on zinc-coated steel plate 961–5
[8] Mathieu A, Matteï S, Deschamps A, Martin B and Grevey D 2006 Temperature control in laser brazing of a steel/aluminium assembly using thermographic measurements *NDT E Int.* **39** 272–6
[9] Ungers M, Fecker D, Frank S, Donst D, Märgner V and Kaierle S 2010 In-situ quality monitoring during laser brazing *Phys. Procedia* **5** 493–503
[10] Quazi M M, Ishak M, Fazal M A, Arslan A, Rubaiee S, Aiman M H, Qaban A, Yusof F, Sultan T, Ali M M and Manladan S M 2020 A comprehensive assessment of laser welding of biomedical devices and implant materials: recent research, development and applications *Crit. Rev. Solid State Mater. Sci.* **0** 1–43
[11] Mittelstädt C, Seefeld T, Radel T and Vollertsen F 2016 Novel approach to increase the energy-related process efficiency and performance of laser brazing *Phys. Procedia* **83** 523–31
[12] Nagatsuka K, Sechi Y, Ma N and Nakata K 2014 Simulation of cracking phenomena during laser brazing of ceramics and cemented carbide *Sci. Technol. Weld. Join.* **19** 682–8
[13] Shapiro A E and Flom Y A 2007 Brazing of titanium at temperatures below 800°C: Review and prospective applications *http://titanium-brazing.com/publications/DVS-Manuscript_1020-Copy2-19-07.pdf* 1–24
[14] Yared W, Chen C Y, Sievers N, Tillmann W, Zielke R and Schimpfermann M 2019 Void

distribution in a brazed cemented carbide steel joint analyzed by X-ray microscopy *Meas. J. Int. Meas. Confed.* **141** 250–7

[15] Quazi M M, Ishak M, Fazal M A, Arslan A, Rubaiee S, Qaban A, Aiman M H, Sultan T, Ali M M and Manladan S M 2020 Current research and development status of dissimilar materials laser welding of titanium and its alloys *Opt. Laser Technol.* **126** 106090

IOP Conf. Series: Materials Science and Engineering 1068 (2021) 012004 doi:10.1088/1757-899X/1068/1/012004

Recent progress on the corrosion characterization of magnesium (Mg) prepared by powder metallurgy technique

M A F Romzi[1]*, J Alias[1] and M I M Ramli[2]

[1]Department of Mechanical Engineering, College of Engineering, Universiti Malaysia Pahang, 26300, Gambang, Pahang, Malaysia

[2]Faculty of Mechanical and Automotive Engineering Technology, Universiti Malaysia Pahang, 26600 Pekan, Pahang, Malaysia

*E-mail: ameerulfikri@ump.edu.my

Abstract. This paper reviews the progress on the corrosion characterization of magnesium (Mg)based material prepared by the powder metallurgy (PM) technique. In recent years, Mg alloys and composites produced by the powder metallurgy technique have been gaining interest in many industries especially in biomedical applications. Mg is also being used to improve battery electrodes and be considered for next-generation secondary batteries with remarkable energy and capacity density. Nevertheless, bare Mg is known to have poor corrosion resistance in most environments with occurrences such as non-uniform corrosion attack, high and fast reaction, microgalvanic corrosion of Mg-matrix and intermetallic particle and formation of non-protective hydroxide film after longer exposure in corrosive environment. New fabrication methods of Mg, such as the PM method are likely to benefit future Mg's mechanical and corrosion properties. Utilizing the latest corrosion characterization method is needed to ensure accurate and precise methods for investigating corrosion behaviour. The recent approach and progress in characterizing Mg's corrosion behaviour, specifically that have been prepared by the PM technique is considered in this paper.

1. Introduction

Magnesium (Mg) is a current lightweight material which is relevant in a wide scope of modern fields, from aviation, automobiles to biomedical applications. Its principle points of interest are good strength to weight proportion, biocompatibility, in blend with biodegradability. In any case, because of the high reactivity of pure Mg in most environment and the poor room temperature strength, making it not so much adequate for building applications. Nevertheless, for the most part, Mg composites and alloys are utilized [1–3]. Unfortunately, even though Mg might have been stable in an atmospheric environment, it is not stable in an aqueous environment since it is more anodic than the other metals, referring to the electrochemical series (EMF) table. Because of that, pitting is likely to occur on Mg due to galvanic corrosion happens on it. Corrosion occurs on Mg when it is in a room temperature environment where a film of oxide occurs on Mg. The moisture in the atmosphere will then converts the MgO to $Mg(OH)_2$ [4]. Mg is known to improve the mechanical properties of materials and it would be a disadvantage if the corrosion resistance of Mg cannot be improved [5–7]. It is proven that the corrosion resistance of pure Mg does not suit some biomedical and technical applications because of its high porosity values [8–14]. High porosity means the materials will corrode faster. Therefore, this paper will be discussing the recent techniques used by the researchers to characterize the corrosion behaviour of Mg prepared by powder metallurgy technique.

2. Powder metallurgy of Mg

Powder metallurgy (PM) is characterised as the pressing of fined powders to bond a strong piece or material. There are five procedures in the powder metallurgy including the powder creation, blending, compaction, sintering and secondary processing. PM technique is used in Mg production to form a layer of thermodynamically stable corrosion products on its surface. The layer produced prevents diffusion processes which is needed for the densification of material during the PM process. Usually, argon and nitrogen are to be used as a protective atmosphere, especially during the sintering process since Mg is highly reacted to oxygen [15].

There have been some discussions regarding the suitability of the PM technique towards Mg alloys production. It is known that the PM technique is effective since it can help to combine metals with different densities and melting points of materials. The PM technique will affect the microstructure of Mg and then will also affect the corrosion behaviour of the alloys produced. Besides, the limitations are also being discussed, for example, the hot pressing process in producing Mg powders caused the powder particles to undergo high plastic deformation which leads to the leakage of the corrosion layer which usually present on the surface of particles [7,16]. When this thing happens, the contact area of the particles will increase, enhancing the diffusion process and then will result in decreased porosity of the materials when processing in bulk production.

Zhao et al. [17] reported that there was improvement in terms of compressive strength when alloying Mg-Zn-Zr-3Y using the PM technique. It was happened due to strain, precipitation and fine crystal strengthening which caused the performance improvements as highlighted by [18,19]. Shi et al. [20] proved that during the compression test which was done according to CMT4305 at ambient temperature, the compressive strength will be improved as well as the matrix alloy which shows the role when suppressing twinning deformation.

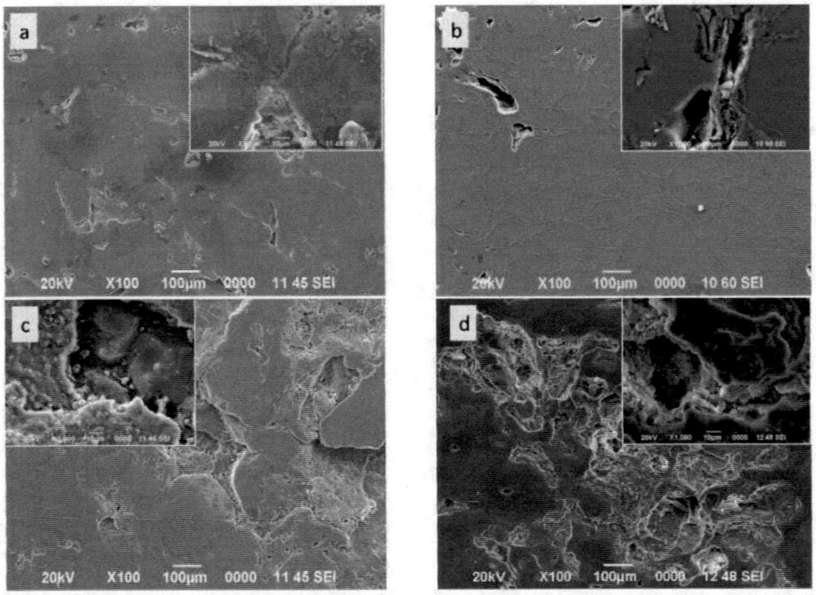

Figure 1. SEM images of a) Mg-Zn-Ca-580, b) Mg-Zn-Ca-630, c) Mg-Zn-Ca10Carb-580, and d) Mg-Zn-Ca-10Carb-630 [21].

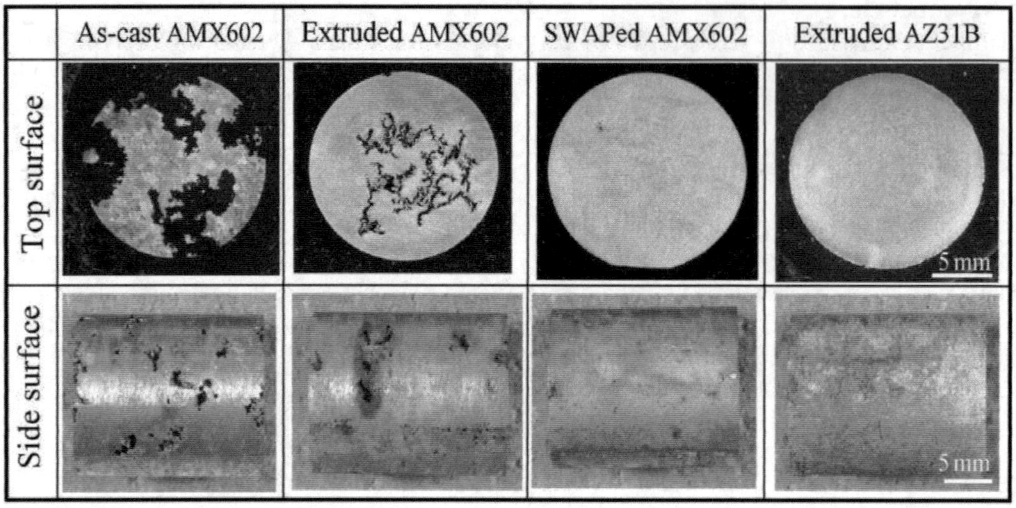

Figure 2. Micrographs of Mg alloys after immersed in 0.1M of NaCl solution for two weeks [22].

Meanwhile, Annur et al. [21] used a space holder agent to create porous Mg-Zn-Ca which was also used by [23–27]. By adding Mg-Zn Ca mixture powder with carbamine particles, the porosity of the Mg alloy was increased even though the mechanical properties were decreased. Figure 1 shows the scanning electron microscopy (SEM) images of the as-sintered alloys. Aghion et al. [24] also found the same effect where the addition of space holder resulted in the increase of porosity, but they also reported that a high porosity resulted decreasing in corrosion resistance when exposed to human body fluid. Liao et al. [22] observed the pitting occurs on the as-cast and extruded AMX602 alloys, whereas for the SWAPed AMX602 and AZ31B alloys, the corrosion is not obvious. Figure 2 shows the micrographs indicating pitting and filiform corrosion occurs in the samples immersed in 0.1M of NaCl solution for two weeks.

Čapek et al. [28] performed a sintering process under two types of atmosphere with 0.1L/min flowrate which are argon gas with 99.996% purity and 55 mm thick of Mg chips purified argon which acted as a getter. The results suggested that sintering under getted argon atmosphere improve the Mg particles diffusion, hence slightly increase the porosity. Rashad et al. [29] added Al-GNP into Mg by mixing Al and Mg using ethanol solution using mechanical agitator. The graphene nanoplatelets (GNP) were ultrasonicated for 60 minutes in ethanol and then being dropped into the mixture of Al and Mg, which later the mixing process will be continued for 60 minutes to obtain homogeneity. The addition resulted in improvements in tensile, strain and yield strength of Mg.

3. Corrosion behaviour of PM Mg and the applied techniques

Some researchers use variation of techniques to characterise the corrosion behaviour of Mg and Mg alloys specifically produced by the PM method. The techniques including the electrochemical characterization, potentiostatic and potentiodynamic polarization, electrochemical impedance spectroscopy (EIS), immersion test, hydrogen evolution test and scanning vibrating electrode technique (SVET). Some of the researchers in the biomedical field use electrochemical tests where the test was carried out in simulated body fluid including Hank's Balanced Salt Solution (HBSS), enriched Hank's Balanced Salt Solution (HBSS) and also in NaCl solution highlighted by [2,30–32].

The EIS provides information about the characteristics of material degradation when going through a corrosion medium. EIS can also be evaluated from other solution such as simulated body fluid (SBF) which was highlighted by [2]. From the data obtained from Nyquist plots, equivalent circuit (EC) is used to determine the characteristic of the material corrosion as shown in Figure 3. Figure 4 shows the EIS plots of the electrochemical behaviour of PM Mg under the different conditions highlighted by [16].

Brezina et al. [16] proved that for hot pressed Mg materials prepared under 500 MPa in a corrosion medium, the highest corrosion resistance was achieved. This was proven by higher polarization resistance after longer corrosion experiment duration compared to other material conditions. More pronounced corrosion attack of samples in HBSS compared to enriched HBSS is shown from experiments. Zhang et al. [33] discussed that when Mg alloy was put into HBSS, the Cl- presence will increase the reaction rate when reacting to Mg(OH)2 and creating MgCl2 which were highlighted by [34]. From their findings, they suggested that the increase of Zn content resulted in the polarization curve becoming longer which also meant that Zn is beneficial as a protective film for Mg. Yan et al. [1] also used Ringer's solution to get the pH values of Mg-Zn against time. The samples were immersed for 24 hours and the pH increased slowly until the peak of stabilized pH values. The increment of pH values means extra OH- is generated and slower increasing of pH values means a lower rate of corrosion as highlighted by [35].

Zheng et al. [36] plotted three Nyquist plots of Mg/Ca composites in DMEM for different immersion times. In the curves obtained, the capacitance loop at high frequency and the other at low frequency are set as constants. For Mg/10Ca composite sample, within 12 hours of immersion, Rf and Rt increased and slowly decreased after being immersed for 72 hours. It shows that the protection effect of the corrosion product layer is enhanced at the early stage of immersion and becomes worse when duration of immersion is increased which was agreed by [37].

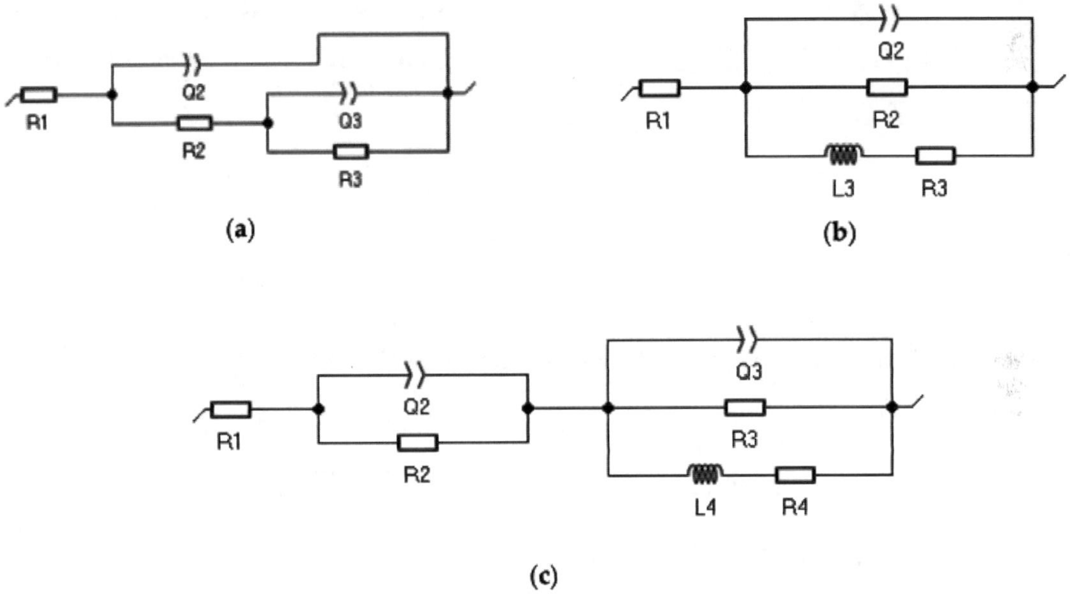

Figure 3. EC used for Nyquist plots evaluation [16].

Liao et al. [22] meanwhile use NaCl solution as a medium to immerse three AMX602 alloys and AZ31B alloy for the electrochemical test. The samples were immersed in 500 ml NaCl solution with ratio of the area of the samples and the volume of NaCl 1:35. After the immersing process, the corrosion products were cleaned away using a solution containing a boiling 10% CrO_3 chromic acid, and later rinsed with water. Then, the samples are cleaned ultrasonically using acetone, dried and then weighed to measure the weight loss. Figure 5 shows the plot of weight loss (mg/cm^2) of AMX602 and AZ31B Mg alloys due to corrosion.

Hydrogen evolution (HE) accompanied Mg dissolution in aqueous solutions. HE controls oxygen reduction at all values of pH where the potentials at Mg dissolves are low. In aqueous solution, Mg and all its alloys exhibited open circuit potential values less than about -1.45 V_{SCE} [38]. Thomas et al. [39] defined that the negative different effect (NDE) is a phenomenon of cathodic hydrogen evolution at anodically polarised Mg which is categorised by hydrogen evolution reaction (HER) rate increasing with anodic polarisation which is mentioned by [36],[40]. Zheng et al [36] found in cathodic polarization current that hydrogen evolution of Mg/1Ca composites is much lower than Mg/5Ca and Mg/10Ca samples. The data was obtained from potentiodynamic polarization curve which is also highlighted by [1,2,22,32,36,40,41]. Figure 6 shows the potentiodynamic polarization curve of Mg/1Ca, Mg/5Ca and Mg/10Ca composite samples in Dulbecco's Modified Eagle Medium (DMEM).

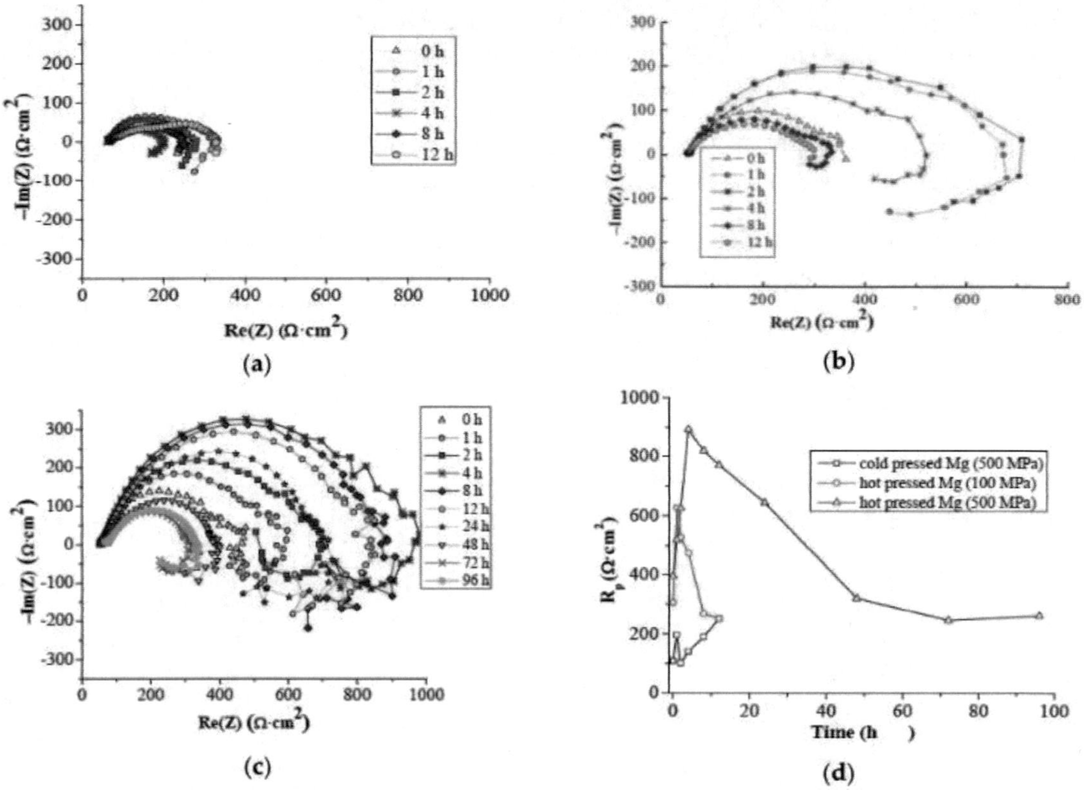

Figure 4. PM Mg EIS behaviour under different condition of pressing temperature and power [16].

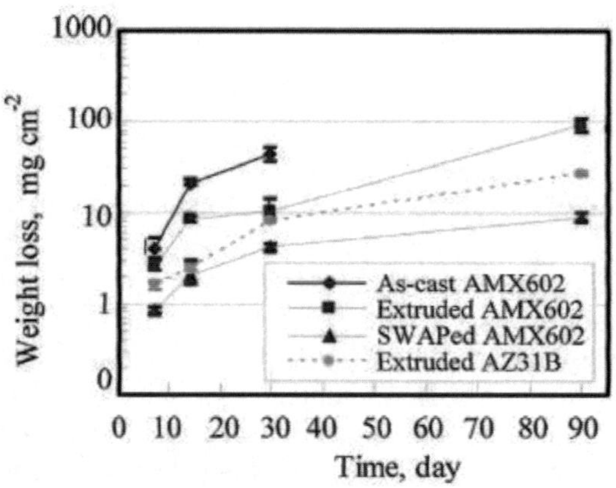

Figure 5. Plot of weight loss against time for three AMX602 and AZ31B alloys during immersion test in NaCl solution [22].

Potentiodynamic polarization curve is also an electrochemical test which is normally done for Mg in SBF. Kubásek et al. [2] performed the potentiodynamic and EIS test under SBF at 37°C which resulted that WE43 PM alloy is characterised by slightly higher corrosion potential than the WE43

IM. Pereda et al. [32] immersed the Mg (PM) in PBS solution for 1hour to 168 hours. The PBS consists of 8gl^{-1} NaCl, 0.2gl^{-1} KCl, 0.2 g l^{-1} KH_2PO_4, 1.15 gl^{-1} Na_2HPO_4. Akinwekomi et al. [40] revealed that materials that exhibit more positive corrosion potential (Ecorr) and lower corrosion current density (icorr) show that there were enhanced corrosion resistance compared with materials that exhibit opposite behaviours as shown on Tafel curves on Figure 7.

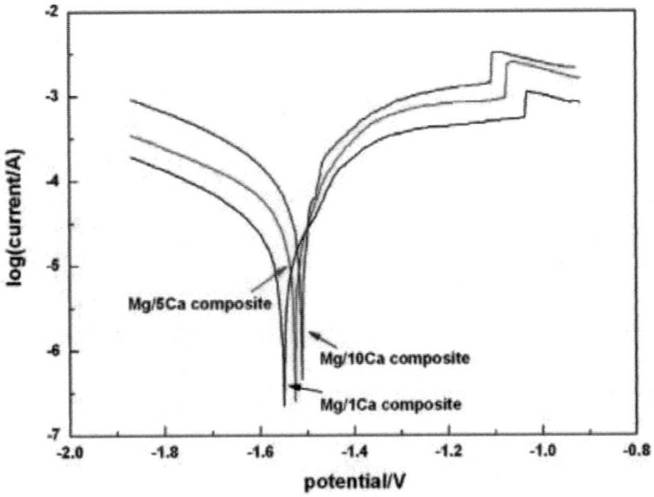

Figure 6. Potentiodynamic polarization curve of Mg/1Ca, Mg/5Ca and Mg/10Ca composite samples in DMEM [36].

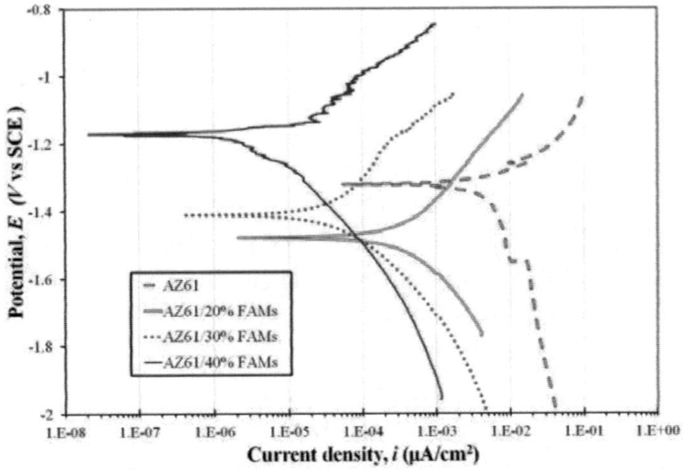

Figure 7. Potentiodynamic polarization curves of AZ61 and AZ61/FAMs syntactic foams [40].

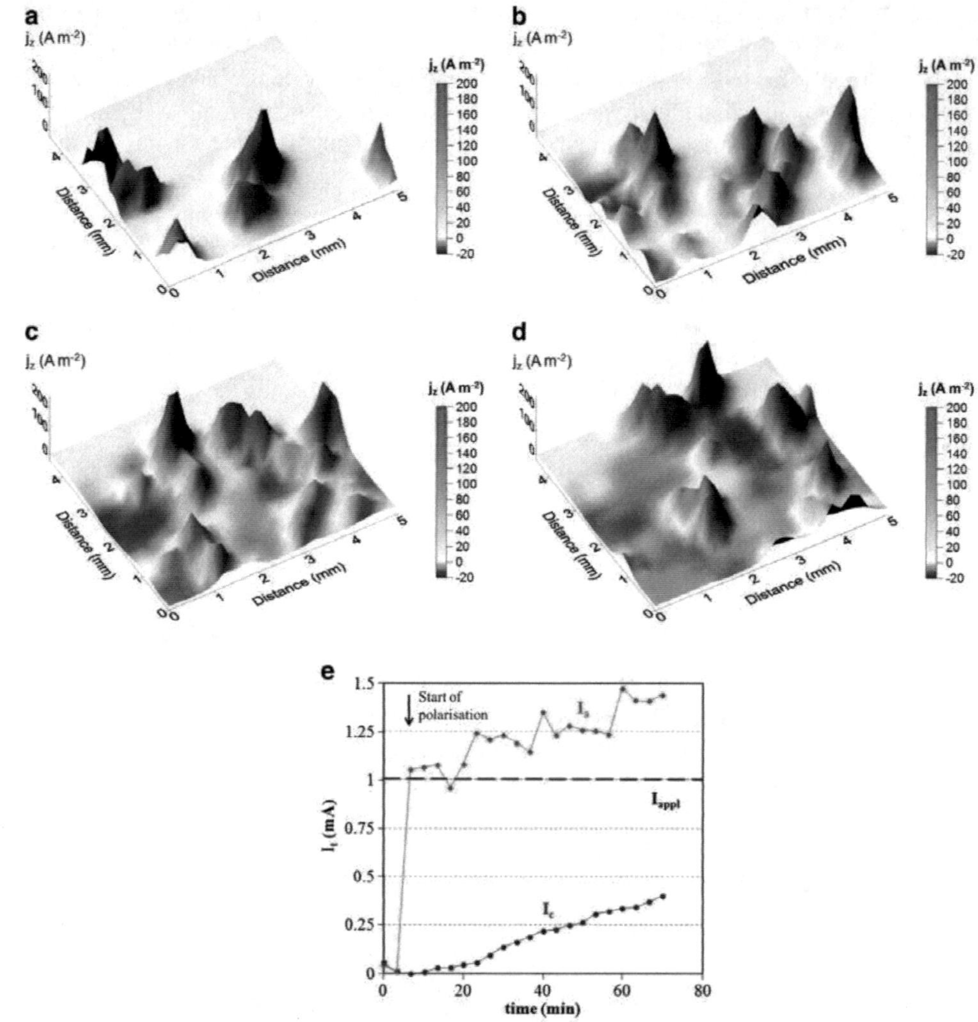

Figure 8. SVET maps on Mg surface immersed in NaCl (aq) electrolyte, a) 7 mins, b) 28 mins, c) 48 mins and d) 68 mins after starting galvanostatic polarisation at +1 mA cm^{-2} and e) SVETderived integrated current (I_t), normalised for a 1 cm^2 area, galvanostatically [42].

Scanning vibrating electrode technique (SVET) is used to characterise the anodes and cathodes distributions on pure Mg during anodic polarisation in NaCl solution. Williams et al. [42] use electrical contact which was made to high purity Mg rod of 10 mm diameter and 8 mm length. The specimen was then cold mounted in two-part epoxy resin so that one of the circular faces of the cylinder was exposed. Based on Figure 8, it was observed that some Mg areas become anodes, which later changes to cathode hence sustaining the HER which is agreed by [43]. Table 1 shows the list of researches of PM Mg and Mg alloys with the corrosion characterization technique.

Table 1. Lists of the corrosion characterization technique applied for PM Mg and Mg alloys.

Author	Year	Alloys	PM routes	Corrosion characterization methods

Balog et al. [44]	2019	Ti-Mg	Compacted by CIP at 200MPa-extruded at 400-500°C	• Immersion test
Akinwekomi et al. [40]	2019	AZ61/fly ash	Sintered for 20 minutes at 550°C	• Electrochemical characterization • Microstructural characterization
Kubásek et al. [2]	2017	Mg–4Y–3RE–Zr	Compacted under HVP at 425MPa and 350°C for 1 hour-extruded at 400°C and 0.2mm/s rate	• Immersion test • Potentiostatic and potentiodynamics test • Open circuit potential measurements
Březina et al. [16]	2017	Pure Mg	Compaction under nitrogen atmosphere-compacted at pressure 100MPa, 200MPa, 300MPa, 400MPa and 500MPa	• Microstructure observation • Potentiostatic • Electrochemical impedance spectroscopy (EIS)
Yan et al. [1]	2016	Mg-Zn	Mix for 8 hours-cold pressed-sintered at 500550°C for 2 hours in vacuum sinter furnace under argon gas	• Immersion test • Electrochemical test
Razak et al. [45]	2013	Al-Mg	Compact using manual hydraulic press at 210 MPa- sintered at 610°C argon gas for 6 hours	• Electrochemical characterization
Liao et al. [22]	2012	Mg-AlMn-Ca	Atomized by spinning water atomization process-compacted at 400MPa-hot extruded at 573-623K	• Immersion test • Microstructural characterization • Electrochemical characterization
Zheng et al. [36]	2010	Mg-Ca	Cold pressed at 400MPahot pressed at 350MPa at 320°C-hot extruded at 300°C	• Immersion test • Electrochemical test • Potentiodynamic test
Pereda et al. [32]	2010	Pure Mg	Cold-pressed 310MPa extruded at 420°C-	• OCP measurements • Electrochemical characterization and EIS • Potentiodynamic and potentiostatic • Surface characterization of corroded Mg (PM)

4. Conclusion

- The review offered here recognize the application of corrosion test to characterize the corrosion behavior of Mg that is specifically produced by PM technique. The use of PM technique provides the route to create a high strength and better properties of Mg, while at the same time pertaining cost effective and less complex production of Mg alloys and composites.
- Accurate results of corrosion behaviour can help to verify the performance of Mg alloys in a corrosive medium, hence producing Mg with the best properties. The most common and reliable strategy to assess the corrosion behaviour is performing electrochemical test such as EIS and potentiodynamic polarization that can tailor the surface properties of Mg to the microstructure developed during the PM route. The use of SVET test also effectively provides important information on the anodic and cathodic reactions, that is best approach to understand the corrosion mechanism of Mg.

Acknowledgement
The authors wish to acknowledge the financial support of Universiti Malaysia Pahang (RDU180333) and Ministry of Higher Education (FRGS/1/2019/TK05/UMP/02/5) research grants.

References

[1] Yan Y, Cao H, Kang Y, Yu K, Xiao T, Luo J, Deng Y, Fang H, Xiong H and Dai Y 2016 Effects of Zn concentration and heat treatment on the microstructure, mechanical properties and corrosion behavior of as-extruded Mg-Zn alloys produced by powder metallurgy *Journal of Alloys and Compounds* **693** 1277-1289

[2] Kubásek J, Dvorský D, Čavojský M, Vojtěch D, Beronská N and Fousová M 2017 Superior Properties of Mg–4Y–3RE–Zr Alloy Prepared by Powder Metallurgy *Journal of Materials Science and Technology* **33** 652–660

[3] Zhou T, Yang M, Zhou Z, Hu J and Chen Z 2013 Microstructure and mechanical properties of rapidly solidified / powder metallurgy Mg – 6Zn and Mg – 6Zn – 5Ca at room and elevated temperatures *Journal of Alloys and Compounds* **560** 161–166

[4] Ma H, Wu L, Liu C, Liu M, Wang C, Li D, Chen X, Dong J and Ke W 2019 First-principles modeling of the hydrogen evolution reaction and its application in electrochemical corrosion of Mg *Acta Materialia* **183** 377-389

[5] Wang X, Hu L, Liu K and Zhang Y 2012 Grain growth kinetics of bulk AZ31 magnesium alloy by hot pressing *Journal of Alloys and Compounds* **527** 193–196

[6] Kayhan S M, Tahmasebifar A, Koç M, Usta Y, Tezcaner A and Evis Z 2016 Experimental and numerical investigations for mechanical and microstructural characterization of micromanufactured AZ91D magnesium alloy disks for biomedical applications *Materials and Design* **93** 397–408

[7] Meng F, Rosalie J M, Singh A and Tsuchiya K 2015 Precipitation behavior of an ultra-fine grained Mg-Zn alloy processed by high-pressure torsion *Materials Science and Engineering A* **644** 386–391

[8] Sumita M, Hanawa T and Ohnishi I 2003 Failure Processes in Biometallic Materials *Comprehensive Structural Integrity* 131–167

[9] Bornapour M, Celikin M, Cerruti M and Pekguleryuz M 2014 Magnesium implant alloy with low levels of strontium and calcium: The third element effect and phase selection improve biocorrosion resistance and mechanical performance *Materials Science and Engineering C* **35** 267– 282

[10] Zhang S, Zheng Y, Zhang L, Bi Y, Li J, Liu J, Yu Y, Guo H and Li Y 2016 In vitro and in vivo corrosion and histocompatibility of pure Mg and a Mg-6Zn alloy as urinary implants in rat model *Materials Science and Engineering C* **68** 414–422

[11] Khalajabadi S Z, Abdul Kadir M R, Izman S and Marvibaigi M 2016 The effect of MgO on the biodegradation, physical properties and biocompatibility of a Mg/HA/MgO nanocomposite manufactured by powder metallurgy method *Journal of Alloys and Compounds* **655** 266–280
[12] Bornapour M, Muja N, Shum-Tim D, Cerruti M and Pekguleryuz M 2013 Biocompatibility and biodegradability of Mg-Sr alloys: The formation of Sr-substituted hydroxyapatite *Acta Biomaterialia* **9** 5319–5330
[13] Yazdimamaghani M, Razavi M, Vashaee D, Moharamzadeh K, Boccaccini A R and Tayebi L 2017 Porous magnesium-based scaffolds for tissue engineering *Materials Science and Engineering C* **71** 1253–1266
[14] Singh Raman R K, Jafari S and Harandi S E 2015 Corrosion fatigue fracture of magnesium alloys in bioimplant applications: A review *Engineering Fracture Mechanics* **137** 97–108
[15] Agrawal D 2013 Microwave sintering of metal powders *Advances in Powder Metallurgy: Properties, Processing and Applications* 361–379
[16] Březina M, Minda J, Doležal P, Krystýnová M, Fintová S, Zapletal J, Wasserbauer J and Ptáček P 2017 Characterization of powder metallurgy processed pure magnesium materials for biomedical applications *Metals* **7** 1–22
[17] Zhao L, Ma G, Jin P and Yu Z 2019 Role of Y on the microstructure, texture and mechanical properties of Mg–Zn–Zr alloys by powder metallurgy *Journal of Alloys and Compounds* **810** 151843
[18] Du B N, Xiao Z P, Qiao Y X, Zheng L, Yu B Y, Xu D K and Sheng L Y 2019 Optimization of microstructure and mechanical property of a Mg-Zn-Y-Nd alloy by extrusion process *Journal of Alloys and Compounds* **775** 990–1001
[19] Peng Q, Sun Y, Ge B, Fu H, Zu Q, Tang X and Huang J 2019 Interactive contraction nanotwinsstacking faults strengthening mechanism of Mg alloys *Acta Materialia* **169** 36–44
[20] Shi X, Luo A A, Sutton S C, Zeng L, Wang S, Zeng X, Li D and Ding W 2015 Twinning behavior and lattice rotation in a Mg-Gd-Y-Zr alloy under ballistic impact *Journal of Alloys and Compounds* **650** 622–632
[21] Annur D, Lestari F P, Erryani A and Kartika I 2018 Study of sintering on Mg-Zn-Ca alloy system *AIP Conference Proceedings 1964* **vol 020029** pp 020029-1-020029-5
[22] Liao J, Hotta M and Mori Y 2012 Improved corrosion resistance of a high-strength Mg-Al-MnCa magnesium alloy made by rapid solidification powder metallurgy *Materials Science and Engineering A* **544** 10–20
[23] Elsayed A, Umeda J and Kondoh K 2011 The texture and anisotropy of hot extruded magnesium alloys fabricated via rapid solidification powder metallurgy *Materials and Design* **32** 4590–4597
[24] Aghion E and Perez Y 2014 Effects of porosity on corrosion resistance of Mg alloy foam produced by powder metallurgy technology *Materials Characterization* **96** 78–83
[25] Annur D, Lestari F P, Erryani A, Sijabat F A, Astawa I N G P and Kartika I 2018 Preparation and characterization of porous Mg-Zn-Ca alloy by space holder technique *AIP Conference Proceedings 1945* vol 1945 pp 020015-1-020015-5
[26] Yang D, Hu Z, Chen W, Lu J, Chen J, Wang H, Wang L, Jiang J and Ma A 2016 Fabrication of Mg-Al alloy foam with close-cell structure by powder metallurgy approach and its mechanical properties *Journal of Manufacturing Processes* **22** 290–296
[27] Singh S and Bhatnagar N 2018 A novel approach to fabricate 3D open cellular structure of Mg10Zn alloy with controlled morphology *Materials Letters* **212** 62–64
[28] Čapek J and Vojtěch D 2014 Effect of sintering conditions on the microstructural and mechanical characteristics of porous magnesium materials prepared by powder metallurgy *Materials Science and Engineering C* **35** 21–28
[29] Rashad M, Pan F, Tang A, Asif M, Hussain S, Gou J and Mao J 2015 Improved strength and ductility of magnesium with addition of aluminum and graphene nanoplatelets (Al+GNPs) using semi powder metallurgy method *Journal of Industrial and Engineering Chemistry* **23** 243-250

[30] Saberi A, Bakhsheshi-Rad H R, Karamian E, Kasiri-Asgarani M and Ghomi H 2020 A study on the corrosion behavior and biological properties of polycaprolactone/ bredigite composite coating on biodegradable Mg-Zn-Ca-GNP nanocomposite *Progress in Organic Coatings* **147** 105822

[31] YANG Z, LI J, ZHANG J, LORIMER G and ROBSON J 2008 Review on Research and Development of Magnesium Alloys *Acta Metallurgica Sinica (English Letters)* **21** 313–328

[32] Pereda M D, Alonso C, Burgos-Asperilla L, Del Valle J A, Ruano O A, Perez P and Fernández Lorenzo De Mele M A 2010 Corrosion inhibition of powder metallurgy Mg by fluoride treatments *Acta Biomaterialia* **6** 1772–1782

[33] Zhang E, Yin D, Xu L, Yang L and Yang K 2009 Microstructure, mechanical and corrosion properties and biocompatibility of Mg-Zn-Mn alloys for biomedical application *Materials Science and Engineering C* **29** 987–993

[34] Song G L 2011 Corrosion of magnesium alloys *Corrosion of Magnesium Alloys* 1–640

[35] Bakhsheshi-Rad H R, Abdul-Kadir M R, Idris M H and Farahany S 2012 Relationship between the corrosion behavior and the thermal characteristics and microstructure of Mg-0.5Ca-xZn alloys *Corrosion Science* **64** 184–197

[36] Zheng Y F, Gu X N, Xi Y L and Chai D L 2010 In vitro degradation and cytotoxicity of Mg/Ca composites produced by powder metallurgy *Acta Biomaterialia* **6** 1783–1791

[37] Song Y, Shan D, Chen R, Zhang F and Han E H 2009 Biodegradable behaviors of AZ31 magnesium alloy in simulated body fluid *Materials Science and Engineering C* **29** 1039–1045

[38] Frankel G S, Samaniego A and Birbilis N 2013 Evolution of hydrogen at dissolving magnesium surfaces *Corrosion Science* **70** 104–111

[39] Thomas S, Medhekar N V., Frankel G S and Birbilis N 2015 Corrosion mechanism and hydrogen evolution on Mg *Current Opinion in Solid State and Materials Science* **19** 85–94

[40] Akinwekomi A D 2019 Microstructural characterisation and corrosion behaviour of microwavesintered magnesium alloy AZ61/fly ash microspheres syntactic foams *Heliyon* **5** 1-14

[41] Zhang E, He W, Du H and Yang K 2008 Microstructure, mechanical properties and corrosion properties of Mg-Zn-Y alloys with low Zn content *Materials Science and Engineering A* **488** 102– 111

[42] Williams G, Birbilis N and Mcmurray H N 2013 The source of hydrogen evolved from a magnesium anode *Electrochemistry Communications* **36** 1–5

[43] Esmaily M, Svensson J E, Fajardo S, Birbilis N, Frankel G S, Virtanen S, Arrabal R, Thomas S and Johansson L G 2017 Fundamentals and advances in magnesium alloy corrosion *Progress in Materials Science* **89** 92–193

[44] Balog M, Ibrahim A M H, Krizik P, Bajana O, Klimova A, Catic A and Schauperl Z 2019 Bioactive Ti + Mg composites fabricated by powder metallurgy: The relation between the microstructure and mechanical properties *Journal of the Mechanical Behavior of Biomedical Materials* **90** 45–53

[45] Razak K A and Derman M N 2013 Corrosion behaviour of anodised powder metallurgy aluminium-magnesium composites *Advanced Materials Research* **795** 469–473

Investigation on Cutting Feasibility on Ni-CBN on HSS Substrates for Aerospace Material

S. N. Azinee[1*], N. Norsilawati [1*], N. Bahiyah Baba [1], C. I. M. Fathil [1], Mohd. H. Ibrahim[1], R. A. N. Izzat[1]
[1]Faculty of Engineering Technology, University College TATI, 24000 Kemaman Terengganu, Malaysia.

[*]Email: nor_azinee@uctati.edu.my, norsilawati@uctati.edu.my.

Abstract. The study involves coating the high-speed steel (HSS) tool with cubic Boron via electroless nickel co-deposition. The surface coating of Ni-CBN was deposited via electroless nickel where the process of involved chemical reactions as the ceramic CBN powders was embedded within the layer. By introducing the electroless nickel process, the coating cost and procedure could be simplified. The electroless deposition has been extensively practised in industry, and the most prevalent coating offers excellent corrosion, wear and abrasion resistance, ductility, lubricity, electrical properties, and hardness. Thus, the introduction of the new electroless nickel (Ni-CBN) tools can be a contribution to the industry. This paper discussed on comparing the performance of Nickel-cubic boron nitride (Ni-CBN) HSS tool substrates and with uncoated HSS tool for its surface tolerance in milling of Aluminium Alloy 7075. The process variables selected during experimentation based on Taguchi L9 are cutting speed, feed rate, and depth of cut with three different levels for each parameter. Result shows, the lowest flank wear for coated measured is 51.34 μm while the highest is 380.26 μm. The highest surface roughness for uncoated is 1.154 μm and lowest 0.42μm. Meanwhile, for coated substrates, the highest Ra is 0.787μm and lowest 0.251μm. At the end of the analysis, the Ni-CBN HSS cutting tools are generally able to slightly reduce the tool wear with the longest tool lifetime 195 minutes compared uncoated 143 minutes. Therefore, it is concluded that the Ni-CBN HSS tool end mill are generally able to reduce the tool wear.

1. Introduction

Nowadays, the productions of coating technology are demanding because of the need for higher productivity rates for industry consumption. Many researchers are developing new technology based on new materials, tools materials, and the evolution method of coatings characterization to maintain the industrial and scientific interest in this field. It shows the growing market of cutting tools has been developed [1]. The application of coated tools has become more critical in the machining process and mostly been made by thermal spraying processes such as Physical Vapor Deposition (PVD) and Chemical Vapor Deposition (CVD). The significant problems of these processes are that they are costly due to high temperature and complex procedures. By introducing the electroless nickel process, the coating cost and method could be simplified [2].

The electroless deposition has been extensively practised in industry, and the most prevalent coating offers excellent corrosion, wear and abrasion resistance, ductility, lubricity, electrical properties, and hardness [3]. Thus, the introduction of the new electroless nickel (Ni-CBN) tools can be used as an alternative and be a contribution to the industry.

Cubic boron nitride (CBN) has already established itself in many areas of machining. CBN being the second hardest material known to man, after diamond, is produced artificially [4]. The CBN tools have excellent mechanical and thermal properties, such as strength at high temperature, abrasion resistance and hardness second only to that diamond. Multilayer CBN coatings represent a new deposition method that can improve adhesion on metal substrates. This multilayer CBN structure showed outstanding adhesion in atmospheric conditions, even with high residual stress. Compared to monolayer CBN coatings, the multilayer CBN films had lower elastic moduli, and their critical loads were twice as high [5]. Many kinds of research had been shown on CBN tools performance in the past few years [6, 7]. The application of CBN as cutting material is a promising approach and could have a significant impact on productivity. Sharad N. Yevad [8] has proven that a composite CBN-TiN coating can be applied to a machine tool. The composite shows excellent CBN-to-TiN and composite coating-to-carbide substrate adhesion. This statement can be supported by [9], which stated that the characterization results show a relatively uniform coating with continuous CBN particle dispersion in an adherent TiN matrix on complex geometry cutting inserts.

In [10] concluded the HSS cutting tool was successfully coated with Ni/YSZ composite via electroless nickel co-deposition process. The average thickness of the coating is 59 μm with a uniform and smooth surface where TiN coated shows the most consistent performance in surface roughness results. In another experiment, it proved that the machining surface roughness is affected by cutting speed in milling Al/SiC metal matrix composites. The best surface roughness is obtained with increasing cutting speed from 300 mm/min to 450 mm/min. The surface roughness of the work materials obtained is gradually decreased from 1.3 μm to 0.6 μm [11].

Milling is the most used machining process for metal cutting purpose. The milling process usually involves a few different operations, but it depends on the shape of the final product and the state of the raw piece. Milling is required for giving a precise finish and adding a few features like slots or threaded holes. The cutting tool quality can determine the performance of the cutting proses. Uhlmann [4] state that the machining of high-performance workpiece materials requires significantly harder cutting materials. In hard machining, the first tool wear occurs due to high process forces and temperatures. There are many cutting tools available in the industry, and it can be generally divided into coated and uncoated. The coated cutting tools are usually performed better than the uncoated ones. Standard commercialized coated cutting tool are AlN, TiN, TiAlN and other [10].

Therefore, in this study, a new approach of Ni-CBN is investigated as this ceramic-metal surface coating is very well known for its high thermal, wear and corrosion resistance. It is also known as an alternative material for the cutting tool coating. Besides, the layer will be processed via electroless nickel co-deposition, which is simple, without consuming electrical power and less costly compared to conventional thermal spraying processes [12].

2. Methodology

The methodology is divided into three parts: Electroless Coating, Machining and Cutting Feasibility.

2. 1 Electroless Coating

The high-speed steel (HSS) cutting tool with standard dimension Ø6 mm was used as a substrate of CBN. The end mill requires sensitising to activate the HSS cutting tool surface by the four pre-treatment processes consecutively. Characters of the substrate sample were treated by chemical etching and mechanical blasting. Reinforcement ceramic particles of Cubic Boron Nitride (CBN) powder will use. CBN powders have exceptional thermal conductivity and provide improved surface integrity in the grinding of hardened alloy steels, tool steels, nickel, and cobalt-based superalloys. Figure 1 below shows End Mill Cutting Tool.

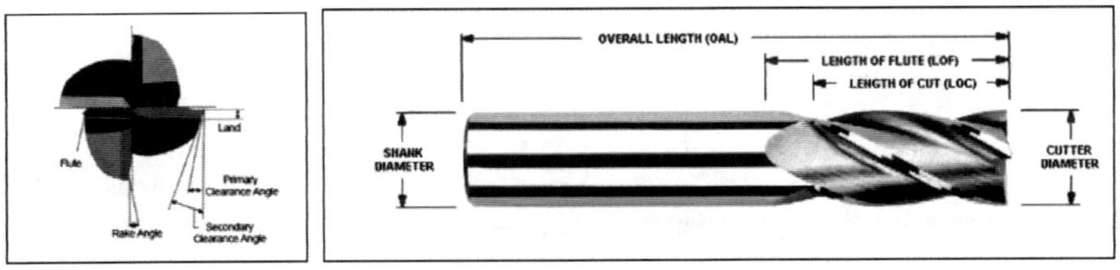

Figure 1. End Mill Cutting Tool

A CBN powder of 50g/l was added to the electroless nickel solution along with the cutting tool. The coating proses were taken in a beaker. The coating time was kept constant at 60 minutes. The bath temperature was kept constant at 89 ± 20 0C. The powder is held in suspension in the nickel solution by either mechanical stirring or air bubbling agitation methods. Mechanical stirring was done by Jenway hotplate with a magnetic stirrer, and air bubbling was performed at 1.2W air pressure. The whole coating process sequence is simplified in Figure 2.

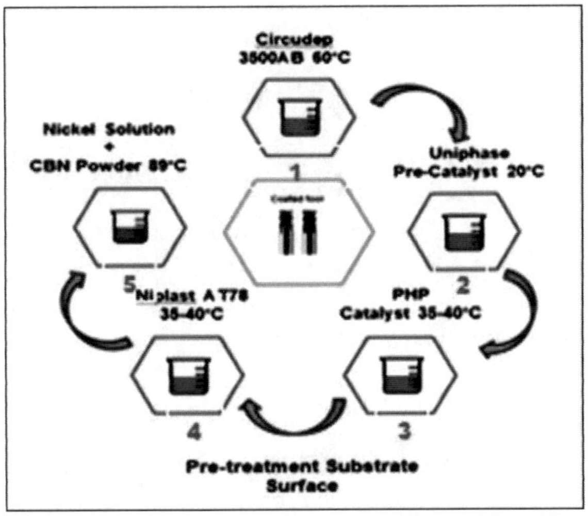

Figure 2. Elecroless Ni-CBN Coated Process

2.2 Cutting Feasibility

The software MINITAB 14 was used for analyze the influence and also the efficiency range of parameters on the magnitude of surface roughness for Aluminium Alloy 7075. The process variables selected during experimentation based on Taguchi L9 are cutting speed, feed rate, and depth of cut with three different levels for each parameter. The details of the machining cutting parameters are provided in Table 1.

Table 1. Machining Cutting Parameter and Levels

Machining Parameters	Units	Values		
		Low	Medium	High
Spindle Speed	(RPM)	1860	2650	3450
Feed Rate	(min/mm)	180	257	334
Depth of Cut	(mm)	1	2	3

The Orthogonal Array of experiment involved three nickel Co-Deposition process parameters shows in Table 2 below. A Taguchi Method of DOE approach with three parameters at three levels gave L9 for each substrate. This experiment uses cutting tool HSS Ni-CBN coated and HSS uncoated which contributed to 18 experiments. The depth of cut and feed rate were designated based on preferences specified by the end mill producers. The selected values of spindle speeds referred to the suggested change of speed to move the experiment into the "high cutting speed" type [13, 14].

Table 2. The Orthogonal Array of Machining Process Parameter and Level for Coated and Uncoated HSS Cutting Tool.

↓	C1	C2	C3
	Cutting Speed, v (rpm)	Feed Rate, F (mm/min)	Depth of Cut (mm)
1	1860	180	1
2	1860	257	2
3	1860	334	3
4	2650	180	2
5	2650	257	3
6	2650	334	1
7	3450	180	3
8	3450	257	1
9	3450	334	2

The cutting feasibility will be tested on aerospace material Aluminium Alloy 7075. The milling pocket dimension for machining is 40 mm x 35 mm. This test method describes a laboratory procedure for determining the flank wear, and surface roughness for nanocoated and uncoated tools using Zeiss Stemi 20000-C Microscope Profile Optical Video Measuring System for tool wear and Surface Tester SJ-210 for surface roughness.

2.3 Machining
After the coating was done for HSS end mills, the diameter of CBN composite coated cutting tool was measured for the thickness using digital micrometer. The machine that used for machining is Deckel Maho DMU 50 CNC Machine. The characteristic of uniform thickness of electroless coating enabled the measurement. The results are obtained by taking three readings for each tool bit to get average thickness. Then, the samples of the HSS substrates are taken for analysis to view the presence of the coating elements and also to observe the coating surface under JSM-7800F Field Emission Scanning Electron Microscope (FESEM).

Later, the cutting tools were tested for its machining performances. The machining profile was done using 18 pockets which using two cutting tools. Nine pockets will be run using cutting tool with coated HSS end mill, and nine pockets will be runs for uncoated HSS end mill. The size of each pocket is 40 mm x 35 mm on material Aluminium Alloy 7075, shown in Figure 3. The interval for each cut is 0.2 mm, while the reading of surface roughness is taken every 0.2 mm.

Figure 3. Machining Profile on Aluminium Alloy 7075 Material

The feed rate, spindle seed and depth of cut was kept constant throughout the end milling process. The machined profile was used as the basis of the surface roughness measurement. The machined surface will be tested using Surftest SJ-210- Series 178-Portable Surface Roughness Tester for surface roughness shown in Figure 4. At the same time, tool wear tool wear will be tested using a Zeiss Stemi 20000-C Microscope Profile Optical Video Measuring System pictured in Figure 5.

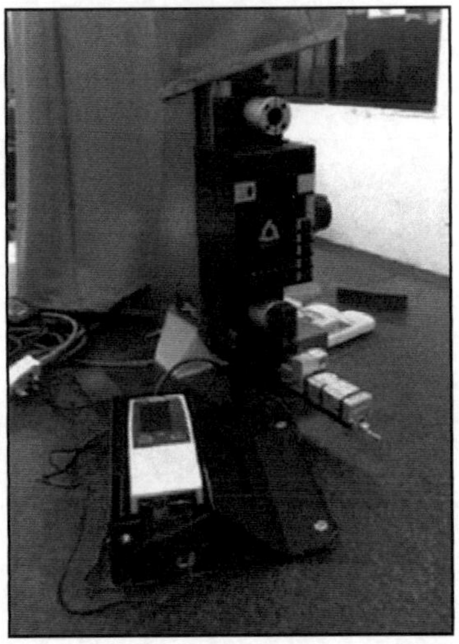

Figure 4. Surftest SJ-210- Series 178-Portable Surface Roughness Tester

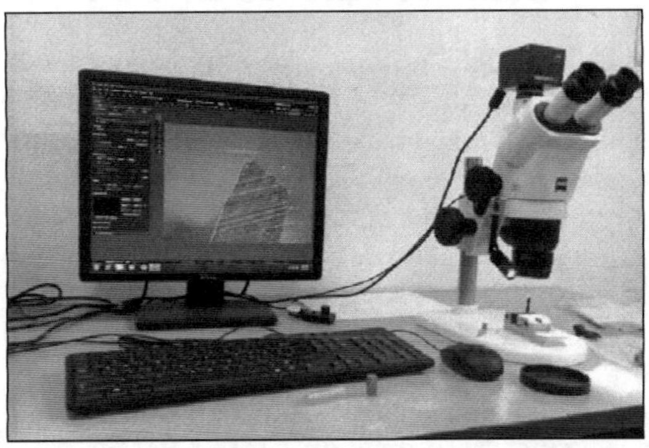

Figure 5. Zeiss Stemi 20000-C Microscope Profile Optical Video Measuring System

3. Results and Discussion

The result will discuss on the surface roughness analysis, tool wear analysis and characterisation for Ni-CBN coated cutting tool.

The HSS end mill cutting tool was successfully coated with CBN composite via electroless coating. The images of HSS cutting tool coated and uncoated were shown in Figure 6 below.

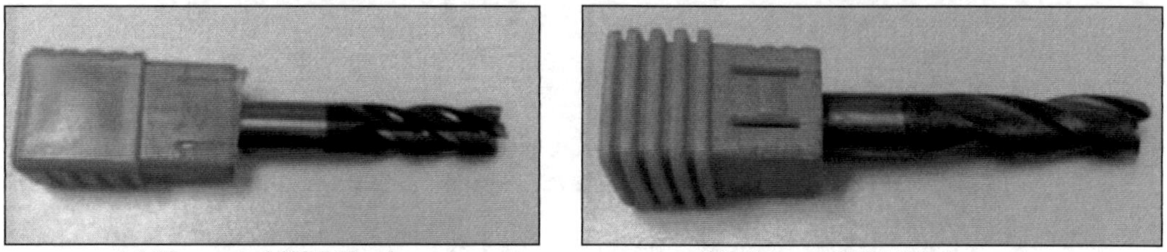

(a) Uncoated HSS Tool (b) Coated HSS Tool

Figure 6. Cutting Tool Image (a) Uncoated HSS Tool (b) Coated HSS Tool

3.1 Surface Roughness Analysis

The cutting tool quality is the main factor to contribute for better surface roughness analysis. The result Ra of machined Alluminium Alloy 7075 for coated and uncoated cutting tool were recorded and compared in Table 3.

Table 3. Data of Design Experiment and Surface Roughness

No of Test	Cutting Speed,V (RPM)	Feed Rate, F (mm/min)	Depth of Cut (mm)	Surface Roughness, Ra	
				Coated (μm)	Uncoated (μm)
1	1860	180	1	0.576	0.695
2	1860	257	2	0.787	1.154
3	1860	334	3	0.890	1.220
4	2560	180	2	0.481	0.534
5	2560	257	3	0.301	0.586
6	2560	334	1	0.412	0.619
7	3450	180	3	0.296	0.485
8	3450	257	1	0.251	0.421
9	3450	334	2	0.527	0.729

Table 3 shows the combination high cutting speed and medium feed rate at Test no. 8 for coated tools where it produce good surface finish Ra 0.251μm. Meanwhile, the combination low cutting speed and high feed rate at Test no 3 for uncoated will resulted bad surface finish with high surface roughness Ra 1.22 μm. It shows that the interaction between high-value feed rate and high-value spindle produce good surface finish [15, 16]. Mohammed [17] also states that the interaction between the cutting speed and feed rate are significantly affecting the surface roughness.

3.2 Tool Wear Analysis

By using Zeiss Stemi 20000-C Microscope Profile Optical, Video Measuring System Tool wear will examine every 0.2 mm of machining. According to ISO 8688-21:1989, as stated previously, it is noted that the lowest tool wear is the best and most durable for end mill cutting tool. Figure 7 shows the tool wear on the cutting tool before and after the machining process.

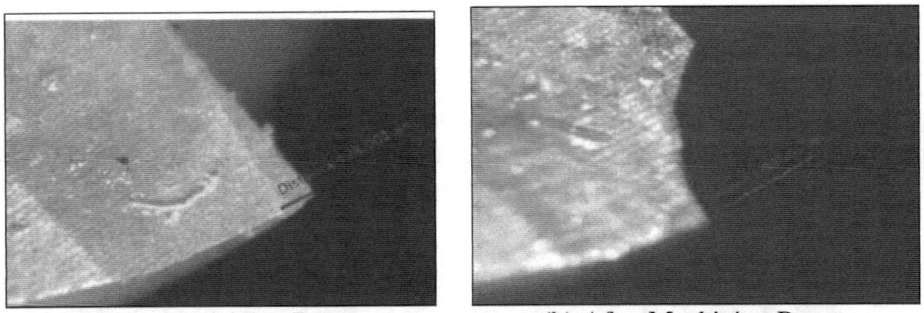

(a) Before Machining Process (b) After Machining Process

Figure 7. Tool Wear on HSS Cutting Tool for Test no 5(a) Before (b) After Machining Process

The flank wear trend in Figure 8 shows that the substrates performed better for coated end mill based on cutting time. Test no 3 and 5 produce longest tool life with 195 minute. Figure 9 shows the flank wear trend for uncoated end mill based on cutting time. The substrates performed scattered than coated end mill. Test no 3 produce longest tool life with 143 minute for uncoated tools.

The highest combination of parameters and time will produce high cutting time. It will cause the occurrence of tool wear. The interaction between the cutting speed and feed rate is significantly affected tool wear [15, 18, 19].

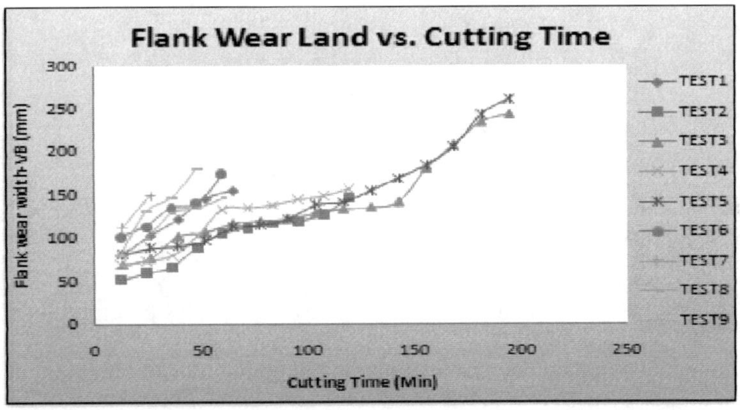

Figure 8. Flank Wear Land vs. Cutting Time for Coated HSS

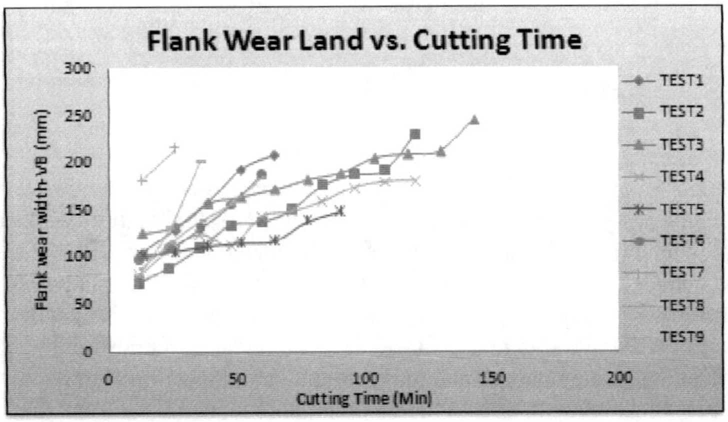

Figure 9. Flank Wear Land vs. Cutting Time for Uncoated HSS

3.3 Characterisation

The samples of the HSS substrates are taken for analysis to view the presence of the coating elements and also to observe the coating surface under JSM-7800F Field Emission Scanning Electron Microscope (FESEM). Figure 10 obtained by morphological image mapping, the scanning image is then processed and matches with the elements. The bubbles in the chart represent the elements present on the surface area, while Table 4 shows the composition of elements.

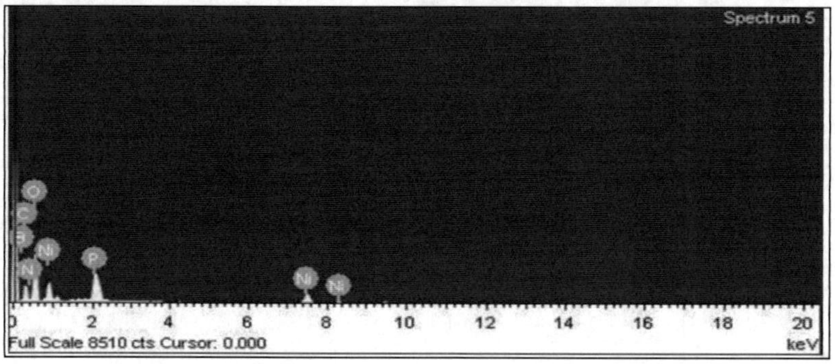

Figure 10. EDX Spectrum Chart with Elements Composition

Table 4. Composition of Elements

Element	Weight %
B	20.69
C	16.71
N	9.16
O	15.01
P	4.20
Ni	34.22
Total	100.00

In this analysis, the goal is to prove that the element of the Ni-CBN is present. As observed in Table 4 and Figures 10, the elements are indeed present, so the electroless coating is fundamentally considered as a success. The finding from this FESEM results is the elements deposition that is present on the surface of the sample, the type of elements are also validated via the EDX spectrum mapping that produces the elements chart and their composition.

4. Conclusion

The electroless Ni-CBN coating has been successfully performed on HSS end mills with a diameter of 6 mm. As a result, the average thickness of the coating HSS cutting tool is 15 μm. Two types of cutting tools were selected and compared which are HSS uncoated and HSS coated. Two types of cutting tools were selected and compared, which are HSS uncoated and HSS coated. The performance of these two types shows the lowest flank wear for coated measured is 51.34 μm, while the highest is 380.26 μm. The highest surface roughness for uncoated is 1.154 μm and lowest 0.42μm. Meanwhile, for coated substrates, the highest Ra is 0.787μm and lowest 0.251μm. At the end of the analysis, the Ni-CBN HSS cutting tools are generally able to slightly reduce the tool wear with the longest tool lifetime 195 minutes compared uncoated 143 minutes. Therefore, it is concluded that the Ni-CBN HSS tool end mill are generally able to reduce the tool wear.

Acknowledgments
The authors gratefully acknowledge to the UC TATI due to the research is fully supported by Short Term Grant (STG) UC TATI 9001-1808 under Advanced Manufacturing Cluster.

References:

[1] Bouzakis K-D, Michailidis N, Skordaris G, Bouzakis E, Biermann D, M'Saoubi R (2012) Cuttingwith Coated Tools: Coating Technologies, Characterization Methods and Performance Optimization. *CIRP Annals* 61:703–723.

[2] Yu D, Wang C, Cheng X, Zhang F. Optimization of hybrid PVD process of TiAlN coatings by Taguchi method. *Applied Surface Science*, (2008), 255(5):1865-1869

[3] Mallory, G.O. and J.B. Hajdu. Electroless plating: *Fundamentals and Applications* (1990). William Andrew.

[4] Uhlmann E, Oyanedel Fuentes JA, Keunecke M (2009) Machining of High-Performance Workpiece Materials with CBN Coated Cutting Tools. *Thin Solid Films* 518(5):1451–1454.

[5] Sung-Tae Park, Jeon-Geon Han, M. Keunecke, Kwangmin Lee (2016). Mechanical and structural properties of multilayer c-BN coatings on cemented carbide cutting tools, Int. Journal of Refractory Metals and Hard Materials

[6] K. Bouacha, M.A. Yallese, T. Mabrouki, J.F. Rigal (2010). Statistical analysis of surface roughness and cutting forces using response surface methodology in hard turning of AISI 52100 bearing steel with CBN tool, *Int. J. Refract. Metals Hard Mater.* 28, 349–361.

[7] Y. Sahin, A.R. Motorcu (2008). Surface roughness model in machining hardened steel with cubic boron nitride cutting tools, *Int. J. Metals Hard Mater.* 26, 84–90.

[8] S.N. Yadave, A.P. Malshe, W.D. Brown, W.C. Russell, Novel composite CBN–TiN coating: synthesis and performance analysis, *Trans. NAMRI/SME* 30 (2002) 485–492.

[9] Wenping Jiang, Abhijeet S. More, W.D. Brown, Ajay P. Malshe, A cBN-TiN composite coating for carbide inserts: coating characterization and its applications for finish hard turning, *Surface and Coatings Technology* 201 (2006) 2443–2449.

[10] NB Baba, MF Omar, S Sharif, SB Mohamed. Processing and properties of Ni/YSZ composite coating on high speed steel cutting tool. *Journal of Fundamental and Applied Sciences*, (2018), 10 (2S), 688-700

[11] Turgut Y, Şahin I, Findik T. Study of cutting force and surface roughness in milling of Al/SiC metal matrix composites. Scientific Research and Essays, (2010), 6(10):2056-2062

[12] Baba N B, Davidson A, Muneer T. Investigation of Ni-YSZ composit manufactured by electroless Ni coating. Applied Mechanics and Materials, (2011), 52:1660-1664

[13] Rama Kotaiah K, Srinivas J, Babu KJ, Srinivas K (2010) Prediction of optimal cutting states during inward turning: an experimental approach. Mater Manuf Processes 25(6):432–441, 1532–2475

[14] Gandarias A, de Luis Norberto Lopez L, Aizpitarte X, Lamikiz A (2008) Study of the performance of the turning and drilling of austenitic stainless steels using two coolant techniques. Int J Mach Mach Mater 3:1–17

[15] J.A. Ghani, I.A. Choudhury, H.H. Hassan, Application of Taguchi method in the optimization of end milling, *J. Mater. Process. Tech.* 145 (1) (2004) 84–92.

[16] Syed Mohd Fadly Bin Syed Hassan , Suzana Binti Shafei , Rokayah Binti A Rashid (2020) Optimization of Machining Parameters in Milling Process for High Speed Machining using Taguchi Method for Best Surface Roughness, IOP Conference Series: Materials Science and Engineering. 864 (2020) 012110

[17] Mohammed T. Hayajneh, Montasser S. Tahat, Joachim Bluhm, A study of the effects of machining parameters on the surface roughness in the endmilling process, *Jordan J. Mech. Ind. Eng*. 1 (1) (2007) 1–5. September 2007; ISSN 1995-6665.

[18] Çagrı Vakkas Yıldırım,Turgay Kıvak, Fehmi Erzincanl (2019) Tool wear and surface roughness analysis in milling with ceramic tools of Waspaloy: a comparison of machining performance with different cooling methods, Journal of the Brazilian Society of Mechanical Sciences and Engineering. 41:83.

[19] N. Norsilawati, B.T.H.T. Baharudin, S.Na'ain, M.A. Joraimee, R.M. Raja Manisa (2014) The Analysis of Coating Performance on Stainless Steel in High Speed Machining, Australian Journal of Basic and Applied Sciences. 528-531

IOP Conf. Series: Materials Science and Engineering 1068 (2021) 012006 doi:10.1088/1757-899X/1068/1/012006

Composites of Polymer Blends and Their Applications Using Natural Fibres: A Review

A. Mohd[1*], N. Bahiyah Baba[1], M. Z. Umor[1] and R. M. Mohamed[1]

[1]Faculty of Engineering Technology, University College TATI, 24000 Kemaman Terengganu, Malaysia.

[*]E-mail: aliasmohd@uctati.edu.my

Abstract. Study on new composite materials in engineering products with promising physical and mechanical properties has been considered as one of the fields of concern in recent decades. Strength, hardness and fatigue properties make engineering structural more flexible. They are extensively used in the aerospace industry, mechanical engineering applications and parts, electronic packaging, vehicle and aircraft structures, process industry equipment, as well as in biomedical equipment. Disposing of composite wastes however, are very difficult because of its structure and compositions. Hence, composite materials recycling has become one of the major measures of the future. This study seeks to analyse the present state of engineering plastics using natural fibers in their properties and manufacturing techniques. The effects of various chemical treatments on natural fibres' mechanical and thermal properties have been studied in strengthening thermosetting and thermoplastics composites. The mix ratio of polymer waste used from the industry sector with natural fiber is expected to rise in the future, thus issues regarding recycling need to be tackled. It concluded that chemically treated natural fibre improved the adhesion between fibre surface and polymer matrix, which gradually increased the properties of natural fibres incorporated composites.

1. Introduction

In the recent years, the awareness of environmental and consumer's awareness towards new products subjected to strictly regulation and law enforcement, the environmentally friendly used materials adopting natural fibers become more popular [1][2]. Natural fiber offered good properties and environmental user friendly compared to synthetic one when developing a new product which aiming quality and sustainable values [3]. The growth of natural fiber applications in composites industries are considerably fast over 5 years (2011 – 2016) by 10% worldwide [4].

The variety of mechanical properties and lack of knowledge for such fibres created major problems for end users in the widespread implementation of methods for collecting, handling, processing and post-processing natural fibres. Natural fibres are clearly defined fibres that are not synthetic or manmade and may be derived from animals or plants. [5]. The use of natural fibre with renewable and non-renewable polymers from both origins, such as oil palm, kenaf and jute, to manufacture composite materials, has attracted considerable attention in recent decades.

Plants producing cellulose fibres can be categorised as fibres of bast (kenaf, hemp, ramie, flax, and jute), fibres of seed (kapok, coir, and cotton), fibres of leaf (abaca, pineapple, and sisal), fibres of grass and reed (rice, corn, and wheat), fibres of core (jute, hemp, and kenaf) and all other types of fibres (roots and wood) [6]. Table 1 indicates the world and world production of the most common and commercially natural fibres.

Table 1: Natural fibers in the world and their world production [4]

Fiber Source	World Production (10^3)
Bamboo	30.000
Sugar cane bagasse	75.000
Jute	2300
Kenaf	970
Flax	830
Grass	700
Sisal	375
Hemp	214
C oir	100
Ramie	100

The problems of natural fibres are the scattered data and the variations in mechanical properties described. Furthermore, the absence of instructions on methods for the collection, storage, processing and post-processing of natural fibres creates a difficulty for both the producers and consumers of these products. These issues are in reality, important deterrents to the widespread use of natural fibres in different applications. The fibre reinforced polymer matrix received considerable attention in terms of its comparatively low weight, low cost, less harm to processing equipment, good relative mechanical properties such as tensile module and flexural module, improved surface finish of moulded composite parts, renewed composite parts due to the good properties and superior benefits of natural fibre over synthetic fibres [7].

Moreover, the mixing of natural fibres and polymers is argued since the chemical compositions of both fibres and the matrix are distinct, it was considered a challenge. Such reasons for the ineffective transferring of stress during interface of the composites produced. Therefore, natural fibre modifications using unique treatments are definitely necessary. Typically, these improvements are focused on the use of groups of functional reagents that have the capacity to react and modify the composition of the fibre structures. As a result, fibre modifications make it possible to minimise the absorption of moisture by natural fibres, resulting in excellent Incompatibility across the polymer matrix and the fibre [8]. Kabir et., al. [9] agreed in a review of chemical treatments of natural fibres that treatment is a main factor to be taken into account mostly in processing of natural fibres. It was found that fibres lose hydroxyl groups due to various chemical treatments, which decreases the hydrophilic characteristics of the fibres and enhances the mechanical strength and dimensional stability of natural fibre reinforced composites.

This paper is organised as follows in four areas: first, we give an overview of natural fibres, focusing on their benefits and practical characteristics. Next, we identify the relevant applications and potential trends. Finally, with some final remarks, we discuss and conclude the reviewed.

2. Natural Fibre Polymer Composites (NFPCs)

Natural fibre polymer composites can be classified as a composite material with a polymer matrix incorporated with natural high-strength fibres such as oil palm, jute, sisal, flax and kenaf [10] (NFPC). Polymers can typically be broken down into two classifications: thermoplastics and thermosets. The thermoplastic matrix material composition tends to make their properties weaker at an elevated heat range and roll back when cooling, consisting of one or two-dimensional molecular. Polymer thermosets can, meanwhile, be recognised as highly cross-linked polymers that recover only using heat, heat and pressure and/or light irradiation. Therefore, its composition gives the thermoset polymer excellent properties such as high flexibility, good strength and modulus [11, 12].

In general, depending on the nature of the constituents, bio-based composites can be graded either as partially eco-friendly or green. Green composite means all the components come from renewable sources, potentially reducing carbon dioxide emissions and depending on petroleum-based materials. On the other hand, if one of the components is partly environmentally friendly, either the fibre or the matrix is not derived from renewable resources [13]. In accordance with length, dimension and orientation, natural fibre reinforcement can be split, as shown in figure 1. The fibre itself is classified as continuous or discontinuous according to its length to diameter ratio (l/d) (i.e. chopped). Generally, the fibre-reinforced phase structure is known as woven or non-woven. A woven fabric is characterised by continuous interlacing of perpendicular yarns in a regular pattern.

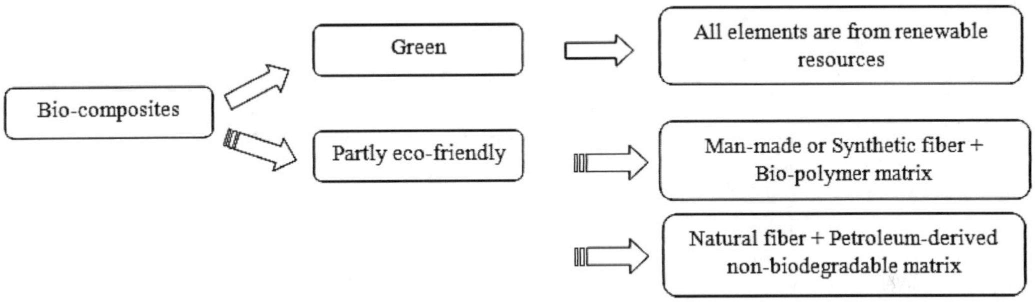

Figure 1. Classification of bio-composite [13–15].

Due to the nonlinear mechanical behaviour of natural fibres, under the influence of tensile shear loads, the fracture activity of composites is also affected [16]. Table 2 demonstrates the mechanical properties of common forms of natural fibres in the world [17].

Composite properties [19] are also affected by the hydrophilic character of the natural fibre [18] and by the mixing of the fibre. In order to achieve good NFPC properties, high fibre loading is typically required [20]. It should be remembered, in general, that the rise in the fibre content contributes to enhancement of the composites' tensile properties. [21].

Table 2. Physicomechanical properties of natural fibers [17].

Fiber	Density (g/cm^3)	Tensile strength (MPa)	Young's modulus (GPa)	Elongation at break (%)
OPEFB	0.7–1.55	2	3.2	2.5
Flax	1.4	88–1500	60–80	1.2–1.6
Hemp	1.48	550–900	70	1.6
Jute	1.46	400–800	10–30	1.8
Ramie	1.5	500	44	2
Coir	1.25	220	6	15–25
Sisal	1.33	600–700	38	2-3
Abaca	1.5	980	—	—
Cotton	1.51	400	12	3–10
Kenaf	1.2	295	—	2.7–6.9
Kenaf	0.21	—	—	—
Bagasse	1.2	20–290	19.7–27.1	1.1
Henequen	1.4	430–580	—	3–4.7
pineapple	1.5	170–1672	82	1–3
Banana	1.35	355	33.8	53

Table 3. Equilibrium humidity content of various natural fibres relative humidity (RH) at 65 percent and 21°C [22].

Fibre	Equilibrium humidity content (%)
Abaca	15
Pineapple	13
Jute	12
Sisal	11
Coir	10
Hemp	9
Ramie	9
Bamboo	8.9
Bagasse	8.8
Flax	7

In the meantime, temperature can also influence the percentage of the composites' water absorption. The equilibrium moisture content of certain natural fibres can be observed in Table 3 at 65 percent humidity at 21 ° C [22].

3. Application

NFPC applications are growing quickly in various engineering fields. In various automotive applications, structural components, packaging and construction components, various types of natural fibres such as oil palm, jute, kenaf, hemp, and bamboo reinforced polymer composites have been given great importance [23, 24]. Society and governments are challenged by many sectors, including automotive, manufacturing, electricity and aerospace, to generate more environmentally sound products and, among others, to reduce their dependency on fossil fuels [25, 26, 27].

In tables 4 and 5, respectively, the application of natural fibres in general and also specifically in the automotive industry is presented. Due to its low specific weight, relatively high strength, relatively low manufacturing costs, corrosion resistance and fatigue resistance, the widespread use of NFPCs in polymer composites.

IOP Conf. Series: Materials Science and Engineering 1068 (2021) 012006 doi:10.1088/1757-899X/1068/1/012006

Table 4. Natural fibre composite applications in industry [28 – 31].

Fibre	Application in building, construction, and others
Hemp fibre	Construction products, textiles, cordage, geotextiles, paper & packaging, furniture, electrical, manufacture bank notes, and manufacture of pipes
Oil palm	fiber Building materials such as windows, door frames, structural insulated panel building systems, siding, fencing, roofing, decking, and other building materials [32]
Wood fibre	Window frame, panels, door shutters, decking, railing systems, and fencing
Flax fibre	Window frame, panels, decking, railing systems, fencing, tennis racket, bicycle frame, fork, seat post, snowboarding, and laptop cases
Rice husk fibre	Building materials such as building panels, bricks, window frame, panels, decking, railing systems, and fencing
Bagasse fibre	Window frame, panels, decking, railing systems, and fencing
Sisal fibre	In construction industry such as panels, doors, shutting plate, and roofing sheets; also, manufacturing of paper and pulp
Stalk fibre	Building panel, furniture panels, bricks, and constructing drains and pipelines
Kenaf fibre	Packing material, mobile cases, bags, insulations, clothing-grade cloth, soilless potting mixes, animal bedding, and material that absorbs oil and liquids
Cotton fibre	Furniture industry, textile and yarn, goods, and cordage
Ramie fibre	Use in products as industrial sewing thread, packing materials, fishing nets, and filter cloths. It is also made into fabrics for household furnishings (upholstery, canvas) and clothing, paper manufacture
Jute fibre	Building panels, roofing sheets, door frames, door shutters, transport, packaging, geotextiles, and chip boards
Coir fibres	Building panels, flush door shutters, roofing sheets, storage tank, packing material, helmets and postboxes, mirror casing, paper weights, projector cover, voltage

Table 5. Applications of natural fibres in automotive industry [33, 34, 35, 36, 37, 38].

Natural Fibres	Component Description	Other Constituents	Reference(s)
Bast fibres (flax, hemp, kenaf, jute, sisal, etc.)	Carrier for covered door panels, covered components for instrument panels, covered inserts, carrier for hard and soft armrests, seat back panels, door panels, door bolsters, headliners, side and back walls, seat backs, rear deck trays, pillars, centre consoles, load floors, trunk trim	Polypropylene (PP) and polyester	[34]
Abaca	Under-floor panel and body panels	-	[33,36]
Coconut	Seat bottoms, back cushions and head restraints, interior trim and seat cushioning, seat surfaces/backrests	Natural rubber	[34,35,38]
Banana	Wrapping paper	-	[37]
Coir	Car seat covers, mattresses, doormats, rugs	-	[38]
Cotton	Soundproofing, trunk panel, insulation	PP/PET	[33,35]
Flax	Seatbacks, covers, rear parcel shelves, other interior trim, floor trays, pillar panels and central consoles, floor panels	Mat with PP (floor panels)	[33,35]
Fibrowood recycled	Plastic retainer for seat back panel	PP granules, thermoplastic	[34]
Flax or Hemp	Carrier for covered door panels	Epoxy resin	[34]

Flax/Sisal	In the interior door linings and panels, door panels	Thermoset resin	[33,35]
Kenaf	Door inner panel	PP	[36]
Kenaf/Flax	Package trays and door panel inserts	-	[33]
Kenaf/Hemp	Door panel, rear parcel shelves, other interior trim, Lexus package shelves, door panels	-	[33]
Wool	Upholstery, seat coverings	Leather	[33,35]
Wood Flour	Carrier for covered door panels, carrier for armrest, carrier for covered inserts	PP or polyolefin (POE)	[34]
Wood	Carrier for covered door panels, carrier for covered door panels, covered or foamed instrument panels, covered inserts and components, covered seat back panels, fibre in the seatback cushions, inserts, spare tire, covers	Acrylic resin and synthetic fibre	[33,34]

4. Future Trends

The broad benefits of natural fiber-reinforced composites, such as a high weight-to-weight ratio, light weight and biodegradability, make them ideal for various applications in the construction industry [39]. To a large degree, the capacity for the application of natural fibres, in particular as composite reinforcement, relies on the increase in regulations and their commercial requirements, including the setting of production and post-treatment.

Advancing the knowledge of natural fibre would make it possible to standardise the fibre type available on the marketplace and enable people to have a top level of trust in the mechanical and chemical properties. The scientific group has an important role to play in the release of specific and extensively characterised studies, for a robust increase in fundamental knowledge of natural fibres [40]. Van de Weyenberg et al. [41] have shown that good properties such as high tensile strength and compression strength, of thin-walled components made of sisal fibre reinforced composite, give it a wide range of applications, such as structural components, permanent shaping, and pipe reinforcement of existing structures. Green buildings are built to be a clean, fitting and environmentally friendly area to live and work. The main materials used at this time as part of green materials are biocomposites [42].

Table 6. Output (in tons) of composites based on wood and natural fibres in 2012 and their forecasts for 2020 [43].

Bio-Composites	Production (in ton)		
	2012	2020 (without Incentives for Bio-Based Products)	2020 (with Strong Incentives for Bio-Based Products)
Wood-Plastic Composites			
Construction, extrusion	190,000	400,000	450,000
Automotive, compression moulding and extrusion/thermoforming	60,000	80,000	300,000
Granulates, injection moulding	15,000	100,000	>200,000
Natural Fibre Composites			
Automotive, compression moulding	90,000	120,000	350,000
Granulates, injection moulding	2000	10,000	>20,000

Carus et al. have shown that the application of natural fibres will dramatically increase in the future as they begin to reach various sectors comparing to automotive sector. In table 6 for 2012 and 2020, the Production Forecast Scenarios are presented. There is a substantial difference in the amount of natural fibre and wood-based composites for this time. So, it can be claimed that in the absence of significant political incentives to increase the proportion of bio-based materials used in products, especially in the automotive sector, there would be no rapid growth [43].

5.Conclusion

When incorporating such natural fibres to the correct loading ratio, composite polymer blends are an efficient way to increase the quality of parts when it comes to cost-effective, technical and environmental characteristics. The benefits of getting the right polymer blend formulations along with the natural fibres give low density and fair price, offering great marketing advantages, such as the automotive and construction industries. It is possible to enhance the physical and chemical properties of a new material by introducing a binding agent as a bonding material between composite polymers with such fibres. This study showed that the properties of fibres are significantly affected by the manner in which they are handled and how the treatments are conducted. The chemical treatment of the fibre used has shown that it is possible to increase the physical and mechanical properties of composite polymers and to regulate the level of moisture by means of alkalization and coupling agents. But the important point that needs to be discussed is that some fibres are very expensive compared to those with the same mechanical properties, such as cotton. Therefore, before withdrawing the rationale and recommendation, the selection of the required fibres along with the polymer blend for such application must be properly carried out using the relevant experimental method and review of the effects.

Acknowledgments

Authors wishing to acknowledge financial support from University College TATI (UC TATI) using Short Term Grant No.: GPJP 1/9001-2010.

References

[1] Paulo Pecas, Hugo Carvalho, Hafiz Salman and Marco Leite 2018 Natural fibre and their application: A Review *J. of Composites Science*, **2**, 66.

[2] Layth Mohammed, M. N. M. Ansari, Grace Pua, Mohammad Jawaid and M. Saiful Islam 2015 A review on natural fiber reinforced polymer composite and its applications *Int. J. of Pol Sci,* **20** pp 1 – 15.

[3] Mitra, B. C. 2014 Environmental friendly composite *materials Def. Sci. J.*, **64,** pp 244 – 261.

[4] N. Uddin 2013, Ed. *Development in Fiber-Reinforced Polymer (FRP) Composites for Civil Engineering*, Elsevier.

[5] A. Ticoalu, T. Aravinthan, and F. Cardona 2010 A review of current development in natural fiber composites for structural and infrastructure applications *Proceedings of the Southern Region Engineering Conference (SREC '10)*, pp. 113–117, Toowoomba, Australia.

[6] O. Faruk, A. K. Bledzki, H.P. Fink, and M. Sain 2012 Biocomposites reinforced with natural fibers: 2000–2010 *Progress in Polymer Science*, vol. **37**, no. 11, pp. 1552–1596.

[7] A. Shalwan and B. F. Yousif, "In state of art: mechanical and tribological behaviour of polymeric composites based on natural fibres *Materials & Design*, vol. **48**, pp. 14–24, 2013.

[8] S. S. Ray and M. Bousmina, "Biodegradable polymers and their layered silicate nanocomposites: in greening the 21st century materials world," *Progress in Materials Science*, vol. **50**, no. **8**, pp. 962–1079, 2005.

[9] M. M. Kabir, H. Wang, K. T. Lau, and F. Cardona 2012 Chemical treatments on plant-based natural fibre reinforced polymer composites: an overview *Composites Part B: Engineering*, vol. **43**, no. **7,** pp. 2883–2892.

[10] H. Ku, H. Wang, N. Pattarachaiyakoop, and M. Trada 2011 A review on the tensile properties of natural fiber reinforced polymer composites *Composites Part B: Engineering* vol. **42**, no. **4**, pp. 856–873.
[11] A. Ticoalu, T. Aravinthan, and F. Cardona 2010 A review of current development in natural fiber composites for structural and infrastructure applications *Proceedings of the Southern Region Engineering Conference (SREC '10)*, pp. 113–117, Toowoomba, Australia.
[12] O. Faruk, A. K. Bledzki, H.P. Fink, and M. Sain 2012 Biocomposites reinforced with natural fibers: 2000–2010 *Progress in Polymer Science*, vol. **37**, no. 11, pp. 1552–1596.
[13] J. Meredith, R. Ebsworth, S. R. Coles, B. M. Wood, and K. Kirwan 2012 Natural fibre composite energy absorption structures *Composites Science and Technology*, vol. **72**, no. **2**, pp. 211–217.
[14] Mohanty, A. Misra, M. Drzal, L. Selke, S. Harte, B. Hinrichsen G. 2005 *Natural Fibers, Biopolymers, and Biocomposites; Mohanty, A., Misra, M., Drzal, L., Eds.*; CRC Press: Boca Raton, FL, USA,; ISBN 978-0-8493-1741-5.
[15] Dicker, M.P.M.; Duckworth, P.F.; Baker, A.B.; Francois, G.; Hazzard, M.K.; Weaver, P.M. 2014 Green composites: A review of material attributes and complementary applications. *Compos. Part A Appl. Sci. Manuf.*, **56**, pp. 280–289.
[16] A. May-Pat, A. Valadez-Gonzalez, and P. J. Herrera-Franco 2013 Effect of fiber surface treatments on the essential work of fracture of HDPE-continuous henequen fiber-reinforced composites *Polymer Testing*, vol. **32**, no. **6**, pp. 1114–1122.
[17] M. Jawaid and H. P. S. Abdul Khalil 2011 Cellulosic/synthetic fibre reinforced polymer hybrid composites: a review *Carbohydrate Polymers*, vol. 86, no. 1, pp. 1–18.
[18] A. Shalwan and B. F. Yousif 2013 In state of art: mechanical and tribological behaviour of polymeric composites based on natural fibres *Materials & Design*, vol. 48, pp. 14–24.
[19] M. A. Norul Izani, M. T. Paridah, U. M. K. Anwar, M. Y. Mohd Nor, and P. S. H'Ng 2013 Effects of fiber treatment on morphology, tensile and thermogravimetric analysis of oil palm empty fruit bunches fibers *Composites Part B: Engineering*, vol. **45**, no. 1, pp. 1251–1257.
[20] I. S. M. A. Tawakkal, M. J. Cran, and S. W. Bigger 2014 Effect of kenaf fibre loading and thymol concentration on the mechanical and thermal properties of PLA/kenaf/thymol composites *Industrial Crops and Products*, vol. **61**, pp. 74–83.
[21] S. Shinoj, R. Visvanathan, S. Panigrahi, and M. Kochubabu 2011 Oil palm fiber (OPF) and its composites: a review *Industrial Crops and Products*, vol. **33**, no. 1, pp. 7–22.
[22] O. Faruk, A. K. Bledzki, H.-P. Fink, and M. Sain 2012 Biocomposites reinforced with natural fibers: 2000–2010 *Progress in Polymer Science*, vol. **37**, no. 11, pp. 1552–1596.
[23] A. Shalwan and B. F. Yousif 2013 In state of art: mechanical and tribological behaviour of polymeric composites based on natural fibres *Materials & Design*, vol. **48**, pp. 14–24.
[24] E. Sassoni, S. Manzi, A. Motori, M. Montecchi, and M. Canti 2014 Novel sustainable hemp-based composites for application in the building industry: physical, thermal and mechanical characterization *Energy and Buildings*, vol. **77**, pp. 219–226.
[25] AL-Oqla, F.M.; Sapuan, S.M. 2014 Natural fiber reinforced polymer composites in industrial applications: Feasibility of date palm fibers for sustainable automotive industry. *J. Clean. Prod.*, **66**, 347–354.
[26] Salazar, V.L.P.; Leão, A.L.; Rosa, D.S.; Gomez, J.G.C.; Alli, R.C.P. 2011 Biodegradation of Coir and Sisal Applied in the Automotive Industry. *J. Polym. Environ.*
[27] Faruk, O.; Bledzki, A.K.; Fink, H.-P.; Sain, M. 2012 Biocomposites reinforced with natural fibers: 2000–2010. *Prog. Polym. Sci.*
[28] A. Ticoalu, T. Aravinthan, and F. Cardona 2010 A review of current development in natural fiber composites for structural and infrastructure applications *Proceedings of the Southern Region Engineering Conference* (SREC '10), pp. 113–117, Toowoomba, Australia.
[29] T. Sen and H. N. Reddy 2011 Various industrial applications of hemp, kinaf, flax and ramie natural fibres *International Journal of Innovation, Management and Technology*, vol. **2**, pp. 192–198.

[30] U. S. Bongarde and V. D. Shinde 2014 Review on natural fiber reinforcement polymer composite," *International Journal of Engineering Science and Innovative Technology*, vol. **3**, no. **2**, pp. 431–436.

[31] L. Mwaikambo 2006 Review of the history, properties and application of plant fibres *African Journal of Science and Technology*, vol. **7**, no. **2**, p. 121.

[32] I. S. M. A. Tawakkal, M. J. Cran, and S. W. Bigger 2014 Effect of kenaf fibre loading and thymol concentration on the mechanical and thermal properties of PLA/kenaf/thymol composites," Industrial Crops and Products, vol. **61**, pp. 74–83.

[33] Huda, M.S. Drzal, L.T; Ray, D. Mohanty, A.K. Mishra M. 2008 Natural-fiber composites in the automotive sector. *In Properties and Performance of Natural-Fibre Composites*; Woodhead Publishing: Oxford, UK, ISBN 9781845692674.

[34] Witayakran, S.; Smitthipong, W.; Wangpradid, R.; Chollakup, R.; Clouston, P.L. 2017 Natural Fiber Composites: Review of Recent Automotive Trends. *In Reference Module in Materials Science and Materials Engineering*; Elsevier Publishing: Amherst, MA, USA,; ISBN 9780128035818.

[35] Food and Agriculture Organization of the United Nations 2012. *Unlocking the Commercial Potential of Natural Fibres; Food and Agriculture Organization of the United Nations*: Rome, Italy.

[36] Holbery, J.; Houston, D. Natural-fiber-reinforced polymer composites in automotive applications. JOM 2006, **58**, pp. 80–86.

[37] Ramdhonee, A.; Jeetah, 2017 P. Production of wrapping paper from banana fibres. *J. Environ. Chem. Eng.*

[38] Food and Agriculture Organization of the United Nations. Common Fund for Commodities 2008 *In Proceedings of the Symposium on Natural Fibres*, Rome, Italy.

[39] A. R. Kakroodi, Y. Kazemi, and D. Rodrigue 2013 Mechanical, rheological, morphological and water absorption properties of maleated polyethylene/hemp composites: effect of ground tire rubber addition *Composites Part B: Engineering*, vol. **51**, pp. 337–344.

[40] Paulo Peças, Hugo Carvalho , Hafiz Salman and Marco Leite 2018 Natural Fibre Composites and Their Applications: A Review, "J. Compos. Sci. 2018, **2, 66**

[41] I. Van de Weyenberg, J. Ivens, A. De Coster, B. Kino, E. Baetens, and I. Verpoest 2003 Influence of processing and chemical treatment of flax fibres on their composites," Composites Science and Technology, vol. **63,** no. **9**, pp. 1241–1246.

[42] Layth Mohammed, M. N. M. Ansari, Grace Pua, Mohammad Jawaid, and M. Saiful Islam 2015 A Review on Natural Fiber Reinforced Polymer Composite and Its Applications *International Journal of Polymer Science*, Vol. 2015, Article ID 243947.

[43] Carus, M.; Eder, A.; Dammer, L.; Korte, H.; Scholz, L.; Essel, R.; Breitmayer, E.; Barth, M. 2015 Wood-PlasticComposites (WPC) and Natural Fibre Composites (NFC): *European and Global Markets 2012 and Future Trends in Automotive and Construction*; Nova-Institute: Hürth, Germany.

Palm Oil Mill Effluent Treatment by Using Copperas as Physicochemical Treatment Agent

Noor Aiman Razak[1], Nor Aisyah Mat Yahya[1*], Nabihah Abdullah[1]
[1]Faculty of Engineering Technology, University College TATI, 24000 Kemaman, Terengganu , Malaysia.

*[1]Email: noraisyah@tatiuc.edu.my

Abstract. The palm oil mill industry in Malaysia are a huge agriculture industry which generates significant amount Palm Oil Mill Effluent (POME). POME is an oily wastewater which consist of various suspended components, the liquid are a combination of steriliser condensate and cooling water. The wastewater is very pollutant to human and environment, the coagulants applied to reduce the suspended materials in POME as following the stringent requirements of biological oxygen demand (BOD), chemical oxygen demand (COD), total dissolved solid (TDS), total solid (TS) and turbidity as stated by the Department of environment. This paper analyses the use of ferrous sulphate heptahydrate as a coagulant as it was applied on the primary treatment and the results improvement upon physicochemical treatment parameters. Series of batch treatment for coagulation and flocculation with ferrous sulphate heptahydrate were conducted. Dosage and pH were conducted to determine the optimum conditions to apply the coagulants. The result of coagulants showed that ferrous sulphate heptahydrate have a good effect on BOD reduction at dosage 1500 mg/l pH of 3.4 before pH adjustment. There were eight different samples used for different concentration of coagulant, the sample of 250 ppm copperas feed till 1000 ppm feed has the COD result difference of 171 ppm. When the copperas feed increased the COD value will decreased gradually. For the second parameter, biological oxygen demands the BOD value are reduced successfully, the first sample with feed concentration 500 ppm of copperas has BOD value of 288 ppm. As we increase the feed of copperas to 2000 ppm the BOD value reduced to 13 ppm. Next the highest value of TSS reduction is at 2000 ppm of copperas feed, at the specific dosage the TSS reading was 7502 ppm

1. Introduction

The palm oil industry in Malaysia is the primary contributor from the rural segment to the national economy, which has contributed income of nearly RM63.5 billion of every 2014 from the export of palm oil-based products. This is regardless of the oil palm plantations just covers 16 percent of the in general planted horticultural zone in Malaysia, or proportionate to about 5.23 million hectares. The palm oil industry prepared 95.4 million ton of oil palm organic products Fresh Fruit Bunches (FFB) in 429 palm oil processes all through Malaysia that created 19.65 million ton of CPO in year 2014.

Higher crude palm oil (CPO) creation has expanded the measure of waste produced from the palm oil extraction process, the two solids and fluid waste. The major waste streams are POME that makes up roughly about 0.5 - 0.8 ton of POME from each ton of FFB handled in the event that the untreated effluent is released into water course, it is sure to cause significant ecological issues because of its high contamination load as biochemical oxygen demand (BOD), chemical oxygen demand (COD), oil and grease, total solids (TS) and suspended solids (SS). The palm oil plant industry in Malaysia has in this way been recognized as the one releasing the biggest contamination load into the streams all through the nation (Asis, Affiq, Arifin, Ngteni, & Tahiruddin, 2016b).

2. Scope of study

This study consists of several lab experiments which utilize copperas (ferrous sulphate) as the coagulant to treat the primary treated palm oil mill effluent POME. The palm oil mill effluent sample was obtained from kilang sawit Panji Alam which is located in Kemaman, Terengganu. The efficiency of the coagulant was assessed based on biological oxygen demand, chemical oxygen demand, total suspended solid and final pH of the treated POME.

The main focus of this research is the application of ferrous sulphate heptahydrate as a potential coagulant for the treatment of POME. Physicochemical properties such as pH, temperature, Biological Oxygen Demand (BOD), Chemical Oxygen Demand (COD), Total Suspended Solid (TSS), Suspended Solid (SS), Oil & Grease (O&G) Fresh POME turbidity from Sludge Pit (Primary POME), secondary anaerobic digester tank (Secondary POME), First tertiary aerobic pond (Tertiary POME) have been identified. The coagulation tests were conducted in accordance with standard coagulation-flocculation testing of industrial wastewater treatment (APHA, 2012) at different coagulation doses 9 (200mg/L-2000 mg/L), dilution factor (DF) at natural pH and temperature for 60 minutes. The coagulation capacity of ferrous sulphate heptahydrate and ferrous sulphate monohydrate were calculated by extracting BOD, COD, TSS, SS, turbidity in primary POME, secondary POME and tertiary POME treated.

The coagulation efficiency of coagulants ferrous sulphate heptahydrate and ferrous sulphate monohydrate has been compared and determined which coagulants and their optimal experimental conditions are most suitable for use as coagulants. POME coagulation was subsequently performed on a pilot scale using the optimum experimental conditions of copperas coagulant (coagulant doses of 750 mg/L-1250 mg/L; natural pH, ambient temperature, reaction time of 3 h). Finally, the economic viability of the copperas coagulation of POME was determined and compared with the existing POME tertiary treatment cost.

3. Materials and Methdology

The sampling method and briefs procedure taken on the sample to be taken , the stage of effluent treatment that to be studied and the safety aspects of running the sampling, the analysis and data collection from sample test , following the parameter the primary effluent should be tested for Total Dissolved Solid (TDS) , biochemical oxygen demand BOD , chemical oxygen demand COD. The jar test method is key to the research and conducted by using the most precise equipment in UC TATI laboratories.

3.1 Jar Test Method

Jar test used the treatment chemicals used in palm oil mill effluent coagulation process. This simulates the coagulation/flocculation process in a water treatment plant and lets operators decide whether they are using the correct amount of treatment chemicals, thus enhancing the performance of the plant. Jar conducted by adjusting the amount of coagulant and the sequence in which they are added to samples of raw water held in jars or beakers. The sample is then stirred in such a way that floc formation, development and settlement can be observed as it would be in the full-scale treatment plant. Flocculates formed when treatment chemicals combine with raw water content and clump together. The operator then conducts a series of tests to measure the effects of different quantities of flocculating agents at different pH values in order to assess the correct floc size for a specific plant. The right floc size varies according to the filter size. Figure 1 shows the setup of jar test equipment.

Figure 1: POME jar testing using setup rpm speed

3.2 Determination of Chemical Oxygen Demand

Chemical oxygen demand was used to determine the quantity of pollution in water after wastewater treatment. The value of COD indicated by testing the water sample using calorimetric analysis. In experiment for COD analysis the 1ml was treated clear solution with the COD reagent were heating for 2 hours at 150 °C using the COD reactor. COD reactor function was used to heat the COD reagent with the samples for 2 hours at 150 °C. After 2 hours, the samples were cooled down for a few minutes before testing the organic matter present in wastewater was oxidized by reagent material in the presence of Sulphuric acid to produce carbon dioxide CO2 and water H2O. The quantity of potassium dichromate used in the reaction is equivalent to the oxygen used to oxidize to organic matter of wastewater. The results of COD under this procedure are described as the mg of O2 consumed per liter of sample. The specimen is tested with Hach dr900 COD calorimeter which utilizes chromatography mechanism to which compared the reading between the blank sample and the wastewater sample that had been tested. 27 The blank can be used twice with the same lot of vials for measurements. Hold it in the night. Decomposition tracking by measuring the absorption at the correct wavelength (420 or 610 nm). Use a vial containing 5 mL of deionized water to zero the device in the absorbance mode and calculate the empty absorbance. The value is registered. Prepare a new blank when approximately 0.01 absorbance units have modified the absorbance.

3.3 Determination of Biochemical Oxygen Demand

Two BOD bottles were completely filled with dilution water. The correct specimen volume was carefully measured. Then the sample was put into the dilution before the containers are filled and the bottle was closed, the initial dissolved oxygen of one of the duplicate bottles are determined, including water dilution blank by one of the approved methods and record data. The bottles are placed in the incubator at 20°C and incubate for five days. The dissolved oxygen content of the incubated bottles was determined at the end of exactly five days. For each dilution, the biochemical oxygen demand was determined by using the formula. From those dilutions that have a depletion of at least 2 mg/L DO and at least 1.0 mg/L residual, the most reliable biochemical oxygen demand will be obtained. If there is more than one dilution that meets these criteria, the biochemical oxygen demand results should be averaged to obtain a final biochemical oxygen value. The blanks of the dilution water are only used to test the dilution water quality. The loss of dissolved oxygen will be less than 0.2 mg/L if the water value is decent and free of impurities. In either case, do not use the obtained depletion as a blank correction. Palm oil mill effluent are known for its high value in biological oxygen demand, as in the presence of dissolved oxygen aerobic bacteria used the organic materials founded in the effluent as a food. The calculation for the five day biochemical oxygen demand must conducted using the formula 3.1 , with $D0$ is the initial dissolve oxygen , $D5$is the bod value after the fifth day , so the difference of dissolved oxygen should be divided by *P* which is the dilution factor of the used volume of sample. 28 The BOD test is used because it measures the treatment plant efficiency in terms of BOD removal and control the plant process. It also used to determine the impact of the effluent discharge on the receiving waters. The final measurement of all oxygen consuming materials in a sample are termed as Total Biological Oxygen Demand (TBOD) and because the test are run in five days it is often referred as Five day BOD (BOD_5)

3.4 Determination OF Total Suspended Solid (TSS)

Very simply, by filtering supernatant from centrifugation (Figure 2) of the specimen through a particular pore size filter, TSS is anything that was captured. Suspended solids can range from silt or sand particles to pieces of plant material like leaves or stems. High amounts of TSS can cause a body of water to look aesthetically undesirable. Either the water's colour or overall turbidity will be adversely affected. There are several things that need to consider maintaining a good sample data. The effluent that are used may contain high concentration of metal waste and sulphates, those presence of impurities would make the sample absorb moisture from the area, so the sample are stored in air thigh container. Next, throughout drying process the sample may lose its constituents, the sample drying time were planned and the sample was cooled. The volume of residue over the whole sample were monitored, so much residue was entrapped water during evaporation.

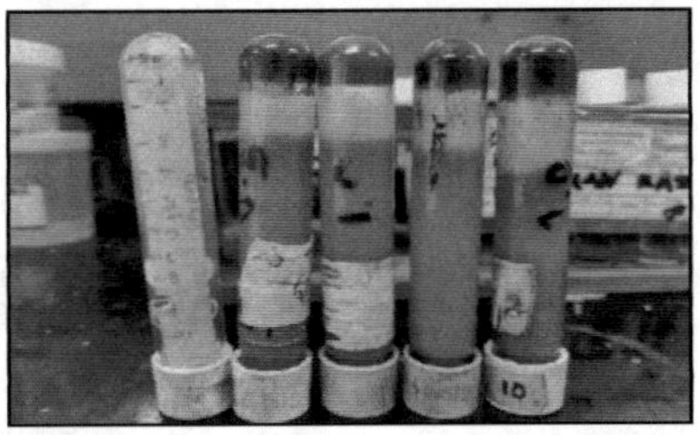

Figure 2: The visual method total suspended solids by using centrifuge

3.5 Salinity Test of Copperas

There are two main methods of determining the salt content of water; Electrical Conductivity (EC), and Total Dissolved Solids (TDS). EC is a measure of how much electrical current can flow through a sample of water while TDS is a measure of the relative weight of dissolved materials in a sample of water. The salinity testing of copperas was conducted as one of the analysis that applied on the material. Copperas was tested using Hach conductivity meter. Electrolytic conductivity is the ion movement in a solution that generates an electrical current and is the reciprocal resistivity of the solution. The ions were derived from inorganic dissolved solids (e.g., chloride, nitrate, anions and sodium, calcium, magnesium, iron, and cations of aluminium). Organic materials such as oils, phenols, alcohols, and sugars were not conductive enough for a reasonable concentration calculation. Conductivity meters calculate the resistance in a solution area specified by the probe's physical design. A voltage was applied between the electrodes and was used to measure the conductivity for each centimetre by the voltage decrease induced by the solution resistance.

3.6 Determination of Moisture Content in Copperas

Since the copperas that used are ferrous sulphate hepta hydrate, the nature of the material is high moisture. By using thermalgravimetric method the percentage of moisture content were determined. During the procedure the samples were taken and by using cubicle, the samples were weighed for the initial measurement, later were being dried in oven.

4. Results and discussion

4.1 Chemical Oxygen Demand

The value of chemical oxygen demand was slightly reduced as every 250 ppm of copperas coagulants were applied (Figure 3). The required dosage of copperas needed can be considered high. To achieve a good state of settling, at 1000 ppm the result of the chemical oxygen demand value just achieved target. Since the copperas were acidic in nature and the treatment that was simulated are as aerobic condition the value of the palm oil mill effluent was reduced. To achieve a neutral condition pH of the effluent, further neutralization should be conducted. As the optimum operating pH for metal-based polymer coagulation are between 4 to 5, adjustment should also be applied during the process, as for the coagulation of the pome from kilang sawit Panji Alam, Kemaman, the pH are adjusted to 5 before it was applied with the copperas.

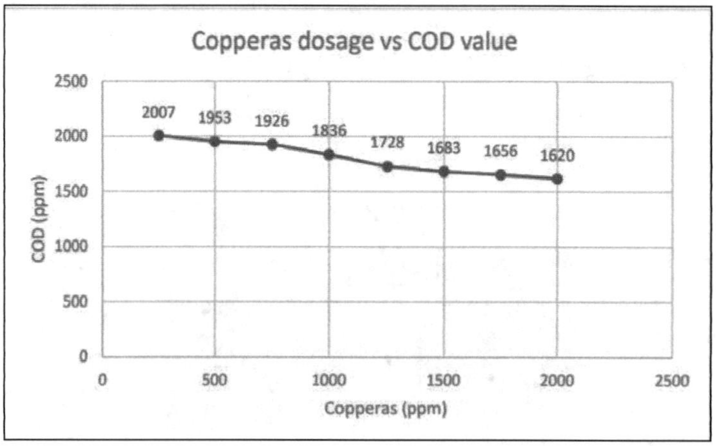

Figure 3: The value of chemical oxygen demand versus the dosage of copperas

4.2 Temperature

It was observed that the value of biochemical oxygen demand (BOD_5) was slightly decreased as we increase the dosage concentration of the copperas (Figure 4). The value was very significant as we compared to the reduction value of chemical oxygen demand (COD). The initial value of biochemical oxygen demand from the diluted palm oil mill effluent of factor of 0.02 give the value of average 400 ppm of BOD_5. At 250 ppm of copperas dosage the value of biochemical oxygen demand (BOD) was dropped to 288 ppm of concentration. The result prove that the copperas coagulant was very effective on biochemical oxygen demand reduction. The trends were almost same to the chemical oxygen reduction as both of the oxygen demand factors showing improvement of quality after the application of coagulants.

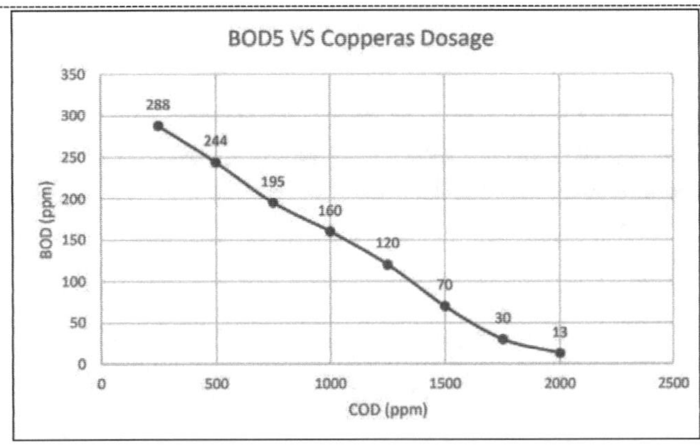

Figure 4: The value of biochemical oxygen demand (BOD_5) versus the dosage of copperas

4.3 Total Suspended Solid

The total suspended solid reduction was significantly reduced as the copperas treatment applied (Figure 5). The nature of the palm oil mill effluent is very high in total suspended solids. The oil and grease composition that exist in the palm oil mill effluent contributes on the high value of suspended solids. The initial reading of the palm oil mill test shows that more than 8000 ppm, for 1500 ppm of copperas dosage the reading of total suspended solid are at 7922 ppm. Total suspended solid was slightly reduced as the treatment was conducted.

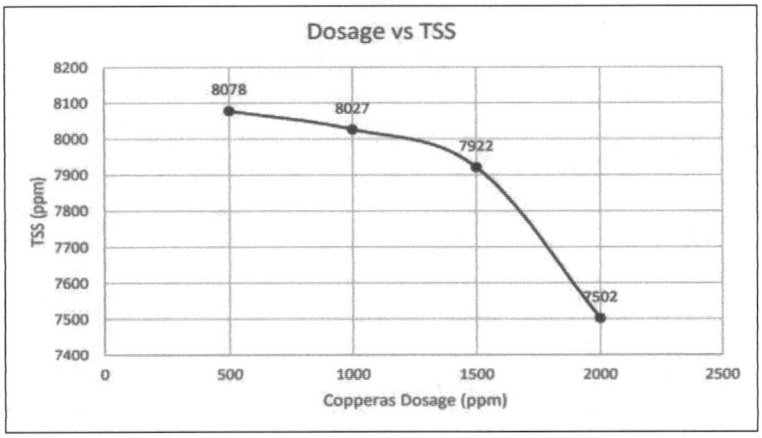

Figure 5: TSS value against copperas dosage value

4.4 Copperas Analysis

From the moisture analysis result in Table 1 we can see the reduction of weight are a very visible hence we can measure the loss of water which is the moisture content that existed in the copperas. The nature for copperas, ferrous sulphate hepta hydrate prove that the material contains high amount of moisture. Table 1 shows the copperas after being overdried. The drying was conducted only to 4 hours at maximum, since the copperas was very reactive to the increased temperature the ferrous sulphate heptahydrate will turn into ferrous sulphate thermal heptahydrate in temperature more than 150°C that has different properties, the physical properties can be seen.

Table 1. Result of moisture analysis of copperas.

Weight of container (g)	Weight of sample (g)	Total of weight (g)	Total weight		Loss of water (%)	
			After 2hrs (g)	After 4hrs (g)	2	4
41.89	20.18	62.07	55.45	51.23	32.82	53.72
47.52	20.66	68.18	61.70	52.13	31.35	77.69
45.74	20.30	66.04	60.74	52.20	26.11	68.18

From the Table 2 we can observe the average salinity value was at 11.6 mS/cm. The salinity value was the conductivity value of the material. In this study 2 gm of copperas was diluted in 50 ml solutions with dilution factor of 0.04 respectively. The concentration of the ferrous sulphate can be determined by using this method as the concentration of soluble salt are proportional to the conductivity value of the test sample. Meanwhile the reading of the pH. value of the copperas was about 2, since the nature of the copperas acidic , the application during the coagulation process does make buffering effects, but this acidic properties of copperas also good in case the palm oil mill effluent pH value was more than five , since the effective pH for this coagulant are between 4-5 pH.

Table 2: The salinity and pH analysis of the copperas

RUN no.	Salinity mS/cm	pH
1	11.72	1.8
2	12.31	1.9
3	12.06	2
4	8.80	1.9
5	12.84	1.8
Average	11.55	1.88

5. Conclusion

Copperas (ferrous sulphate hepta hydrate) was considered as a weak coagulant compared to commercial coagulant such as alumina sulphate. The research proven the ability of the copperas to perform observable reduction on the value of biochemical oxygen demand (BOD_5).To gain minimum 50 ppm discharge limit of the biochemical oxygen demand (BOD) the required dosage is 1500 ppm or 1.5 gram per liter of palm oil mill effluent (pome). The first parameter used in this research, chemical oxygen demand, there were eight different samples used for different concentration of coagulant, the sample of 250 ppm copperas feed till 1000 ppm feed has the COD result difference of 171 ppm. When the copperas feed increased the COD value will decreased gradually. For the second parameter, biological oxygen demands the BOD value was reduced successfully, the first sample with feed concentration 500 ppm of copperas has BOD value of 288 ppm. As we increased the feed of copperas to 2000 ppm the BOD value reduced to 13 ppm. Next the highest value of TSS reduction was at 2000 ppm of copperas feed, at the specific dosage the TSS reading was 7502 ppm. The application of copperas as primary effluent treatment agent supposed to help reducing the biochemical oxygen demand (BOD_5) primarily and also improve others parameters which is in the studies, chemical oxygen demand (COD), and total suspended solids. With the improvement level gained in primary stage of the wastewater treatment the final discharge should be cleaner than before, hence making the palm oil mill effluent treatment more efficient.

Acknowledgement

This work was financed by University College TATI and laboratory works conducted at department of chemical, Faculty of Engineering Technology, University College TATI.

References

[1] Abdul Latif Ahmad, Norliza Ibrahim, S. I. and S. B. 2018. Coagulation Sedimentation Extraction Pretreatment Methods for The Removal of Suspended Solids and Residual Oil from Palm Oil Mill Effluent (Pome). *IIUM Engineering Journal*, **3(1),** 1–9

[2] Ahmad, A., & Krimly, M. (2014). Palm Oil Mill Effluent Treatment Process Evaluation and Fate of Priority Components in an Open and Closed Digestion System. *Current World Environment*, **9(2),** 321–330.

[3] Asis, A. J., Affiq, M., Arifin, M., Ngteni, R., & Tahiruddin, S. 2016a. Palm Oil Mill Effluent Tertiary Treatment by Physicochemical Treatment Using Ferrous Sulphate. *Iranica Journal of Energy & Environment* **7(2),** 163–168.

[4] AremuMuyibi, S., Tajari, T., Jami, M. S., & KoladeAmosa, M. 2014. Removal of organics from treated palm oil mill effluent (POME) using powdered activated carbon (PAC). *Advances in Environmental Biology*, **8**(3 SPEC. ISSUE), 590– 595.

[5] Foo, K. Y., & Hameed, B. H. 2010. Insight into the applications of palm oil mill effluent: A renewable utilization of the industrial agricultural waste. *Renewable and Sustainable Energy Reviews,* ***14*(5),** 1445–1452.

[6] Iskandar, M. J., Baharum, A., Anuar, F. H., & Othaman, R. 2018. Palm oil industry in South East Asia and the effluent treatment technology—A review. *Environmental Technology and Innovation*, **9**, 169–185

[7] Khairuddin, M. N., Zakaria, A. J., Isa, I. M., Jol, H., Nazri Wan Abdul Rahman, W. M., & Salleh, M. K. S. 2016. The potential of treated palm oil mill effluent (Pome) sludge as an organic fertilizer. *Agrivita*, **38(2),** 142–154.

[8] Kheang Loh, S., Mei, Lai, E. E., Muzzammil Ngatiman, Weng, Lim, S., Salimon, J. 2013. Zero Discharge Treatment Technology of Palm Oil Mill Effluent. *Journal of Oil Palm Research,* **25(3),** 273–281.

[9] Madaki, Y. S., & Lau Seng. 2013. Palm oil mill effluent (POME) from Malaysia. International Journal of Science, *Environment and Technology*, **2(6),** 1138–1155.

[10] Parmar, K. A., Prajapati, S., Patel, R., & Dabhi, Y. (2011). Effective use of ferrous sulfate and alum as a coagulant in treatment of dairy industry wastewater. *Journal of Engineering and Applied Sciences,* **6(9),** 42–45.

[11] Rupani, P. F., Singh, R. P., Ibrahim, M. H., & Esa, N. (2010). Review of Current Palm Oil Mill Effluent (POME) Treatment Methods: Vermicomposting as a Sustainable Practice, **11(1),** 70–81.

[12] Ullah, A., & Rathnasiri, P. G. (2015). Optimization of Adsorption-Coagulation Process for Treatment of Palm Oil Mill Effluent (Pome) Using Alternative Coagulant

IOP Conf. Series: Materials Science and Engineering 1068 (2021) 012008 doi:10.1088/1757-899X/1068/1/012008

Development of *Penambang* Boat Driving Cycle to Evaluate Energy Consumption and Emissions

E.A.E.S Shahiran[1], I.N. Anida[1], J.S Norbakyah[1,2,3], A.R Salisa[1,2,3]*

[1]Faculty of Ocean Engineering Technology and Informatics, Universiti Malaysia Terengganu, 21030 Kuala Nerus, Terengganu, Malaysia.

[2]Energy Storage Research Group (ESRG), Universiti Malaysia Terengganu, 21030 Kuala Nerus, Terengganu, Malaysia.

[3]Renewable Energy and Power Research Interest Group (REPRIG), Universiti Malaysia Terengganu, 21030 Kuala Nerus, Terengganu, Malaysia.

*E-mail: nazri.dagang@umt.edu.my

Abstract. *Penambang* Boat is a service that has been developed in the Kuala Terengganu River to boost the public transport network in Terengganu. As far as real-world cities are concerned, the *Penambang* Boat process is important for each running waterway to increase energy efficiency and reduce pollution. This research study is to define and validate the parameters of this driving cycle and to develop the *Penambang* Boat driving cycle on the Kuala Terengganu River throughout its operation path, to analyze fuel economy and emissions using Advance Vehicle Simulator (ADVISOR) software. Through the analysis of energy consumption and emissions, the last two of the micro trips were chosen and the real driving cycles are developed. The results show that the fuel consumption is 8.7 l/100km, the CO emission is 3.479 g/km, the HC emission is 3.718 g/km and lastly the NO_x emission is 0.003 g/km. *Penambang* Boat has been shown the needs to reduce the power consumption and thus minimize the impacts on the environment and economy.

1. Introduction

Many studies show that every conventional engine type emission rates of carbon dioxide and fuel consumption were high. To develop the driving cycle which satisfies the pollution regulatory requirements and high energy performance, the priority needed to improve the modern motor driving cycle. Nowadays, issue on fuel price, fuel consumption and emissions have attracted society's attention to alternative vehicles. It has become increasingly serious issues, thus greatly promoting research on and applications of energy-saving and emission reducing technologies in the automobile industry according to [1]. The Penambang Boat service has invested a certain amount on fuel annually. The solution was therefore addressed to reduce fuel usage. Driving cycle is a data collections of vehicle speed against time collected from the real data situation. This cycle is commonly developed as for few specific region or city, certain road and routes and vehicle types. Driving cycle study is essential to the vehicle developers as well as researchers to study the performance of the proposed transport and vehicle through simulations. However, the driving cycles are not equivalent when comes to different countries or cities, even though they may look identical according to [2].

With the development of this driving cycle, it represents a typical driving graph for the specific area or region either free flow or saturated traffic. Driving cycle also based upon operating factors situations such as the idle state, acceleration and deceleration also in steady state to portray the pattern type in a place of the region according to [3]. Analysis has been conducted on this driving cycle by using collection of data trip from the vehicles in real life operation. With the driving cycle analysis, the vehicles performance can be assessed. Furthermore, this study used three major tasks that are data collection, driving cycle development and route selection. Thus, the real driving cycle simulation characteristics must be executed. Few methods were proposed for the development of the driving cycle such as micro-trip-based cycle construction and modal cycle construction.

For this driving cycle in Kuala Terengganu a micro-trip method is used to develop driving cycle. Based on the driving cycle of the Kuala Terengganu river route driving cycle was developed using *k*-means clustering method and Global Positioning System (GPS) according to [4]. Furthermore, this driving cycle has been compared with another research results. There are two methods for obtaining and tabulating driving cycle data, chasing boat or embarking techniques using GPS or both, that can be emphasized. In this analysis, the GPS technique is used to collect the speed-time data of a boat. It has been analyzed and characterized once the data is obtained. Driving speed data has been obtained using GPS technique. This technique involves the processing in the real world of a series of speed-time data, where the time-related speed of the ship is registered. This process has been repeated in order to obtain large amounts of data.

2. Methodology

There are few main tasks to be tackled in order to complete this project, which are; route selection for *Penambang* Boat driving cycle development, data collection method, characterization and validation of 'Bot Penambang' driving cycle and also development of a driving cycle using *k*-means clustering method. Also, the analysis of energy consumption and emissions using ADVISOR software has been discussed.

2.1 Route Selection

Determining the representative routes throughout the interested area is the most important part for the development of the driving cycle. Kuala Terengganu (KT) River is located between Kuala Terengganu and Kuala Nerus between the Sultan Mahmud Bridge and Drawbridge. Figure 1 shows the river map.

Figure 1. Route of *Penambang* Boat

2.2 Data Collection

The selected route for *Penambang* Boat driving cycle started from Seberang Takir jetty of Kuala Terengganu. The data were collected on the selected route and the data has been collected by using Speedometer GPS android application. The data series of *Penambang* Boat driving cycle are speed in second by second. Collections of data have been held with 10 runs of data under 2 commute routes combined as one complete cycle. In this research, the on-board measurement method has been used using a Global Positioning System.

2.3 Penambang Boat cycle assessments

Cycle assessment is a crucial component in order to characterize and validate the driving cycle. For this project, the parameters involved are average speed, average running speed, average acceleration and average deceleration, and the root mean square of acceleration. Each driving cycle has its own characteristics in term of the cycle assessments or parameters. Before developing a real driving cycle, it is a must to characterize the driving cycle first. For *Penambang* Boat driving cycle, the selection of the cycle assessment has to be made and must be validated using existing data for standard driving cycle such as LA92, EUDC, NYCC, HWFET and IM240. Table 1 shows the equation that be used for every parameter calculating in *Penambang* Boat driving cycle. Nine parameters have been chosen as the assessment parameters since they are the fundamental assessment in order to characterize the driving cycle.

Table 1: Equation for parameters

Parameters	Unit	Equation
Average speed	Km/h	$V_1 = 3.6\frac{\mathrm{dist}}{T_{\mathrm{total}}}$
Average running speed	Km/h	$V_2 = 3.6\frac{\mathrm{dist}}{T_{\mathrm{drive}}}$
Average acceleration of all acceleration phase	m/s^2	$a = \left(\sum_{i=1}^{n}\begin{cases}1 & (a_i > 0)\\ 0 & (else)\end{cases}\right)^{-1}\left(\sum_{i=1}^{n}\begin{cases}a_i & (a_i > 0)\\ 0 & (else)\end{cases}\right)$
Average deceleration of all deceleration phase	m/s^2	$d = \left(\sum_{i=1}^{n}\begin{cases}1 & (a_i < 0)\\ 0 & (else)\end{cases}\right)^{-1}\left(\sum_{i=1}^{n}\begin{cases}a_i & (a_i < 0)\\ 0 & (else)\end{cases}\right)$
Time proportion of idling	%	$\%\ \mathrm{idle} = \frac{T_{\mathrm{idle}}}{T_{\mathrm{total}}}$
Time proportion of cruising	%	$\%\ \mathrm{cruise} = \frac{T_{\mathrm{cruise}}}{T_{\mathrm{total}}}$
Time proportion of acceleration	%	$\%\ \mathrm{acc} = \frac{T_{\mathrm{acc}}}{T_{\mathrm{total}}}$
Time proportion of deceleration	%	$\%\ \mathrm{dec} = \frac{T_{\mathrm{dec}}}{T_{\mathrm{total}}}$
Root mean square acceleration	m/s^2	$\mathrm{RMS} = \sqrt{\frac{1}{T}\int_0^T (a)^2 dt}$

2.4 Driving cycle development

The development of *Penambang* Boat driving cycle used the combination of micro-trips-based approach and pattern classification approach by using *k*-means clustering method. The inputs of *Penambang* Boat driving cycle are second-by-second speed. The data gathered has been divided into several micro-trips and from each micro-trip, the features such as average speed and percentage idle be calculated. The clustering of the micro-trips using *k*-means method took place in order to find the final driving cycle of *Penambang* Boat along selected route.

2.5 K-mean clustering

K-means is one of the simplest unsupervised learning algorithms that solve clustering issue. The *k*-means algorithm endeavors to solve the clustering issue by optimizing a given metric. There is a parcel of benefits by using this *k*-means strategy such as cluster centers can minimize conditional change (a great representation of data), straight forward and quick strategy, and simple to actualize. In this algorithm, mostly Euclidean distance is used to find distance between data points and centroids. Figure 2 shows the micro-trips clustering and its centroid.

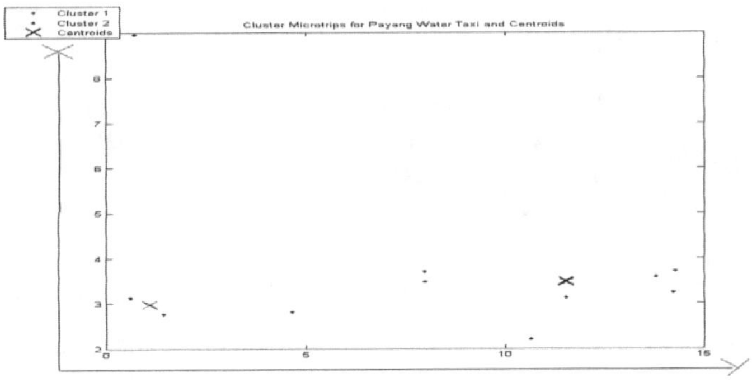

Figure 2: Micro-trips clustering

2.6 Penambang Boat Driving cycle development

The micro-trips have been clustered into two groups. The representatives of micro-trips are determined in order produce the driving cycle for each cluster. The closest micro-trips to the cluster centre or centroid have been considered as the representative micro-trips. The selected micro-trips for each group are presented as in Figure 3 and Figure 4. The micro-trips then combined in order to produce the final driving cycle of *Penambang* Boat along the selected route.

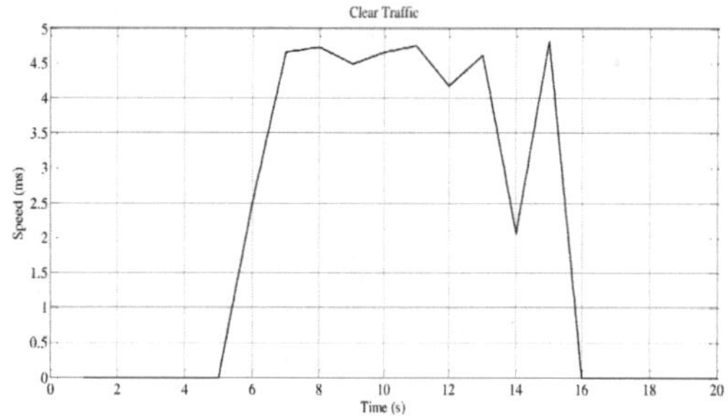

Figure 3: Clear flow

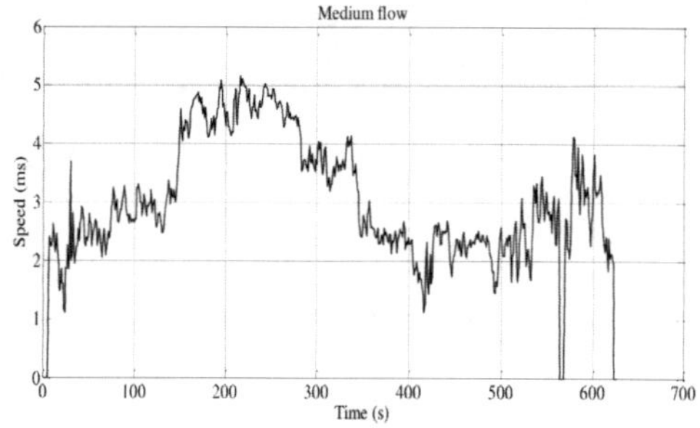

Figure 4: Medium flow

2.7 Fuel rate and emissions analysis using ADVISOR software

The primary goal is to develop a viable driving cycle to assess exhaust emissions, which are Hydrocarbon (HC), Carbon monoxide (CO), Nitrogen oxide (NOx), and Particulate matter (PM) and fuel expenditure of boats that travel in Terengganu Rivers. After the driving cycle has been developed, the data of energy consumption and emissions has been analyzed by using ADVISOR software.

3. Result and Discussion

The validation of driving cycle assessment has been discussed. Also, the development driving cycle has been analyzed. Furthermore, the analysis of the fuel rate and gas emission determined and the comparison between *Penambang* Boat driving cycle with another conventional vehicle has also been compared. The overall route analysis also discussed.

3.1 Validation of cycle assessments

Table 2 and Table 3 shows the percentage difference between standard data retrieved from a few published journals and calculated data of driving cycle using MATLAB code for LA92, EUDC, NYCC and SFTP US06. As in the tables, the average percentage different of error is 0.4% and the minimum error different percentage is 0%. Thus, the percentage error data are lower than 10% of percentage difference so it can be concluded that the MATLAB code is acceptable for further analysis as clarified by [5].

Table 2: Percentage difference of NYCC and EUDC

Parameters	NYCC			EUDC		
	Journal	**Calculated**	**Different (%)**	**Journal**	**Calculated**	**Different (%)**
Distance (m)	1902.76	1898.44	0.22	6955.07	6954.93	0.002
Total time (s)	599	599	0	400	401	0.25
Average speed (km/h)	11.4	11.4	0	62.6	62.6	0.32
Average running speed (km/h)	17.5	17.5	0	69.6	69.7	0.14
Average acceleration (m/s^2)	0.616	0.621	0.81	0.378	0.377	0.27
Average deceleration (m/s^2)	0.601	0.605	0.66	0.926	0.926	0

Percentage idle (%)	31.90	32.05	0.47	10.0	10.2	2.0
Percentage cruise (%)	2.0	2.0	0	53.8	53.6	0.37
Percentage acceleration (%)	32.60	32.55	0.15	25.8	25.7	0.39
Percentage deceleration (%)	33.40	33.38	0.05	10.5	10.5	0

Table 3: Percentage difference of SFTP US06 and LA92

	SFTP US06			LA92		
Parameters	**Journal**	**Calculated**	**Different (%)**	**Journal**	**Calculated**	**Different (%)**
Distance (m)	12893.77	12887.58	0.05	12893.77	12887.58	0.05
Total time (s)	596	601	0.84	596	601	0.84
Average speed (km/h)	76.90	77.19	0.38	76.90	77.19	0.38
Average running speed (km/h)	83.0	83.44	0.53	83.0	83.44	0.53
Average acceleration (m/s^2)	0.666	0.670	0.60	0.666	0.670	0.60
Average deceleration (m/s^2)	0.724	0.728	0.55	0.724	0.728	0.55
Percentage idle (%)	6.5	6.6	1.54	6.5	6.6	1.54
Percentage cruise (%)	5.5	5.5	0	5.5	5.5	0
Percentage acceleration (%)	45.8	45.8	0	45.8	45.8	0
Percentage deceleration (%)	42.2	42.1	0.24	42.2	42.1	0.24

3.2 Penambang Boat Driving Cycle

The final development of the *Penambang* Boat driving cycle can be achieved. Figure 5 shows the final cycle of the *Penambang* Boat. The total distance is 0.541 km and the total micro-trips is 2. The characteristics of the *Penambang* Boat driving cycle in terms of nine assessment parameters are shown in Table 4

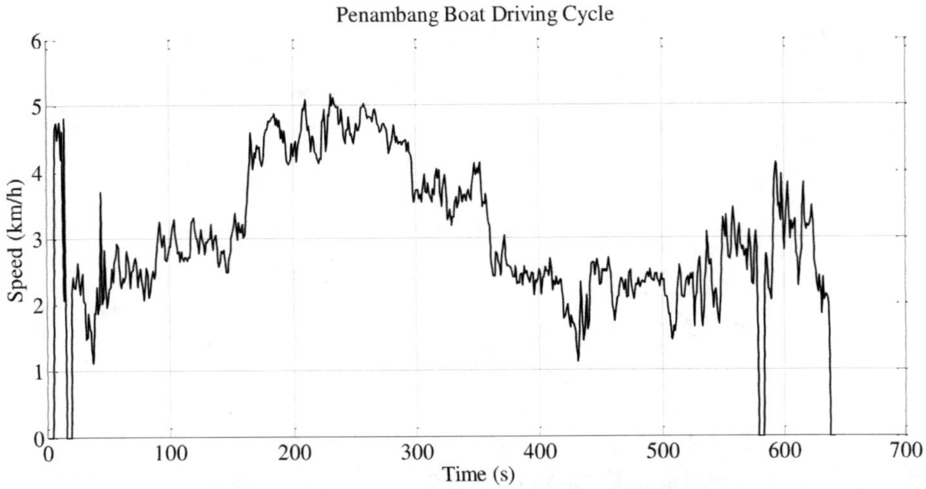

Figure 5. *Penambang* Boat Driving Cycle

Table 4. Driving cycle parameter assessments

Parameters	*Penambang* Boat driving cycle
Distance (km)	0.541
Total time (s)	636
Average speed (km/h)	3.06

Average running speed (km/h)	3.14
Average acceleration (m/s^2)	0.030
Average deceleration (m/s^2)	0.029
Percentage idle (%)	2.04
Percentage acceleration (%)	48.90
Percentage deceleration (%)	49.06
Percentage cruise (%)	0

3.3 Comparison of Penambang Boat driving cycle with existing driving cycle

To evaluate the developed Bot Penambang driving cycle, a comparison was made between the existing cycles as listed in Table 5. The *Penambang* Boat driving cycle data is near to the NYCC driving cycle for average system speed and average running speed data. The average acceleration and deceleration with the EUDC, NYCC and US06 SFTP driving cycle is below 1.0. While the percentage idle, acceleration and deceleration is near to the US06 SFTP driving cycle.

Table 5. Comparison *Penambang* Boat driving cycle and existing driving cycle

Parameters	***Penambang* Boat**	**EUDC**	**NYCC**	**SFTP US06**
Distance (km)	0.541	6.955	1.903	12.894
Total time (s)	636	400	599	596
Average speed (km/h)	3.06	62.6	11.4	76.90
Average running speed (km/h)	3.14	69.6	17.5	83.0
Average acceleration (m/s^2)	0.030	0.378	0.616	0.666
Average deceleration (m/s^2)	0.029	0.926	0.601	0.724
Percentage idle (%)	2.04	10.0	31.9	6.5
Percentage acceleration (%)	48.09	25.8	32.6	45.8
Percentage cruise (%)	0	53.8	2.0	5.5

3.4 Fuel rate and emissions analysis using ADVISOR software

Table 6 shows the result of final *Penambang* Boat driving cycle fuel consumption and emission analysis retrieved from ADVISOR software. From the energy consumption and emissions analysis of *Penambang* Boat driving cycle, it can be concluded that this boat gives the best result in order to prevent the over use of fuel and prevent the increasing various pollution in Kuala Terengganu city and river.

Table 6. Energy consumption and emissions of *Penambang* Boat driving cycle

Parameter	**Value**
Distance (km)	0.9
Fuel consumption (l/100km)	8.7
CO emission (g/km)	3.479
HC emission (g/km)	3.718
NO_x emission (g/km)	0.003
PM emission (g/km)	0

4. Conclusion

The primary objective of this work, which is to develop a driving cycle of *Penambang* Boat at Seberang Takir Jetty, Terengganu along operational routes using *k*-means method, had been successfully accomplished. The nine parameter that commonly used has been applied to assess the driving characteristics for *Penambang* Boat driving cycle. In fact, the exclusivity of the methodology employed in this study is that the driving cycle was developed using a combination method of micro-trips base approach and pattern classification approach using *k*-means clustering method. *Penambang* Boat has been shown the needs to reduce the power consumption and thus minimize the impacts on the environment and economy.

Acknowledgments
The authors would like to be obliged to Ministry of Education Malaysia for providing financial assistance under FRGS 2020 (59623) grant and Faculty of Ocean Engineering Technology and Informatics, UMT for all their technical and research support for this work to be successfully completed.

References

[1] Lee T, Adornato B, 2011. Synthesis of real-world driving cycles and their use for estimating PHEV energy consumption and charging opportunities: *Case study for midwest U.S.* **60(09)** 4153–4163.
[2] Anida I N, Norbakyah J S, Zulfadli M, Norainiza M H, and Salisa A R 2019 Driving cycle development of BAS KITe in Kuala Terengganu city to optimize the energy consumption and emissions. *Materials Science and Engineering* **469** 012112.
[3] Liu Y, Li J and Shen B 2014 Research on Driving Cycle of Long-distance Passenger Vehicles Based on Principal Component Analysis and Cluster Algorithm. *International Journal of Control & Automation* **7(3)**.
[4] Salisa A R, Atiq W H, Norbakyah J S 2015. Development of a KL River Driving Cycle for PHERB Powertrain. *Jurnal Teknologi (Sciences & Engineering)* **76(8)** 101–106.
[5] Ruoyun Ma, Xiaoyi He b, Yali Zheng c, Boya Zhou d, Sheng Lu a and Ye Wu a 2019 Real world driving cycles and energy consumption informed by large-sized vehicle trajectory data. *Journal of Cleaner Production* **223** 564-574.

An Experimental Evaluation of SiO_2 nano cutting fluids in CNC Turning of Aluminium Alloy AL319 via MQL Technique.

A.M. Syafiq, A.A.M. Redhwan *, A.A. Hazim, A.R.M. Aminullah*, S. Zainal Ariffin, W. Nughoro, A. Arifuddin and A.B.S. Hawa

Faculty of Engineering Technology, University College TATI (UC TATI), 24000 Kemaman Terengganu, Malaysia. Tel./Fax +6-09-8601000/+6-09-8635863,

Email: *redhwan323@gmail.com, maminullah@uctati.edu.my.

Abstract. The present study evaluate the performance of SiO_2 nano cutting fluid in CNC turning of AL319 Aluminium Alloy combined with MQL technique. The cutting performance namely surface roughness, cutting temperature and tool wear were investigated against feed rate and spindle speed. The spindle speed of 1000 to 1800 RPM and feed rate of 0.10 to 0.20 mm/rev were used as the input parameter. While the MQL nozzle pressure is kept constant at 0.5 MPa. The cutting performances of SiO_2 nano cutting fluid with three volume concentrations (0.5, 1.0 and 1.5%) were then compared to the conventional CNC cutting fluid. Beforehand, the suspended form of SiO_2 is dispersed in CNC conventional coolant base. One-step and dilutions methods were used in preparing nano cutting fluid. Response surface method (RSM) via Face centered design (FCD) was employed in designing the experimental work. The stability of SiO_2 Nano cutting fluid is observed via visual sedimentation. The experimental outcome depicts lowest cutting temperature of 28.4°C and surface roughness of 1.08902 Ra when SiO_2 nano cutting fluid with 1.5% volume concentration is employed. It can be concluded that the higher the SiO_2 Nano fluid volume concentration, the better the surface roughness quality and lowering the cutting temperature. While no significant result shown in tool wear due to short period of experimental cutting time and relative soft material coupled with coated insert carbide used. Longer cutting time with higher concentration of SiO_2 nano cutting fluid is recommended in future work in order to obtain significant result in tool wear.

1. Introduction

Coolant is one of alternative when come to reducing the temperature and provide proper lubrication any mechanical equipment or machinery, continuous movement of mechanical part between metal to metal generate excessive heat and can cause mechanical failure during operation. Many researcher and scientist develop various of coolant/lubrication to enhances the thermal performance during machining. For the past decade, nanofluid are used by dispersing nanoparticle into base coolant. The benefits of nanofluid is amplification in term of thermal performance [1-3], viscosity, tribology [4, 5], and heat transfer coefficients compared to the base fluid. Nano sized particles (1-100 nm) suspended inside which are metal or metal oxide. The base fluid can be distilled water, ethylene glycol, lubricant or another type of lubrication.

Preparation nanofluid or nano-lubricant can be done by 2 methods, either one-step method or two-step method. Most commonly used to prepare nanofluid is two-step method because economical cost compares to one-step method, which nanoparticle are producing in a massive scale by-way-of the two-step preparation. The first step is to create the nanomaterials dry powders by thermal decomposition and photochemical routines, salt reduction of transition metal methods metal vapour synthetization and electrochemical synthetization techniques, and ligand reduction. Meanwhile one-step preparation method is a process in which the nanoparticles are directly developed by physical vapour deposition (PVD) techniques or methods of fluid chemicals for synthetization nanofluid. Even though two-step method is economical, but nanoparticle have liability to aggregate due to high surface.

Nanoparticle Silicon oxide, SiO_2 are used in the making of nanofluid by using one step method which nanoparticle added into the base coolant. SiO_2 nanoparticle characteristics of high strength and hardness and inexpensive make SiO_2 nanoparticle widely used in turning, drilling, milling and grinding application [6].

MQL/NDM technique is the process sprayed the very small amount of lubricant/coolant mixed with air to aerosol at high pressure in the cutting zone by the help nozzle. MQL offers a safer, greener alternative to traditional fluid supply, allowing manufacturers to follow sustainability strategies and incorporate the capabilities of greener production. The MQL system incorporate of cutting fluid, compressor, fluid reservoir, air-oil mixing chamber, external pipes and nozzle. Main product used for machining with MQL system are fatty alcohols and synthetic esters (vegetable oil) [7]. Researcher [8] investigated the performance parameter of MQL grinding and found that MQL provide an extraordinary quality of lubrication which it turns can improve surface quality. This is supported by [9] whom studied the influence of the auxiliary wheel cleaning jet (WCJ) under MQL using aluminium oxide grinding wheel and obtained MQL + WCJ method showed outstanding performance compare to traditional MQL in term of cooling capacity in dissipate heat from grinding zone.

2. Methodology

The method of preparing the nanofluid is major aspect and it vital to be handled properly according to formulation. Solid nanofluid can contribute to numerous properties enhancement once apply to selected application and it can be achieved by proper preparation method. In this paper, one-step method were used to prepare single nanofluid which is SiO_2, the formulation for this single nanofluid need to be calculate using different volume concentration for different concentration of nanofluid as be done by following previous researcher. This nanofluid SiO_2 was create specially in the machining application.

$$\Delta v = v2 - v1 = v1(\frac{\varphi 1}{\varphi 2}\text{-}1) \tag{1}$$

2.1. Preparation of Nano Cutting fluid

Synthesis of nanoparticle is the first step of good nanofluid. Nanoparticle were used in the preparation of nanofluids is SiO_2 nanoparticle suspended in water base with concentrations 25 wt% from US research nanomaterials Inc. SiO_2 nanoparticle were 50 and 22 nm, respectively. In this paper, nanofluid in base fluid mixture of distilled water and base coolant with volume ratio of 95:5(W: BC). The characteristic of SiO_2 nanoparticle and water/base coolant mixture are presented in table 1. Method of preparation are the SiO_2 nanoparticle were dispersed into base fluid mix with base coolant for 30minute using magnetic stirrer, then the water ultrasonic bath were used to assist the breakdown of the agglomeration between the nanoparticle from 30 minute. There sample were prepared by volume concentration 1.5%, 1% 0.5%. During time period of sonication time, after passing optimal time sonication process the sample will enter the phase which it causes drawback in term agglomeration and fouling emanate to rapid sedimentation [10].

Table 1: The Characteristic of SiO_2 Nanoparticle and Water/Base Coolant Mixture

Characteristic	SiO2	Water/BC
Purity	99.99	99.5
Colour	Colourless	Colourless
Size (nm)	22	-
Concentration (wt%)	25	-
Density (kg/m^3)	2220	-

2.2. Stability Evaluation

Stability evaluation is vital step to ensure the best performance of the nanofluid as done by previous researchers [11, 12]. Nanofluid which had be expose to sonication process will outgoing amplification of its stability and to mitigate the agglomeration size [13]. After sonication process, the sample will be left for sedimentation for 30 days and visual sedimentation where be observed in Fig. 1. The nanofluid is consider to be stable when the concentration or particle size of supernatant particles were kept constant throughout 30 days [14].

Figure 1. Visual sedimentation of SiO_2 Nano Cutting fluid

2.3. Experimental setup

The experiment was conducted using a Mori Seiki NL2500 CNC machine. The parameters of the turning are presented in Table 2. During data collection, Mitutoyo SJ-210-NATA Surface Mitutoyo SJ-210-NATA Surface The surface roughness of each workpiece was measured by using Mitutoyo SJ-210-NATA Surface. The Cutting temperature was measured by Non-Contact Laser Infrared Thermometer Gun (DT-8280) and Ziess Stemi 2000-C Microscope Profile Optical Microscope and computer were used to record and measure data for tool wear the experimental setup is shown in Fig. 2 while the experimental system is shown in Fig. 3.

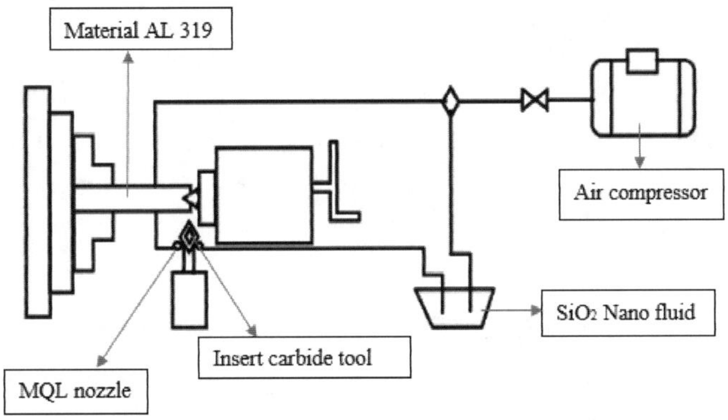

Figure 2. Experimental scheme for CNC Turning

This experiment was conducted at CNC turning where experiment were setup as shown in the figure 3. Laser thermometer and nozzle MQL were setup and attached to the machine using magnet to make sure there is no vibration during measuring data process. The experiment use AL 319 aluminium alloy as the work piece. The MQL nozzle were setting at 0.5Mpa air pressure to give the best and minimum use of nano-cutting fluid. By using the formula (2), RPM of the spindle will change relative to the diameter metal cutting process. The actual diameter for the starting workpiece is 110mm, cutting speed of the aluminium is 400 mm/min referring to the material data sheet. Each 20mm of metal been remove, the spindle speed will increase 400 mm/min, incremental spindle speed change in proportion to level 1 to level 3. RSM is employed for design of experiment and data analysis as done by previous researcher [15].

$$\text{Spindle Speed } (S) = (V*1000)/(\pi*D) \tag{2}$$

Figure 3. Experimental CNC Turning Setup

Table 2: Experimental Parameter of CNC Turning Machining

Control Factor	Symbol	Units	Level 1	Level 2	Level 3
Spindle speed	V	RPM	1000	1400	1800
Feed rate	F	Mm/rev	0.10	0.15	0.20
Concentration Nano fluid	Np	Vol%	0.5	1.0	1.5

3. Results and Discussion

For the result of parameter which consists of temperature, surface roughness and tool wear will measure and record using the specific equipment and machine. The result will show the highest and lowest data of each parameter after running the experiment.

Table 3: Result of Experiment RSM

STD	RUN	BLOCK	Volume Concentration (vt%)	Spindle speed (RPM)	Feed rate (mm/rev)	Temperature (C)	Surface roughness(Ra)	Tool wear (µm)
5	1	Block 1	1.5	1000	0.10	28	0.825	922.599
19	2	Block 1	1.0	1400	0.15	29.5	2.052	842.099
11	3	Block 1	1.0	1400	0.10	29.2	1.06	1143.5
9	4	Block 1	1.0	1000	0.15	29.5	2.104	1191.49
2	5	Block 1	0.5	1800	0.10	30.1	1.346	1023.96
14	6	Block 1	1.5	1400	0.15	29.2	2.285	946.411
1	7	Block 1	0.5	1000	0.10	31	0.835	1058.7
6	8	Block 1	1.5	1800	0.10	29	1.376	997.012

20	9	Block 1	1.0	1400	0.15	29.5	2.924	1011.94
7	10	Block 1	1.5	1000	0.20	28.5	2.778	1074.99
15	11	Block 1	1.0	1400	0.15	28	2.261	1175.5
4	12	Block 1	0.5	1800	0.20	30.5	2.21	768.93
3	13	Block 1	0.5	1000	0.20	29.5	2.863	760.282
8	14	Block 1	1.5	1800	0.20	28.5	2.712	922.738
18	15	Block 1	1.0	1400	0.15	27.5	2.15	975.672
17	16	Block 1	1.0	1400	0.15	28.5	2.196	1067.31
10	17	Block 1	1.0	1800	0.15	28.6	2.038	1197.02
12	18	Block 1	1.0	1400	0.20	29.5	1.873	1058.7
16	19	Block 1	1.0	1400	0.15	28.8	2.202	1320.6
13	20	Block 1	1.5	1400	0.15	30.2	2.439	779.104

3.1. Surface Roughness

Surface roughness was measured by using profilometer, Mitutoyo SJ-210 Surface Roughness Tester. From the graph shown the comparison surface roughness between three different concentrations 0.5, 1.0, 1,5 (vt%). The lower value is the best surface roughness produce by the higher contain nanofluids concentration 1.5 (vt%) which is produce for Ra is 0.0.825 µm. while for the 1.0 (vt%) nanofluids concentration the Ra was measured is 1.06 µm. For the 0.5 (vt%) nanofluids concentration the Ra was measured 1.346 µm and for the true conventional coolant was give the higher Ra is 1.886 µm. It justifies the higher nanofluids concentration will give the lowest and the best Ra for this experiment. The experiment them compared to the pure MQL coolant and it prove that the higher concentration of nanofluid will produce the better result of surface roughness such shown in Fig 4. Previous researcher [16] investigated the optimum uses of SiO_2 nano lubrication parameters in hard turning AISI4140 steel and obtaining Surface roughness enhance with 0.5%wt concentration in mineral oil, less air stream and a 30degree nozzle orientation angle.

From the table 3 shown the comparison surface roughness between feed rate (mm/min). The lowest value of feed rate will produce best surface roughness which is the lower surface roughness Ra. For the lowest feed rate 0.1, the surface roughness Ra was average at 0.825(µm). Then for the medium feed rate which 0.15 the surface roughness Ra was average at 2.038 (µm). Lastly for the higher feed rate 0.2 and get the higher surface roughness Ra was average at 2.21 (µm). This graph shows that feed rate plays important role in measuring surface roughness.

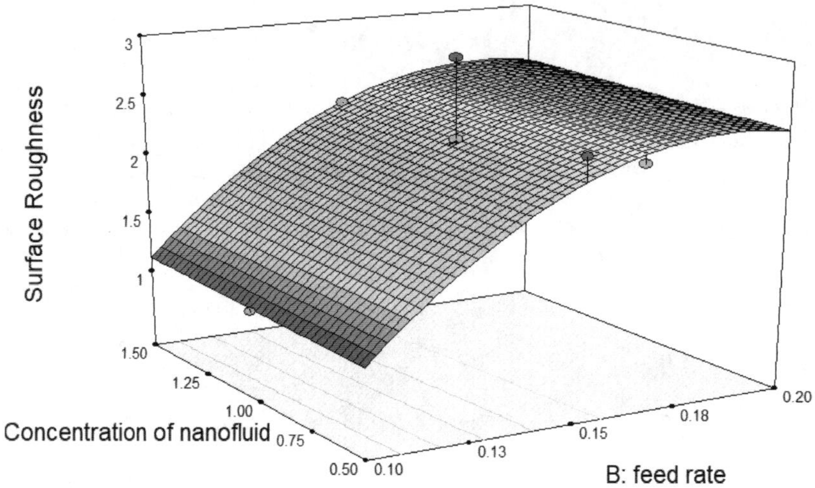

Figure 4. Surface Roughness Result Between Concentration and Feed Rate

3.2. Cutting Temperature

Cutting Temperature was measured by using Non-Contact Laser Infrared Thermometer Gun (DT8280) at distance 5 centimetres. This measurement can read a temperature with range (-50°C to 280°C). From the graph shown the comparison cutting temperature between three different concentrations 0.5, 1.0, 1.5 (vt%). The higher value of nanofluids concentration (vt%) produce the lower of cutting temperature. For the higher contain nanofluids concentration 1.5 (vt%) which is produce for cutting temperature is 29.2 (°C). while for the 1.0 (vt%) nanofluids concentration the cutting temperature was measured is 29.2 (°C). For the 0.5 (vt%) nanofluids concentration the cutting temperature was measured 31. (°C) and for the pure conventional coolant was give the higher cutting temperature is 32.5(°C). It justifies the higher nanofluids concentration will give the lowest cutting temperature for this experiment [17].

From the Figure 5 shown the comparison cutting temperature between three feed rate 0.1, 0.15, 0.2, feed rate (mm/min-1). The higher value of feed rate produces the higher of cutting temperature. For the higher contain nanofluids concentration 1.5 (vt%) which is produce for cutting temperature is 29.1, 29.5, 29.7 (°C). while for the 1.0 (vt%) nanofluids concentration the cutting temperature was measured is 30.2, 30.4, 30.7 (°C). For the 0.5 (vt%) nanofluids concentration the cutting temperature was measured 31.2, 31.5, 31.8 (°C) and for the 0.0 (vt%) nanofluids concentration or true conventional coolant was give the higher cutting temperature is 32.3, 32.8, 33.1 (°C). it vindicates the lowest feed rate will give the lowest cutting temperature for this experiment.

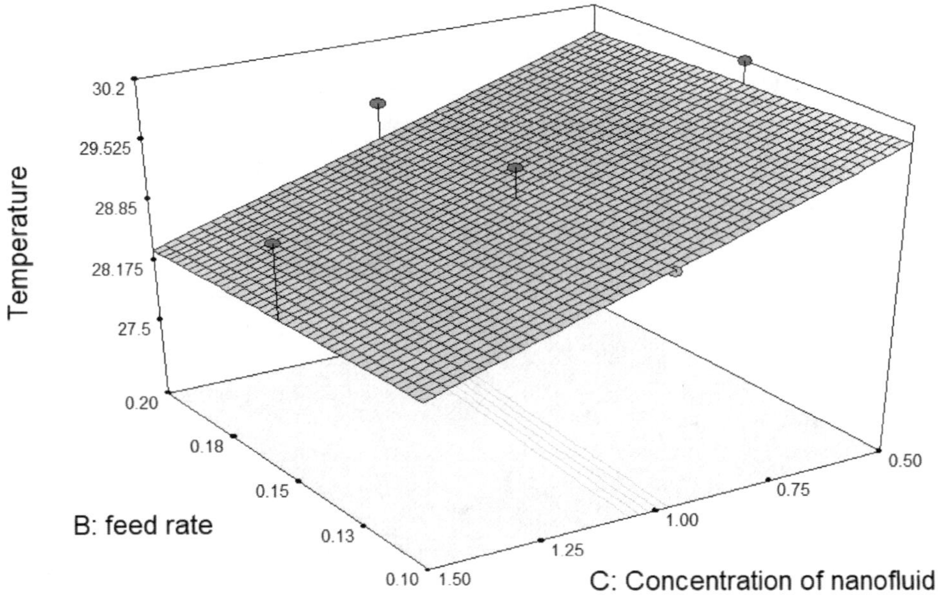

Figure 5. Cutting Temperature Between Feed Rate and Concentration of Nanofluid

3.3. Tool Wear

Figure 6. Lowest and highest tool wear

For the result of tool wear, Figure 7 the lowest result was recorded is 779.104um which is using 1.5 volume concentration of Nano fluid, 1400 RPM of spindle speed and 0.15 feed rate. The highest result is 1320.6um which is using 1.0 volume concentration of Nano fluid,1400 RPM of spindle speed and 0.15 feed rate.

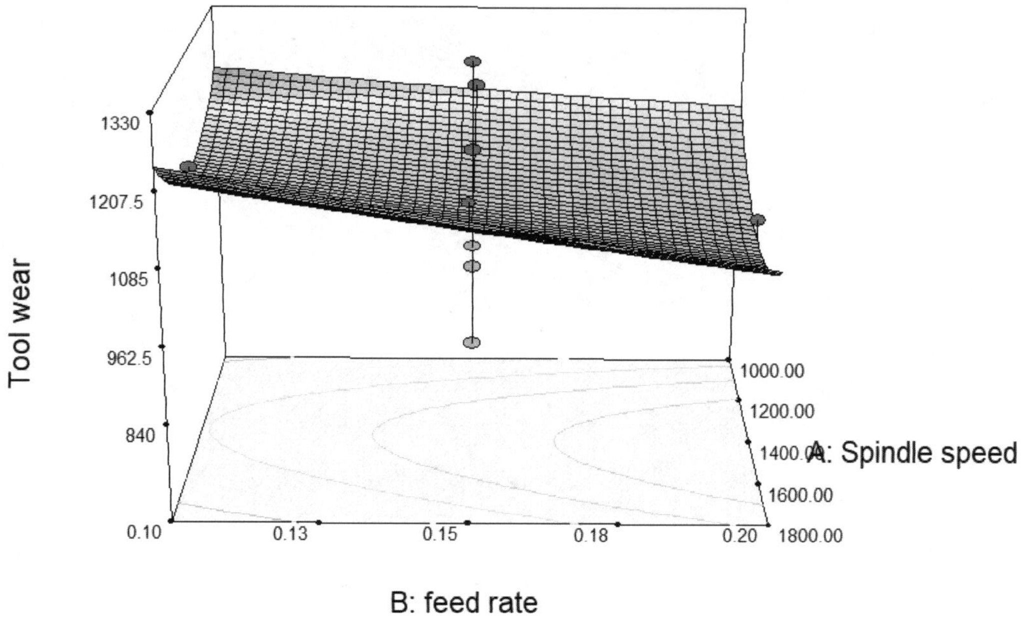

Figure 7. Tool Wear Result Feed Rate Vs Spindle Speed

4. Conclusions

According to the experiment, the main objective of this project is to investigation was conducted to characterize the SiO_2 nano-cutting fluid to be use on CNC turning lathe machine and to study the SiO_2 nano-cutting fluid performance in Turning of Silicon Oxide (SiO_2) with MQL technique.
In this investigation, the aim was to assess the objective experiment is achieving and the result showed the containing huge amount of volume concentration SiO_2 nanofluid improve the surface quality and cutting temperature but cannot justify the tool wear because of the factor that have been told. The following conclusion can be drawn to describe the performance of different concentration of SiO_2 at AL319 in CNC Milling.

- SiO_2 nanofluids improved surface roughness quality and cutting temperature.
- By using higher volume concentration of SiO_2 nanofluid the minimum surface roughness and cutting temperature was obtained.
- The best volume concentration of SiO_2 nanofluid is 1.5 will result the best value of surface roughness Ra=1.08902 μm and cutting temperature 28.395°C.

Acknowledgments

The authors would like to thanks to the UC TATI for sponsoring the work study under Short Term grant (STG) with grant STG 9001-1902

References

[1] N. Zawawi, W. Azmi, A. Redhwan, and M. Sharif, "Thermo-physical properties of metal oxides composite nano lubricants," *J Mech Eng,* no. **1**, pp. 28-38, 2018.

[2] A. Redhwan, W. Azmi, M. Sharif, and N. Zawawi, "Thermal conductivity enhancement of Al_2O_3 and SiO_2 nanolubricants for application in automotive air conditioning (AAC) system," in *MATEC web of conferences*, 2017, vol. **90**: EDP Sciences, p. 01051.

[3] A. Redhwan, W. Azmi, M. Sharif, R. Mamat, and N. Zawawi, "Comparative study of thermo-physical properties of SiO_2 and Al_2O_3 nanoparticles dispersed in PAG lubricant," *Applied Thermal Engineering,* vol. **116**, pp. 823-832, 2017.

[4] N. Zawawi, W. Azmi, A. Redhwan, and M. Sharif, "Coefficient of friction and wear rate effects of different composite nanolubricant concentrations on Aluminium 2024 plate," in *IOP Conference Series: Materials Science and Engineering*, 2017, vol. **257**, no. **1**.

[5] A. Aminullah *et al.*, "Tribology investigation of automotive air condition (AAC) compressor by using Al_2O_3/PAG nanolubricant," *J Mech Eng,* pp. 49-61, 2018.

[6] S. Khatai, R. Kumar, A. K. Sahoo, A. Panda, and D. Das, "Metal-oxide based nanofluid application in turning and grinding processes: A comprehensive review," *Materials Today: Proceedings,* 2020.

[7] Z. Said *et al.*, "A comprehensive review on minimum quantity lubrication (MQL) in machining processes using nano-cutting fluids," *The International Journal of Advanced Manufacturing Technology,* vol. **105**, no. **5-6**, pp. 2057-2086, 2019.

[8] M. Bhuyan, A. Sarmah, K. Gajrani, A. Pandey, T. Thulkar, and M. Sankar, "State of art on minimum quantity lubrication in grinding process," *Materials Today: Proceedings,* vol. **5**, no. **9**, pp. 19638-19647, 2018.

[9] R. L. Rodriguez *et al.*, "Evaluation of grinding process using simultaneously MQL technique and cleaning jet on grinding wheel surface," *Journal of Materials Processing Technology,* vol. **271**, pp. 357-367, 2019.

[10] N. A. C. Sidik, M. M. Jamil, W. M. A. A. Japar, and I. M. Adamu, "A review on preparation methods, stability and applications of hybrid nanofluids," *Renewable and Sustainable Energy Reviews,* vol. **80**, pp. 1112-1122, 2017.

[11] M. Sharif, W. Azmi, A. Redhwan, R. Mamat, and G. Najafi, "Energy saving in automotive air conditioning system performance using SiO 2/PAG nanolubricants," *Journal of Thermal Analysis and Calorimetry,* vol. **135**, no. **2**, pp. 1285-1297, 2019.

[12] A. Redhwan, W. Azmi, M. Sharif, R. Mamat, M. Samykano, and G. Najafi, "Performance improvement in mobile air conditioning system using Al_2O_3/PAG nanolubricant," *Journal of Thermal Analysis and Calorimetry,* vol. **135**, no. **2**, pp. 1299-1310, 2019.

[13] M. Sharif, W. Azmi, A. Redhwan, N. Zawawi, and R. Mamat, "Improvement of nanofluid stability using 4-step UV-vis spectral absorbency analysis," *Journal of Mechanical Engineering,* pp. 233-47, 2017.

[14] M. Sharif, W. Azmi, A. Redhwan, and N. Zawawi, "Preparation and stability of silicone dioxide dispersed in polyalkylene glycol based nanolubricants," in *MATEC web of conferences*, 2017, vol. 90: EDP Sciences, p. 01049.

[15] A. A. M. Redhwan, W. Azmi, M. Sharif, N. Zawawi, and S. Z. Ariffin, "Utilization of Response Surface Method (RSM) in Optimizing Automotive Air Conditioning (AAC) Performance Exerting Al_2O_3/PAG Nanolubricant," in *Journal of Physics: Conference Series*, 2020, vol. **1532**, no. **1**: IOP Publishing, p. 012003.

[16] M. Sayuti, A. A. Sarhan, and F. Salem, "Novel uses of SiO_2 nano-lubrication system in hard turning process of hardened steel AISI4140 for less tool wear, surface roughness and oil consumption," *Journal of Cleaner Production,* vol. **67**, pp. 265-276, 2014.
[17] Y. Su, L. Gong, B. Li, Z. Liu, and D. Chen, "Performance evaluation of nanofluid MQL with vegetable-based oil and ester oil as base fluids in turning," *The International Journal of Advanced Manufacturing Technology,* vol. **83**, no. **9-12**, pp. 2083-2089, 2016.

Effects of aircraft noise on residents nearby a Malaysian airport

Athirah, B.[1, *], **and Nurul Shahida M.S.**[1]

[1]Faculty of Mechanical Engineering, Universiti Malaysia Pahang, 26600 Pekan, Pahang, Malaysia

[*]Corresponding author: trahbadrulhisham@gmail.com

Abstract. In this study, the effects of aircraft noise on the residents living nearby Sultan Ismail Petra Airport, Kelantan, Malaysia, were investigated. A questionnaire survey involving 60 participants revealed that 46.67% agreed that aircraft noise is a source of disturbance in their daily routine. On-site noise level measurements were carried out over a 2-week period at seven locations (P1–P7) surrounding the airport, where each location represents a different distance from the noise source. The daily average noise levels were within 50–65 dB(A), which exceeded the World Health Organization's recommended maximum indoor noise level and recommended maximum outdoor noise level of 35 and 55 dB(A), respectively. However, the daily average noise levels were lower than permissible exposure limit of 90 dB(A). Based on the results, it can be concluded that exposure to the aircraft noise may affect the quality of life of nearby residents in the long term.

1. Introduction

With the escalating population growth over the years, more and more lands are being developed into cities, which include landed houses, condominiums, low-cost flats, offices, retail stores, restaurants, hotels, motels, medical centers, and transportation facilities (bus and railway stations, and airports). With urbanization, environmental noise becomes a major problem where the people are exposed to various types of noise such as leisure noise (e.g. noise from sporting events, concerts, and night clubs), transportation noise (road, railway, and aircraft noise), impulse noise (noise from guns and firecrackers), noise in the working environment (noise from office machines, telephone ringing, ventilation systems, machines in intensive care units, industrial machinery (lathe machine, punch press, cutting machine, jack hammer, furnace), and other types of noise (wind turbine noise, grass cutting noise, noise from construction and renovation sites [7, 11, 13, 23]. The development of airports and aircraft technologies induced high demand of airplanes activities. Even by improved the quality and the technology of the aircraft system, the level of noise emitted still cannot being controlled to the maximum range of human hearing. This aircraft noise is a noise impact that is generated during multiple stages of flight by any aircraft or its parts. The production of noise may trigger health issues due to this high level of noise.

As defined by the Oxford University Press (2020) noise is 'a sound, especially one that is loud or unpleasant or that causes disturbance'. According to Cerletti et al. (2020), environmental noise is one of the factors detrimental to public health and guidelines have been developed by the World Health Organization (WHO) to address this issue. Indeed, numerous studies have been carried out to investigate the effects of environmental noise on people's health with varying results. Some of the common complaints as a result of environmental noise are sleep disturbance with frequent awakenings, difficulties to fall asleep, and sleepiness and fatigue during daytime [7]. Sleep deprivation in turn, impairs the ability to think, understand, concentrate, and make decisions in both schools going children and working adults. Acute exposure to environmental noise can also lead to higher stress levels, annoyance, irritability, as well as speech intelligibility. Prolonged exposure to high levels of noise can also lead to more serious health problems such as tinnitus (sensation of sound even without external sound), hearing loss, hypertension (high blood pressure), ischemic heart disease (condition in which the heart does not receive sufficient blood and oxygen because the arteries are narrowed) as well as psychological health problems (emotional instability, anxiety, and nervousness) [5, 22].

Aircraft noise is no exception, especially to residents living within proximity of an airport and those working in an airport or aircraft hangar. This has been proven in independent studies performed at different parts of the world. Araghi and Yagobhi (2015) investigated the effects of aircraft noise on the physical and mental health of residents living nearby Birjand airport by questionnaire survey. Four domains were assessed: sleep disorders, nervousness, stress and mental illness, and speech intelligibility. The results showed that for residents in Zone 2 (Mehrshahr and Doulat Street, located farther from the airport) suffered from sleep disorders, but they did not suffer nervousness, stress and mental illness, and speech intelligibility. They suggested that both office and residential buildings should be sound insulated to minimize the impact of aircraft noise. In addition, the number of aircraft departures and arrivals should be limited from 10.00 p.m. to 8.00 a.m.

Besides, in Malaysia a study about the community noise at commercial business area was conducted in vicinity of Penang International Airport. Through this assessment, they would like to observe the noise pollution that occur and from this information, it can give idea or guidance for future building or area development for a more viable city in the terms of noise pollution [15].

Bartels, Rooney, and Müller (2018) investigated the effects of aircraft noise on annoyance of 1262 residents living nearby Cologne/Bonn Airport, Germany (which is particularly busy during night time) through telephone survey. The results revealed that the majority of the residents surveyed had high aircraft noise-induced annoyance. Likewise, Gasco, Asensio, and de Arcas (2017) highlighted that aircraft noise-induced annoyance is still high even though efforts have been made by the aviation industry to reduce noise emissions by producing quieter aircraft. Fujiwara, Lawton, and MacKerron (2017) quantified the relationship between aviation (airport location, aircraft noise, and airport activities) and the well-being of the people in and around airports in the United Kingdom. Those in areas with excessive aircraft noise levels had lower happiness and relaxation levels. Nassur et al. (2019), in their study involving 92 residents within vicinity of Paris Charles de Gaulle Airport and Toulouse-Blagnac Airport, France, found that exposure to the maximum sound pressure level during aircraft overflight increased the heart rate of the residents during sleep. Tezel et al. (2019) analysed the noise levels for the residents within proximity of two largest airports in Turkey (Izmir Adnan Menderes Airport and Ankara Esenboga Airport). The noise levels were assessed according to the European Noise Directive as well as annoyance and sleep disturbance indices. Those with acute exposure to excessive noise levels experienced an increase in blood pressure and heart rate, all of which can induce hearing impairment, hypertension, ischemic heart disease, annoyance, and sleep disturbance.

Ibhadode et al. (2018) assessed the aircraft noise exposure levels of people living/working within proximity of four airports in Nigeria (Ibadan, Benin-City, Warri, and Owerri). Four parameters were evaluated: (1) ambient noise level (ANL), (2) sound pressure level (SPL), (3) aircraft take-off noise level (ATNL), and (4) aircraft landing noise level (ALNL). They performed 120 periodic noise sampling surveys from January to December 2017 at 30 randomly-selected study locations according to the ISO 3891, ISO 1996-1, and ISO 1996-2 standards. The results showed that the SPL, ANL, ATNL, and ALNL were within a range of 103–115 dB(A), 52.3–64.1 dB(A), 69.6–87.7 dB(A), and 66.2–82.7 dB(A), respectively. The values exceeded the maximum noise levels recommended by the WHO: (1) 35 dB(A) (indoor), (2) 55 dB(A) (outdoor), and (3) 90 dB(A) (permissible noise limit for 8-h of exposure). The results revealed that the effects of aircraft noise were most pronounced at the neighborhood of Ibadan Airport, followed by Benin Airport, Owerri Airport, and least of all, Warri Airport. Those living within proximity of Ibadan Airport were at high risk of suffering from headaches, sleep disturbance, noise annoyance, and speech intelligibility in the long term.

A few studies have also been carried out in Malaysia. Ismail et al. (2010) investigated the effects of aircraft noise on the residents living nearby a Malaysian airport, where the noise levels were measured less than 3 km from the airport. The noise levels were measured for 24 h over a 30-day period to determine the following parameters: L_{eq} (equivalent noise level, defined as the noise level with the same energy as the original fluctuating noise for the same period), L10 (dominant noise level, defined as the noise level that exceeds 10% of the time during the whole measurement period), L90 (background noise level, defined as the noise level that exceeds 90% of the time during the whole measurement period), and L_{max} (maximum sound level). The overall noise level exceeded the Federal Aviation Administration day-night average sound level requirement of 65 dB(A). Most of the peaks in the Leq and L10 occurred from 12.00 p.m. to 5.00 p.m., consistent with an aircraft event (aircraft overflight). The results revealed that the aircraft events mostly contributed to the spike in noise levels of more than 65 dB(A), which will likely interfere with communication outside of the building, and lead to fatigue and vocal strain in the long term. However, the airport was not specified.

Abdul Aziz et al. (2017) investigated the noise levels inside a Lockheed C130-H Hercules aircraft hangar of the Royal Malaysian Air Force and the hearing status of the aircraft maintenance crew (63 members) in the No. 20 Squadron, Subang Air Base. They found that the highest noise levels at the center of the hangar were 92.2 and 94.2 dB(A) for daytime and night time, respectively, when all four turboprop engines were started (engines operated at 69.0–75.5% rpm). The noise levels were slightly higher during engine ground run (engines operated at 100.0% rpm), with a value of 95.3 and 97.3 dB(A) for daytime and night time, respectively. Moreover, 41.2% of the maintenance crew suffered from hearing impairment resulting from prolonged exposure to such high noise levels. Even more alarming, because of insufficient resources, only 8.57% of the maintenance crew were supplied with ear defenders. Lack of education regarding occupational noise hazard, arbitrary use of ear defenders, and reluctance to own a personal ear defender despite lack of resources were identified to be factors that led to the hearing impairment.

Recently, the authors [4] evaluated the effects of aircraft noise on 10 workers (comprising ground handling staff, aviation security officers, and general workers) working at the landing area of Sultan Ahmad Shah Airport, Kuantan, Malaysia. Questionnaire survey and noise level measurements were both conducted. The noise levels were measured over a 3-day period around the Boeing 737-800 aircraft with two turbofan engines. The results revealed that the workers were exposed to high levels of noise, which exceeded the permissible exposure limit (100 dB(A) for 2 h of exposure) stipulated by the Occupational Safety and Health Act. Three workers suffer from hearing problems, and one already received compensation from the Social Security Organization (SOCSO), Malaysia, for hearing loss (ruptured eardrum) due to noise exposure. This indicates that prolonged exposure to high levels of noise can lead to hearing loss.

Figure 1. Sultan Ismail Petra Airport, Kota Bharu, Kelantan.

All of the aforementioned studies indicate the harmful effects of environmental noise, particularly aircraft noise. In Malaysia, there is a lack of awareness concerning the effects of aircraft noise on the physical and mental health of the workers working in an airport or aircraft hangar, as indicated by the authors' previous study [4] and the work of Abdul Aziz et al. (2017), which is a cause for concern. Only one study [16] has been carried out on how aircraft noise affects residents living within proximity of a Malaysian airport, but even then, it cannot be ascertained which airport was the focus of their investigation. With this in mind, this study was conducted to investigate the effects of aircraft noise on the residents living nearby a Malaysian airport, where Sultan Ismail Petra Airport (SIPA), located in Kota Bharu, Kelantan (east coast of Peninsular Malaysia), was chosen as the study location. Kota Bharu is the capital city of the Kelantan state, with an estimated total population of 1.89 million people in 2019 [9]. According to newspaper articles [2, 16], the SIPA would be extended and upgraded to an international airport to accommodate 4 million passengers annually from its current capacity of 1.5 million passengers annually. The expansion and upgradation project are expected to begin in 2020 and complete in 2023. The SIPA served 2.06 million passengers in 2015, which exceeded its current capacity [2], indicating an increasing demand for flight travel. Hence, it can be reasonably assumed that the residents nearby the SIPA will be exposed more frequently to aircraft noise, which will affect their daily routine. For this reason, SIPA was chosen as the study location. This study will shed some light on how aircraft noise affects these residents and from here, some precautionary measures can be proposed for the stakeholders (government policymakers, airport operator, airport construction firms, and residents) to mitigate the effects of aircraft noise.

2. Methodology

The methodology adopted in this study consisted of two phases (Phase 1 and Phase 2), which are described in the following sub-sections.

2.1 Phase 1

Literature survey was first carried out by searching for journal articles, conference papers, Internet articles, newspaper articles, and scholarly books pertaining to aircraft noise published from 2010 to 2020. The literature survey was focused on environmental noise, the types of environmental noise, the effects of environmental noise on public health, aircraft noise, the effects of aircraft noise on residents living nearby airports, and the effects of aircraft noise in Malaysia. The articles were selected based on the criteria for formal meta-analysis [23]: the type of subjects/participants/respondents, the type of exposure, the parameters used to measure the response, and findings. The literature survey was conducted to lay the foundation to investigate the effects of aircraft noise on the residents living nearby the SIPA, Malaysia.

A questionnaire survey was carried out to elicit information on the effects of aircraft noise exposure on the residents living nearby the SIPA. The questionnaire was designed in English and Malay based on the 'Noise Exposure History Interview Questions' [17]. Observations were carried out at residential areas nearby the SIPA, and there were a school, mosque, houses, and stalls. The questionnaire included questions such as how often the participants were exposed to high noise levels and their daily activities that may produce noise, which will negatively affect themselves. The questionnaires were distributed to 60 participants (n = 60) at different locations within proximity of the airport. Informed consent was obtained from each participant before the questionnaire survey.

Figure 2 shows the photograph of a participant answering the questionnaire. All of the participants cooperated willingly in the interview and gave their best effort. Most of the villagers or permanent residents living nearby the airport were aware of the aircraft noise, but they were accustomed to the noise in their daily routine. The results of the questionnaire survey were analyzed.

Figure 2. Photograph of a participant during the questionnaire survey and interview session

2.2 Phase 2

Phase 2 was conducted after the results from Phase 1 revealed that aircraft noise had a significant impact on the residents living nearby the SIPA. The noise levels were measured using RS PRO sound level meter (Model: RS-95, RS Components, China) with a resolution of 0.1 dB(A) and accuracy of ±0.5 dB(A). Phase 2 was conducted over a 2-week period to ensure consistency in the data. The locations selected for noise level measurements were within the range of human hearing.

The noise level measurements were conducted at seven locations (P1–P7), comprising villages, residential areas, a university, and a school, as shown in Figure 3. Each location represents a different distance from the noise source in the airport (i.e. point of aircraft take-off and landing, designated as

IOP Conf. Series: Materials Science and Engineering 1068 (2021) 012010 doi:10.1088/1757-899X/1068/1/012010

CP) to determine the noise levels exposed to those within proximity of the airport. The noise levels were recorded during aircraft take-off and landing. The coordinates of Locations P1–P7 are presented in Table 1.

Table 1. Description of the points chosen for noise level measurements.

Point	Description	Coordinates	Measured distance from the airport (km)
P1	Kampung Kemumin (village)	6°10'03.4"N, 102°18'41.5"E	2.20
P2	Kampung Pengkalan Chepa (village)	6°10'12.7"N, 102°16'42.1"E	1.64
P3	Kota Bharu (village)	6°10'04.5"N, 102°16'36.2"E	1.82
P4	Kota Bharu (village)	6°09'58.8"N, 102°16'28.7"E	2.20
P5	Taman Kurnia Jaya (residential area)	6°09'41.0"N, 102°18'24.0"E	2.52
P6	Universiti Malaysia Kelantan, Taman Bendahara (university)	6°09'54.2"N, 102°17'01.0"E	0.95
P7	Sekolah Kebangsaan Parang Putting (school)	6°10'04.3"N, 102°18'14.8"E	1.67

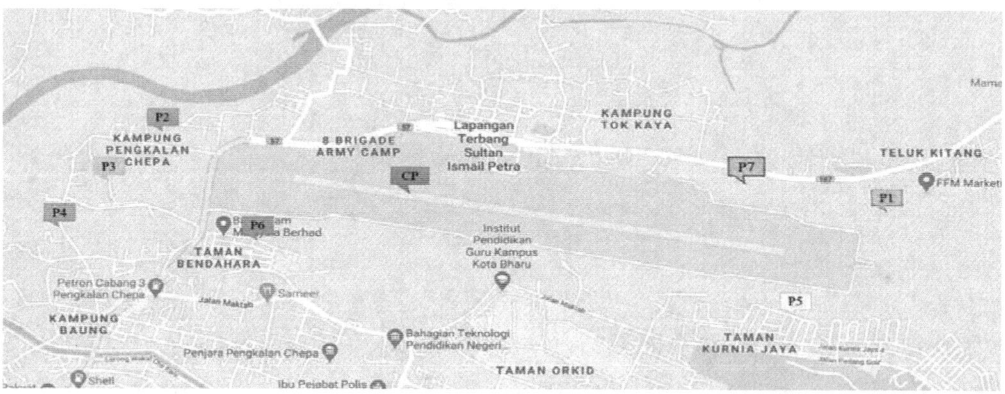

Figure 3. Locations selected for noise level measurements at Pengkalan Chepa, Kelantan (Source: Google Maps

The distance of each location from the airport was measured to assess the differences in the noise level readings. The data also included other on-site environmental factors such land transport noise; car, lorries, motorcycle, human talking, climate change, construction and others related nature noise. The sound level meter was also placed in different set-ups in relation to the individuals exposed to the noise such as at the school assembly hall (in which student activities were conducted) and residential areas where aircraft noise exposure was prevalent. The methodology adopted in this study is summarized in a flow chart, as shown in Figure 4.

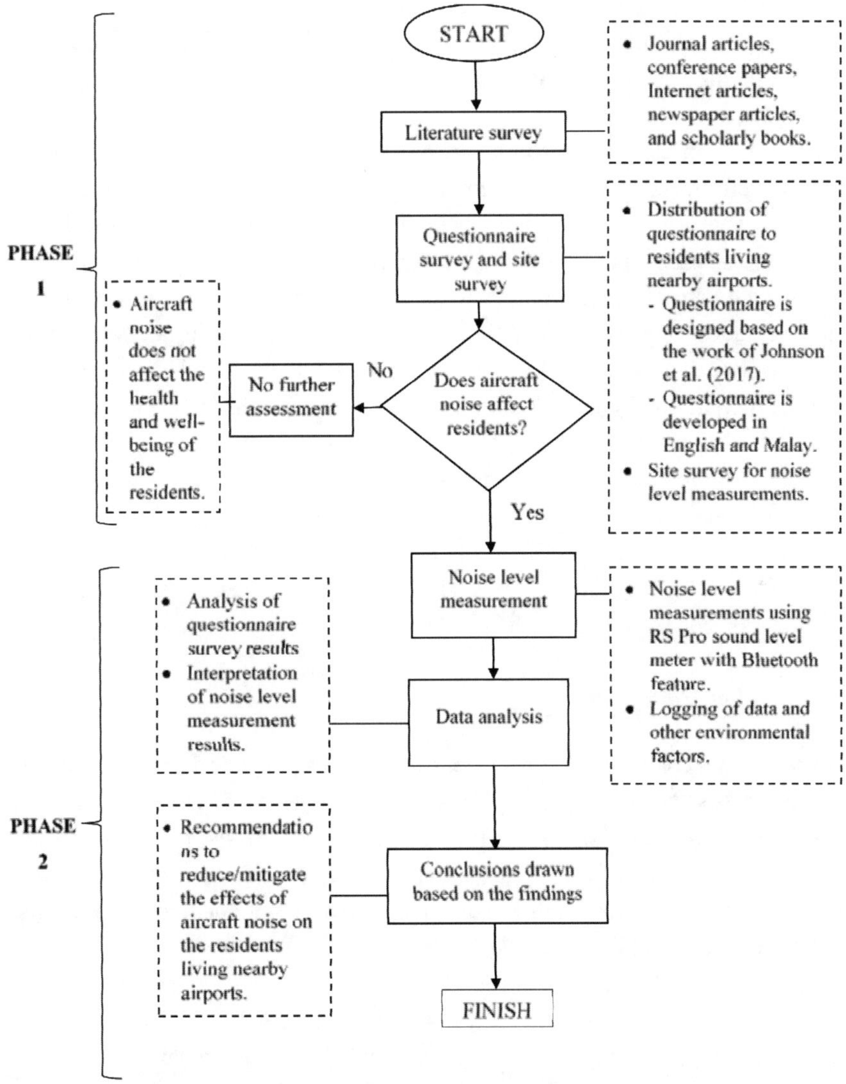

Figure 4. Flow chart of the methodology adopted in this study.

3. Results

3.1 Questionnaire result

Based on the literature survey, it is evident that prolonged exposure to high levels of noise has adverse effects on an individual's physical and mental health, where the most notable effect is the individual's hearing. To obtain the residents' opinions regarding aircraft noise exposure nearby SIPA, a questionnaire survey was carried out with 60 participants (n = 60) during the study. From these participants, 28 (46.67%) agreed that aircraft noise exposure causes disturbance to their hearing. The sampling size is deemed sufficient for this study, and nearly 50% of the participants agreed that aircraft noise affected them. The participants were chosen randomly. This number of responses amicable to say that people still concern about the aircraft noise exposure.

Besides, based on the results, some participants were not concerned regarding the aircraft noise because they were accustomed to the noise. Interviews were also conducted with the students at Universiti Malaysia Kelantan, which is located less than 1 km away from the airport, and the results revealed that the students were disturbed by the aircraft noise. The high noise levels emitted by the aircraft may adversely affect the students' psychological health in the long term.

During the questionnaire survey, three (3) teachers from Sekolah Kebangsaan Parang Puting agreed that the aircraft noise causes disturbance to the students. The disturbance usually occurs during the aircraft arrival and departure. The high noise levels emitted by the aircraft cause disturbance to both the students and teachers, and interfere with the teaching and learning process at school. The high noise levels also disrupt the students' concentration and are a source of distraction during classes.

Based on the questionnaire survey, the high noise levels do not only cause disturbance to hearing, but may also lead to other health problems such as annoyance, increased stress levels, sleep disturbance with awakening, and other physical/psychological health problems. These health problems will affect the individuals' quality of life.

3.2 Noise level measurements

The noise levels were recorded at the selected locations (P1–P7) during aircraft take-off and landing over a 2-week period. According to Mr Hasdy Yufaais, Head of Engineering of Malaysia Airports Holdings Berhad, an average of 29 flight activities were recorded daily. Based on the 2018 airport statistics [19], 24,481 aircraft operations were recorded at the SIPA throughout the year. The number of aircraft operations vary depending on consumer demands and the number of flights, which typically spikes during festive or holiday seasons.

Table 2 shows the noise levels recorded from each location. The data were analyzed based on the minimum, maximum, and average values of noise for each aircraft operation. The noise level measurements at each location were repeated six times to obtain the daily average noise level for each location. The measurements were repeated to ensure that the data were consistent. It shall be noted that the measured noise levels included environmental noise present in the absence of aircraft take-off and landing.

The results are presented in Table 2 with different aircraft landing points. The results were categorized based on the number of noise readings, minimum noise reading, maximum noise reading, average noise reading, distance of the location from the noise source, and daily average noise level.

The last two columns represent the difference between the maximum noise reading recorded and the WHO recommended maximum outdoor noise level (WHO RONL) [55 dB(A)] as well as the corresponding percentage difference.

Based on the percentage difference between the maximum noise reading recorded and the WHO RONL, the percentage increase in noise level was more than 50% for 7 out of 42 measurements. For these measurements, the maximum noise level was within a range of 80–100 dB(A), which is classified as hazardous noise level, and may lead to serious hearing impairment for prolonged exposures. The measured noise level was also influenced by the distance between the location and noise source. For locations with a distance of less than 2 km from the aircraft runway (e.g. P2 and P3), higher noise levels were recorded. Likewise, for locations with a distance of a more than 2 km from the aircraft runway, lower noise levels were recorded, but they still cause disturbance to the residents. The daily average noise level was within a range of 50–65 dB(A).

Table 2. Data obtained from the on-site noise level measurements.

Time (Location)	Noise level readings [dB(A)]				Distance between the location and noise source		Daily average noise level [dB(A)]	Difference between the maximum noise reading and the WHO RONL[a] [dB(A)]	Percentage difference between the maximum noise reading and the WHO RONL[a] [%]
	No.	Min	Max	Average	< 1.5 km from runway	> 1.5 km from runway			
Day 1 (P1)	1	46.1	69.3	54.4883		√		14.3	26.00
	2	46.0	82.5	59.4150		(2.20	55.4755	27.5	50.00
	3	45.4	71.2	52.5233		km)		16.2	29.45
Day 2 (P1)	1	43.4	82.2	56.9333		√		27.2	49.45
	2	46.5	72.4	56.3917		(2.20	57.4656	19.5	35.45
	3	52.4	74.5	59.0717		km)		17.4	31.64
Day 3 (P2)	1	50.5	84.9	57.0650	√			29.9	54.36
	2	51.9	80.4	60.0167	(1.64		58.0895	31.3	59.91
	3	46.1	86.3	57.1867	km)			25.4	46.18
Day 4 (P2)	1	47.3	80.0	55.3517	√			25.0	45.45
	2	46.8	70.8	58.9717	(1.64		56.3984	15.8	28.73
	3	45.2	85.6	54.8717	km)			30.6	55.64
Day 5 (P3)	1	45.0	85.8	56.9683	√			30.8	56.00
	2	43.7	85.3	55.9100	(1.82		56.0811	30.3	55.09
	3	44.8	85.2	55.3650	km)			30.2	54.91
Day 6 (P3)	1	53.3	80.0	65.4567	√			25.0	45.45
	2	53.2	74.2	59.2600			62.6506	19.2	34.91

	3	45.4	79.7	63.2350	(1.82 km)			24.7	44.91
Day 7 (P4)	1	46.3	71.6	54.7883		√ (2.00 km)	54.4200	16.6	30.18
	2	45.8	69.3	56.9217				14.3	26.00
	3	42.3	62.0	51.5500				7.0	12.73
Day 8 (P4)	1	44.3	76.2	58.3500		√ (2.00 km)	56.2594	21.2	38.55
	2	44.0	73.9	54.8800				18.9	34.36
	3	44.9	74.9	55.5483				19.9	36.18
Day 9 (P5)	1	46.0	72.1	56.6450		√ (2.52 km)	55.2605	17.1	31.09
	2	45.2	81.1	57.6983				26.1	47.45
	3	44.7	71.2	51.4383				16.2	29.45
Day 10 (P5)	1	43.8	70.8	52.7350		√ (2.52 km)	51.8455	15.8	28.73
	2	43.4	75.0	47.7033				20.0	36.36
	3	44.6	74.0	55.0983				19.0	34.55
Day 11 (P6)	1	51.9	73.8	56.6700	√ (0.95 km)		57.9056	18.8	34.18
	2	52.5	77.4	57.6250				22.4	40.73
	3	53.3	75.7	59.4217				20.7	37.64
Day 12 (P6)	1	53.6	80.0	61.3783	√ (0.95 km)		59.3811	25.0	45.45
	2	53.6	68.2	57.2450				13.2	24.00
	3	53.0	67.3	59.5200				12.3	22.36
Day 13 (P7)	1	57.2	78.4	64.8100	√ (1.67 km)		65.1567	23.4	42.54
	2	58.7	74.6	67.1117				19.6	35.64
	3	57.4	70.2	63.5483				15.2	27.64
Day 14 (P7)	1	53.9	76.4	60.8183	√ (1.67 km)		60.1511	21.4	38.91
	2	53.6	74.4	59.6483				19.4	35.27
	3	53.3	75.6	59.9867				20.6	37.45

[a] WHO RONL: Recommended maximum outdoor noise level by WHO

In general, most of the measured minimum noise levels did not exceed the WHO RONL [55 dB(A)]), where the values were within a range of 42–54 dB(A). The lowest and highest minimum noise levels recorded were 42.3 and 58.7 dB(A), respectively. Figure 5 shows the minimum noise levels recorded in this study. The WHO RONL [55 dB(A)] is also included for comparison, as indicated by the red line. The WHO RONL represents the recommended outdoor noise level and it can be seen that most of the minimum noise levels did not exceed this limit. However, three minimum noise levels exceeded the WHO RONL, all of which were recorded at Sekolah Kebangsaan Parang Puting. This is indeed expected because the school is located next to the airport runway, with a distance of 1.67 km away from the point of aircraft take-off and landing.

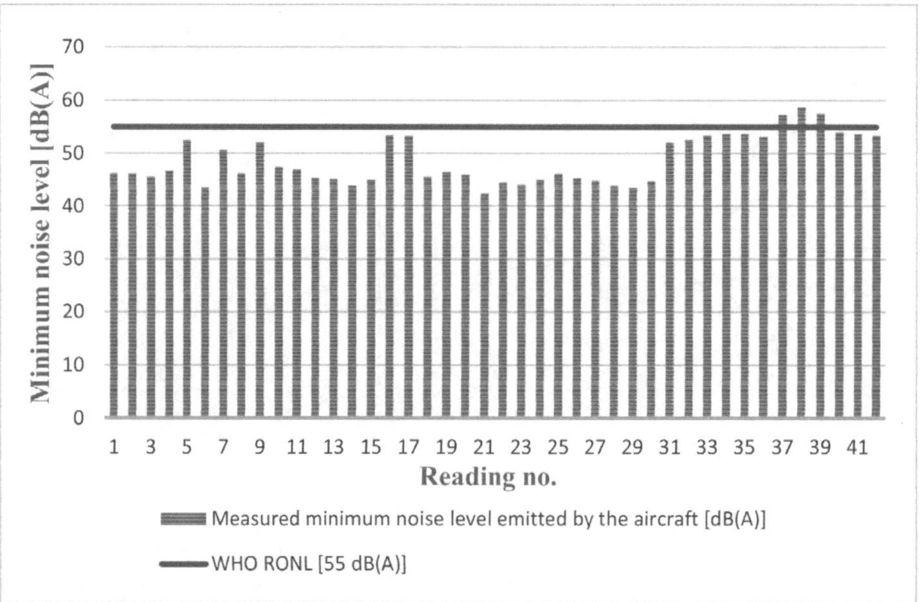

Figure 5. Comparison between the minimum noise levels emitted by the aircraft (measured at P1–P7) and the WHO RONL [55 dB(A)].

During the measurements, the maximum noise levels were recorded when the aircraft approached the sound level meter, which was the nearest point of noise level exposed to human hearing. Such high noise levels occur frequently daily, which leads to annoyance and disturbance to human hearing. In addition, such high noise levels are detrimental for individuals with sensitive hearing such as newborn babies, students, and the elderly. This will affect their mental health.

Figure 6 shows the maximum noise levels emitted by the aircraft measured over a 2-week period. It is evident that all of the maximum noise levels exceeded the WHO RONL [55 dB(A)]. Overall, the maximum noise levels were within a range of 60–90 dB(A). The maximum noise level was highest at P2, with a value of 86.3 dB(A). The measured maximum noise levels vary depending on the type of aircraft taking off and landing at the airport. The lowest maximum noise level was recorded at P4, with a value of 62 dB(A).

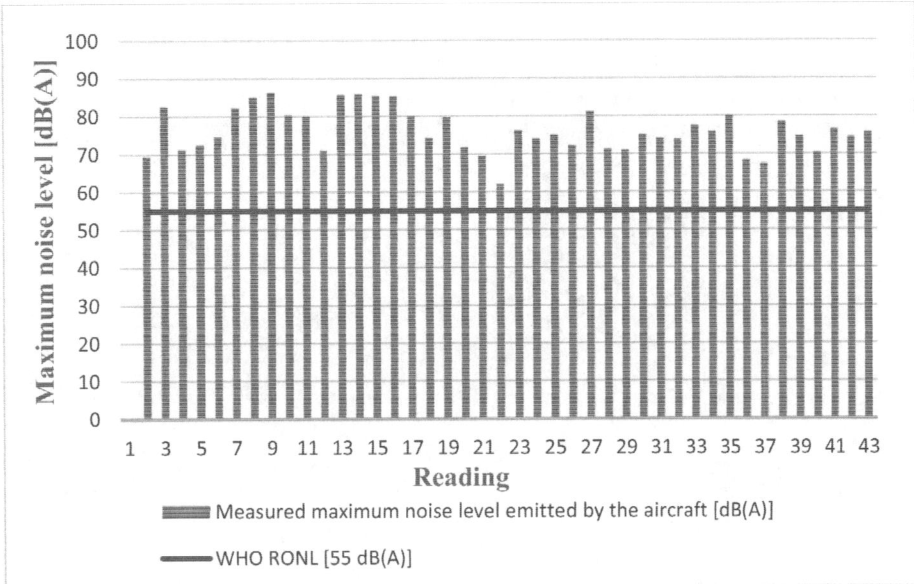

Figure 6. Comparison between the maximum noise levels emitted by the aircraft (measured at P1–P7) and the WHO RONL [55 dB(A)].

Figure 7 shows the comparison between the maximum noise levels emitted by the aircraft measured at P1–P7 and the WHO SRL [90 dB(A)], which is the permissible limit for 8-h of daily noise exposure. In general, the results were positive since the maximum noise levels did not exceed the WHO SRL. Despite this, safety precaution measures are needed to reduce the effects of aircraft noise exposure on those living within vicinity of the airport. However, some individuals seemed to be negligent regarding aircraft noise exposure, which is a concern since this will adversely affect their health and well-being in the long term.

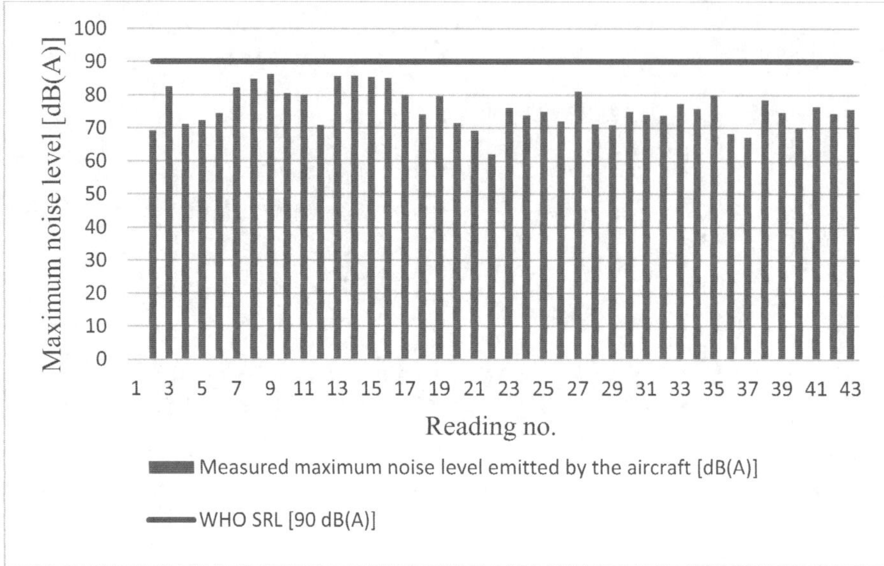

Figure 7. Comparison between the maximum noise levels emitted by the aircraft (measured at P1–P7) and the WHO SRL [90 dB(A)].

Conclusion

The effects of aircraft noise exposure on the residents living nearby Sultan Ismail Petra Airport, Kelantan, Malaysia, have been investigated in this study by questionnaire survey with 60 participants and on-site noise level measurements. Based on the results, it can be concluded that aircraft noise exposure causes disturbance and annoyance to the residents nearby the Sultan Ismail Petra Airport. About 46.67% participants agreed that aircraft noise exposure is a source of disturbance in their daily life. Based on the on-site noise level measurements, the lowest and highest noise levels recorded were 42.3 and 86.3 dB(A), respectively. The daily average noise level was within a range of 50–65 dB(A). The results showed that the measured noise levels exceeded the WHO recommended maximum indoor noise level [35 dB(A)] and recommended outdoor noise level [55 dB(A)]. However, the total noise levels did not exceed 90 dB(A), which is the permissible noise limit for daily 8-h exposure of noise. Based on the results, it is recommended that individuals should be more aware regarding aircraft noise exposure and reduce their time on outdoor activities especially during aircraft take-off and landing. Some of the actions that can be taken are improving the architectural designs and building orientation in relation to the local wind direction. In addition, most developed countries should consider adopting an aero polis concept instead of merely building airports. Finally, the use of anechoic and soundproof wedges as wall-claddings or soundproof material are also recommended.

Acknowledgment
This work was funded by Universiti Malaysia Pahang (grant no.: RDU180388). The authors wish to acknowledge Mr Mohd. Hasdy Yufaais bin Mat Sidi, Head of Engineering, MABH, for the helpful discussions.

References

[1] Abdul Aziz, A.R., Abd Aziz, S., Hassan, M.Z., 2017. Noise exposure among maintenance crews of C130H aircraft lead to the hearing impairment. Malays. *J. Public Health Med. Special Volume* (**1**), 89–94.

[2] Abdullah, S.M., Zainuddin, M.Z., 2018. RM483m upgrade to finish in 3 years. In: New Straits Times, 3 April, viewed 24 July 2020.

[3] Araghi, M., Yaghobi, M.M., 2015. The study of noise pollution caused by Birjand airport on the surrounding residents. *Indian J. [3Sci. Technol. 8* (**11**).

[4] Athirah, B., Nurul Shahida, M.S., 2019. Occupational noise exposure among airport workers in Malaysia: An ergonomic investigation. *J. Phys.: Conf. Ser.* **1262**, 012010.

[5] Australian Academy of Science, 2017. Health effects of environmental noise pollution, viewed 19 July 2020.

[6] Bartels, S., Rooney, D., Müller, U., 2018. Assessing aircraft noise-induced annoyance around a major German airport and its predictors via telephone survey – The COSMA study. *Transp. Res. Part D: Transp. Environ.* **59**, 246–258.

[7] Basner, M., McGuire, S., 2018. WHO environmental noise guidelines for the European region: A systematic review on environmental noise and effects on sleep. *Int. J. Environ. Res. Public Health 15* (**3**), 519.

[8] Cerletti, P., Eze, I.C., Schaffner, E., Foraster, M., Viennau, D., Cajochen, C., Wunderli, J.-M., Röösli, M., Stolz, D., Pons, M., Imboden, M., Probst-Hensch, N., 2020. The independent association of source-specific transportation noise exposure, noise annoyance and noise sensitivity with health-related quality of life. *Environ. Int.* **143**, 105960.

[9] Department of Statistics, Malaysia, 2020. Kelantan, viewed 24 July 2020.

[10] Fujiwara, D., Lawton, R.N., MacKerron, G., 2017. Experience sampling in and around airports. Momentary subjective wellbeing, airports, and aviation noise in England. *Transp. Res. Part D: Transp. Environ.* **56**, 43–54.

[11] Foo, K.Y. 2014. A vision of the environmental and occupational noise pollution in Malaysia. *Noise Health* 16 (**73**), 427–436.

[12] Gasco, L., Asensio, C., de Arcas, G., 2017. Communicating airport noise emission data to the general public. *Sci. Total Environ.* **586**, 836–848.

[13] Gerges, S.N.Y., Sehrndt, G.A., Parthey, W. (n.d.) Noise sources, viewed 23 July 2020.

[14] Ibhadode, O., Oyedepo, O.S., Ogunro, A.S., Azeta, J., Solomon, B.O., Umanah, I.I., Apeh, E.S., Ayoola, A.R., 2018. An experimental-assessment of human exposure-levels to aircraft noise-hazards in the neighbouring-environments of four Nigerian airports. *IOP Conf. Ser.: Mater. Sci. Eng.* **413**, 012080.

[15] Haliza, M. Z. N., Syazwan, N. A. M., Suriani, J. N., Hafiidz, J. M., Shaharudin, S. M., & Sarah, I. Z. (2018). Assessment of community noise at commercial business area (government buildings) in the vicinity of Penang International Airport. *AIP Conference Proceedings, 2030.*

[16] Idris, S.R., 2018. Kelantan's Sultan Ismail Petra Airport gets RM500m upgrade. In: New Straits Times, 27 September, viewed 24 July 2020.

[17] Ismail, A.R., Tahir, M.F.M., Nor, M.J.M., Haniff, M.H.M., Zulkifli, R., 2010. Assessment of noise impact: A case study due to aircraft activities. *In: 2nd Natl. Conf. Mech. Eng. Res. Postgrad. Stud. Universiti Malaysia Pahang, Pahang, Malaysia.*

[18] Johnson, T.A., Cooper, S., Stamper, G.C., Chertoff, M., 2017. Noise exposure questionnaire: A tool for quantifying annual noise exposure. *J. Am. Acad. Audiol. 28* (**1**), 14–35.

[19] MABH. (2018). Airports statistic 2018.

[20] Nassur, A.-M., Léger, D., Lefèvre, M., Elbaz, M., Mietlicki, F., Nguyen, P., Ribeiro, C., Sineau, M., Laumon, B., Evrard, A.-S., 2019. Effects of aircraft noise exposure on heart rate during sleep in the population living near airports. *Int. J. Environ. Res. Public Health 16* (**2**), 269.

[21] Oxford University Press, 2020. Noise, viewed 23 July 2020.

[22] Schreckenberg, D., Benz, S., Belke, C., Möhler, U., Guski, R., 2017. The relationship between aircraft sound levels, noise annoyance and mental well-being: An analysis of moderated mediation. In: *12th Int. Comm. Biol. Eff. Noise (ICBEN) Congr. Noise Public Health Probl.* ICBEN, Zurich, Switzerland.

[23] Śliwińska-Kowalska, M., Zaborowski, K., 2017. WHO environmental noise guidelines for the European region: A systematic review on environmental noise and permanent hearing loss and tinnitus. *Int. J. Environ. Res. Public Health 14* (**10**), 1139.

[24] Tezel, M.N., Sari, D., Erdol, M., Hamamci, S.F., Ozkurt, N., 2019. Evaluation of some health impact indices in two airports' domain. *Appl. Acoust.* **149**, 99–107.

Tribology of Electroless Nickel and Interlayer Coatings: A Review

A. Haniff Bharim[1], M. Zakwan Sarian[1] and N. Bahiyah Baba[1*]

[1]Faculty of Engineering Technology, University College TATI (UC TATI), 24000 Kemaman Terengganu, Malaysia. Tel./Fax +6-09-8601000/+6-09-8635863,
*Email: bahiyah@uctati.edu.my

Abstract. This review emphasizes on the development of electroless nickel composite coatings by incorporating different types of material and interlayers to improve the mechanical and tribological properties of the coatings. Electroless nickel coatings are the main focus of this review as it possesses excellent mechanical and tribological properties such as, hardness, wear and corrosion resistance. Besides, electroless nickel coating may behave as lubricating as well as wear resistance coating compared to other interlayer coatings. Tribology aspect is concern on the friction and wear phenomena which involved the moving parts that measures the overall systems' life, efficiency, and reliability. Interlayer coatings are applied to improve the adhesion between the substrate and the outer coating materials. Various materials are used as interlayers either pure or compound metallic elements.
Keywords: Tribology, Electroless Nickel, Composite coatings, Interlayer

1. Introduction

Electroless is an in-situ chemical process where a reduction reaction of a metallic ions in the solution oxidised by itself. The process was founded by Wurtz in 1844 then Brener and Riddle developed the electroless coating in 1946 [1]. Electroless process includes electroless nickel, electroless copper, electroless palladium, electroless gold [2]. The most common and widely used electroless process is electroless nickel process. Electroless nickel has been used in chemical, textile, printing, electrical, food processing and automotive to aerospace industries.
Electroless nickel process can be divided into conventional and composite coating. The conventional electroless nickel is the traditional electroless process without addition of particles. On the other hand, electroless nickel composites is combination of deposition of metallic nickel with various particle together. The conventional electroless nickel usually denoted as Ni-P where the main composition of the deposits are nickel and phosphorus. Electroless Ni-P is relatively cheap, easy to control, good wear and corrosion resistance as well as high hardness and good lubrication [3]. The focus subject of the review is on electroless nickel composite or sometimes called electroless nickel co-deposition.
Most of electroless nickel coating is homogeneous and uniform onto the substrate. The adhesion of the substrate and the coating are depending on the types of substrate. In order to improve the adhesion

between substrate and coating, an interlayer is introduced. There are various types of interlayer coating were investigated in the previous studies such as chromium [4], titanium nitride [5], copper [6] and others.

Electroless nickel is very well known for its high hardness, good wear and corrosion resistance which categorise them as having splendid tribological properties. The friction and wear characteristics are of the electroless nickel has been well investigated. The tribological properties of conventional electroless nickel has been enhanced by electroless nickel composites where the addition of particles either metallic, ceramics or polymers. The new variants electroless composites such as Ni–W–P, Ni–Cu–P, Ni–P–SiC, Ni–P–TiO_2 had shown improving their tribological properties [7].

2. Electroless nickel composite

Electroless nickel composite coating is a deposition of nickel together with particles either ceramics or polymers. The overview of electroless division is illustrate in Figure 1 below where the conventional electroless is a medium range phosphorus content and the composite is in low range phosphorus and the composite with metallic particles known as ternary alloys is in high range phosphorus content.

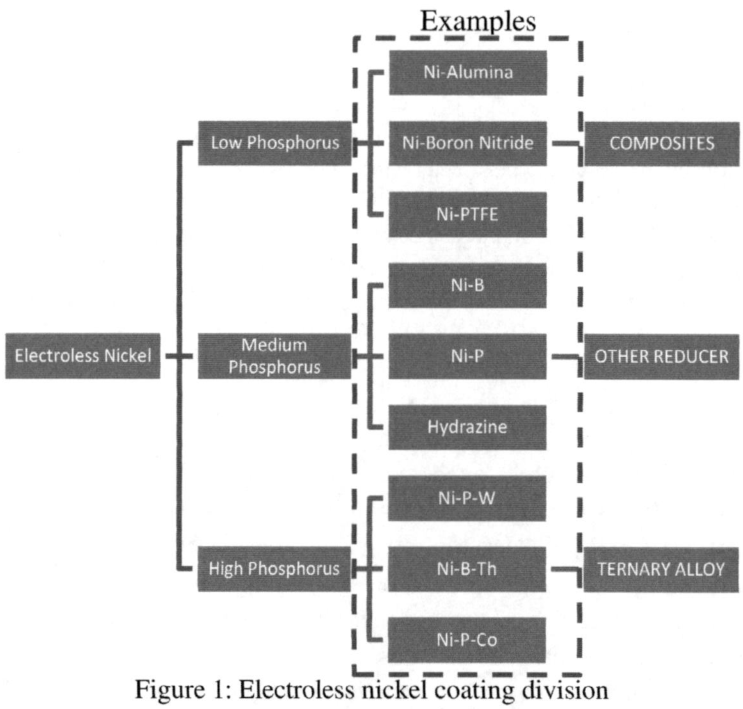

Figure 1: Electroless nickel coating division

The electroless nickel composite coating is deposited simultaneously between metallic nickel and the particles onto the surface of the substrate. The particles are wrapped by the matrix material that is the nickel, which are deposited onto the substrate [8]. There are many variables involved in ensuring a successful co-deposition such as, inert nature of particles, size of particles, bath composition, suspension of particles, method of incorporation of particles in bath, bath reactivity, plating rate, stirring rate, pH, bath temperature and the compatibility of the particles with metal matrix [9].

3. Electroless nickel composite coating

Electroless nickel composite coatings in the field of tribology, can be divided into lubricating composite coatings and wear-resistant composite coatings. Electroless composite coating acts as lubricating composites since electroless coatings are known to have smoothening effect above a critical substrate roughness therefore they are in general very smooth and lubricious [10]. The surface morphology of the coating similar to a cauliflower and the smooth and lubricious properties are due to their unique nodular microstructure [11].

High wear-resistant coating can be enhanced by incorporated of high hardness particles and different surface treatment lead to change in conventional microstructure of coating that modified friction behaviour. The incorporation of particle in electroless nickel composite is highly depend on the dispersion of co-deposited particle in the coating where the stability of particle dispersion with high surface energy and activity in the plating bath control the segregation and agglomeration of the particles [12]. The particle dispersion in the bath can be conveniently realized by capping the particles with appropriate surface-modifying agents [13]. Besides incorporation of inert particles, the wear properties of a coating are affected by numerous other parameters such the applied stress and the surface morphology the wear resistance of electroless nickel deposits depends on both phosphorus content and the type of post heat treatment applied.

The purpose of tribological based application of electroless coating is to impart both smoothness and hardness onto the surface as to reduce friction and wear [14]. Surface treatment such as heat treatment able to reduce the friction coefficient of various electroless coatings and minimization of the friction of the coating is a necessary [15]. Previous studies found that heat treatments of the coatings able to decrease the average and maximal roughness [16-17].

Heat treatment increases hardness and wear resistance however at higher temperature, the grain coarsening gives negative impact to the wear resistance [18]. Surface treatment such as laser irradiated onto the coatings induced low wear than that of as-deposited amorphous alloy and furnace annealed coatings [19].

Poor adhesion between substrate and coating do not provide good protection, because the adhesion evaluation of electroless nickel are still unsolved [20]. The existence of a stronger metal-to-metal bonding makes the electroless nickel coatings having better adhesion compared to the nickel electrodeposition.

4. Interlayer coating

Interlayers are coating introduced between substrate and the outer coating layer. It acts as intermediate to enhance the adhesion of interfacial bonding between the substrate and the outer coating. Other than electroless nickel coating as interlayers, there are many other intermetallic coatings that had been investigated. Other interlayers can be divided into copper-based, and titanium-based.

Copper (Cu)-based interlayer is used either in its pure composition or intermetallic. A self lubricating coating of Cu/Cu-MoS_2 was investigated where Cu is the interlayer [6]. It was found that Cu interlayer on high speed steel give better wear resistance than the uncoated ones. Another study used copper in term of transient liquid phase (TLP) bonded joints with nickel at 580°C which is comparatively low temperatures and it showed that the shear strength of joints increase with holding time [21].

Titanium (Ti)-based interlayers has widely used such as Ti, TiN, TiC, TiAlN and others. A study done on Ti deposition onto 316L stainless steel substrate with thickness 200 nm by magnetron sputtering in a closed field arrangement showed that the adhesion increased with increase in the interlayer thickness due to crystalline tribology superelastic property [22]. Interlayers of TiC and TiC/Ti(C,N)/TiN were investigated as an effective interlayer for diamond film onto high speed steel substrate [5]. This is because coated diamond directly onto high speed via chemical vapour deposition induced high diffussion carbon into steel, thus by adding TiC interlayer the adhesion was improved and no delamination was found.
Chromium (Cr)-based interlayers are another type of interlayers used to enhanced adhesion bonding between coating and substrate. CrZrN was coated onto AISI H13, high speed steel and tungsten carbide substrate with Cr interlayer in between using unbalanced magnetron sputtering [4]. The adhesion strength is proportional to the interlayer thickness as similar trend found for TiC interlayer [22]. It is shown that maximum value at 300 nm thickness of the interlayer, and then the adhesion strength showed decrement as the interlayer increased further to 450 nm. In another study, CrZrN coated on H13 steel substrate using DC magnetron PVD method used Cr as interlayer due to its close lattice parameter and cubic crystal structure to H13 substrate [23]. The suitability of the interlayers should consider the crystal structure that will give better adhesion in between. Bouzakis et. al used Cr/CrN-nanointerlayer onto carbide cutting insert via high power pulse magnetron sputtering (HPPMS) showed the adhesion bond is strongly depends on the substrate surface integrity and on the interlayer thickness [24]. It concludes that if interlayer thickness is adapted to the substrate roughness, the interlayer adhesion and cutting performance can be appreciably improved.

5. Conclusions

Tribological aspect of the conventional electroless nickel, electroless nickel composite and several interlayers were reviewed. The focus was given to materials, mechanical properties, deposition method, and friction and wear characteristics of these types of coatings. The extensive literature evaluated on the electroless nickel coating either with and without incorporated inert particles. The speciality of electroless nickle coatings are their ability to act as lubricating as well as wear resistance coatings. Whereas other types of interlayer coatings mostly act as wear resistance coating. Moreover, either electroless nickel or other interlayers coatings, show the thickness of the coating plays an important criteria for good adhesion between substrate and outer coating. Surface integrity and treatment affected the tribology friction, corrosion behaviour, and wear resistance in which improve the properties.

Acknowledgments
The author acknowledge the financial support of Ministry of Higher Education Malaysia for the fund under Fundamental Research Grant Scheme FRGS1/2018/STG07/TATI/01/1.

6. References

[1] Mallory G O, Hadju J B 1991 Electroless Plating: Fundamentals and Applications AESF, Orlando

[2] Azmah Hanim M A 2017 Chapter 3.15 Electroless Plating as Surface Finishing in Electronic Packaging in Reference Module in Materials Science and Materials Engineering **3** 220-229

[3] Baudrand D W 1978 *Metals Hand book* 8th edition

[4] Grips V K W, Selvi V E, Barshilia H C and Rajam K S 2006 *Electrochimica Acta* Effect of electroless nickel interlayer on the electrochemical behavior of single layer CrN , TiN , TiAlN coatings and nanolayered TiAlN / CrN multilayer coatings prepared by reactive DC magnetron sputtering **51(17)** 3461–3468

[5] Polini R, Pighetti Mantini F, Braic M, Amar M, Ahmed W, Taylor H & Jackson M J 2006 *Journal of Materials Engineering and Performance* Effects of Ti- and Zr-based interlayer coatings on hot-filament chemical vapor deposition of diamond on high-speed steel **15** 201–207

[6] Tongkun C, Shuting L, Meng Z 2015 *Surface and Coatings Technology* The friction and wear behavior of Cu/Cu-MoS_2 self-lubricating coating prepared by electrospark deposition **270(25)** 2015 24-32

[7] Sahoo P 2009 *Mater. Des.*Wear behavior of electroless Ni-P coatings and optimization of process parameters using Taguchi method **30** 1341–1349

[8] Agarwala R C & Agarwala V 2003 *Sadhana* Electroless alloy/composite coatings: A review **28(3-4)** 475–493

[9] Delaunois F, Lienard P. Heat treatments for electroless nickel–boron plating on aluminium alloys. Surf Coat Technol 2002;**160(2–3)**:239–48.

[10] Balaraju J N, Sankara Narayanan T S N & Seshadri S K 2003 *Journal of Applied Electrochemistry* Electroless Ni-P composite coatings **33(9)** 807–816

[11] Alirezaei S, Monirvaghefi S M, Salehi M, Saatchi A 2007 *Wear* Wear behavior of Ni–P and Ni–P Al_2O_3 electroless coatings **262** 978–85

[12] Dong D, Chen X H, Xiao W T, Yang G B, Zhang P Y 2009 *Appl Surf Sci* Preparation and properties ofelectroless Ni–P–SiO_2 composite coatings **255(15)** 7051–5

[13] Wu Y, Shen B, Liu L, Hu W 2006 *Wear* The tribological behaviour of electroless Ni–P–Gr–SiC composite **261(2)** 201–207

[14] Palaniappa M, Seshadri S K 2008 *Wear* Friction and wear behavior of electroless Ni–P and Ni–W–P alloy coatings **265(5–6)** 735–740.

[15] Hou K H, Jeng M C, Ger M D 2007 *J. Alloys Compd.* The heat treatment effects on the structure and wear behavior of pulse electroforming Ni-P alloy coatings **437** 289–297

[16] Vitry V, Delaunois F, Dumortier C 2008 *Surf Coat Technol* Mechanical properties and scratch test resistance of nickel–boron coated aluminium alloy after heat treatments **202(14)** 3316–3324.

[17] Tsujikawa M, Azuma D, Hino M, Kimura H, Inoue A 2005 *J Metastab Nanocryst Mater* Friction and wear behavior of laser irradiated amorphous metal 24–25 375–8

[18] Staia M H, Puchi E S, Castro G, Ramirez F O, Lewis D B 1999 *Thin Solid Films* Effect of thermal history on the microhardness of electroless Ni-P 355–356 472–479

[19] Riedel W 1991 Electroless nickel plating Ohio: ASM International

[20] Yan J, Xu Z, Wu G, Yang S 2004 *Scripta Materialia* Interface structure and mechanical performance of TLP bonded joints of Al_2O_3p/6061Al composites using Cu/Ni composite interlayers **51(2)** 147–150

[21] Abubakar T, Rahman M, Dowling D P, Stokes J, Hashmi M S J 2010 *Surface Engineering* Adhesion performance of TiN coating with amorphous NiTi alloy interlayer onto 316L stainless biosteel deposited by sputtering process **26(7)** 499–505

[22] Kim K, Kim H, La J and Lee S 2016 *Japanese Journal of Applied Physics* Effects of interlayer thickness and the substrate material on the adhesion properties of CrZrN coatings 2 1–6

[23] Taweesup K, Lertsakwanit R, Jiravanichkul J, Yongvanich N, Lothongkum G, and Visuttipitukul P, 2015 *Key Eng. Mater.* Effect of type and coating time of interlayer on properties and morphology of Cr-Zr-N film prepared by DC magnetron sputtering PVD on H13 steel 658 111–119

[24] Bouzakis K, Makrimallakis S, Katirtzoglou G, Skordaris G, Gerardis S, and Bouzakis E 2010 *Surf. Coat. Technol.* Surface & Coatings Technology Adaption of graded Cr / CrN-interlayer thickness to cemented carbide substrates' roughness for improving the adhesion of HPPMS PVD fi lms and the cutting performance **205(5)** 1564–1570

IOP Conf. Series: Materials Science and Engineering 1068 (2021) 012012 doi:10.1088/1757-899X/1068/1/012012

Electric Powered Wheelchair Trajectory Planning on Artificial Potential Field Method

I.M Sollehudin[1], P.M. Heerwan[12]* M.I Ishak[3], M.A. Zakaria[2,4]

[1]Faculty of Mechanical and Automotive Engineering Technology, University Malaysia Pahang, 26600 Pekan, Pahang, Malaysia

[2]Automotive Engineering Centre, University Malaysia Pahang, 26600 Pekan, Pahang, Malaysia

[3]Department of Mechanical Engineering, College of Engineering, University Malaysia Pahang, 26600 Pekan, Pahang, Malaysia

[4]Faculty of Manufacturing and Mechatronic Engineering Technology, University Malaysia Pahang, 26600 Pekan, Pahang, Malaysia

*Email: mheerwan@ump.edu.my

Abstract. An addendum of People with Disabilities (PWDs) worldwide quite worries the World Health Organization in which they are consist of people that incapability of doing their own daily routine without the aid of a third party. The third party may consist of a person or a specific equipment or tools that could aid for their daily mobility. This paper tends to help PWDs with their daily routine by introducing a path planning method that recognize the trajectory of the route by using the Artificial Potential Field Method (APF) that will act as an algorithm in a control system of an Electric Powered Wheelchair (EPW). The EPW will be modelled into the MATLAB and Simulink according to the parameters such as inductance, resistance, and driving torque that is exactly as an actual model. Artificial Potential Field Method (APF) are consist of two main force field, Repulsive Force Field (RFF) and Attractive Force Field (AFF). The Repulsive Force Field (RFF) is designated as obstacle on a path whereas influencing the EPW to navigate away from the obstacle. While the Attractive Force Field (AFF) act as goal or the destination that designated for the EPW in which the EPW will be navigated in a way by approaching towards the AFF. The RFF consists of spacing headway that defines the distance from the obstacle to the EPW and while the limit distance is the maximum distance for the EPW to navigate away from the obstacle before the EPW collides with the obstacle. The strategy is to control the trajectory of the EPW during the obstacle avoidance with yaw angular velocity of the EPW. With spacing headway, limit distance and yaw angular velocity results, the trajectory of the EPW during the obstacle avoidance could be determined.

1. Introduction

The rapid increment of population leads to increase of People with Disabilities (PWDs) worldwide. Researchers tirelessly developing the all sorts of device to improve the daily life of the PWDs. The Electroencephalogram-based brain–computer interfaces (BCIs) is a machine interactive technology that allows the user to interact with the wheelchair without using the peripheral muscles[1]. Moreover, researchers nowadays have actively developed numerous system that could aid the PWDs. The researcher also tends to build the smart wheelchair from the conventional wheelchair [2] that allows the system to be in customized condition suits with the type of impairment of the PWDs faced. Integrating head gesture navigation is also another approach that help PWDs for mobility. The user only uses the head gesture as the control input to navigate the wheelchair that comes with various speed for different direction depends on the head gesture applied. [3]

The development of autonomous wheelchair is also actively developed among the researcher. Some of the breakthrough is the semi-autonomous wheelchair that uses hand motor imagery (MI) and jaw clench to navigate the wheelchair[4]. The selection of control system in order to navigate the wheelchair is also important in which [5] uses the MPC control scheme to address the motion problems of an autonomous wheelchair in realistic environment. The problem arises with autonomous system is the path planning for the wheelchair. For recent years, the most used approaches to solve the path planning are Genetic Algorithm (GA), Particle Swamp Optimization (PSO), Artificial Potential Field (APF), and Ant Colony Optimization [6]. This method of path planning is widely used among the researchers to solve the path planning problems. The most used path planning is GA compared to the others. GA is able to find a valid and feasible path between two positions while avoiding obstacle and could optimize the route and providing the optimal path for the mobile robot [7].

Particle Swamp Optimization is a method used to search an optimum path to the target or goal of the destination. This method is taken into another step in which using the Self-Adaptive Learning Particle Swamp Optimization (SLPSO) in order to search more feasible path for the mobile robot [8]. Ant Colony Optimization (ACO) is also an another method to search an optimal path for the mobile robot. This method uses pheromone trail laying and following the real behaviour of ants[9]. In order to achieve the optimum ACO, improvement has been made to the ACO such as implementation of fuzzy logic with the ACO in minimizing the iterative learning error [10]. Another improvement made for the ACO is based on research from [11] which increase the accuracy of the and accelerate the convergence speed and to enlarge the search space.

The Artificial Potential Field (APF) uses the concept of potential field that consist repulsive and attractive field. Commonly researchers tend to choose this method due to the simple algorithm structure, the concise mathematical description and convenience for real time control [12]. The issues faced for utilizing the APF method is that the subject influenced by the local minima and affecting the trajectory of the mobile robot or the subject. Some researchers developed an improved version of APF to avoid the local minima by changing the repulsive potential function [13]. Method that also being used to avoid the local minima us route planning based on [13]. The main idea for the APF is mobile robot or vehicle is being surrounded with charged potential field that makes the mobile robot either attracted or repelled away [14].

The development of the wheelchair is crucial to help the PWDs facing their daily life and selection of the path planning method is based on the criteria that of the research itself.

2. Methodology

This paper purposely to implement the Artificial Potential Field Method (APF) on the Electric Powered Wheelchair. The application of APF on the EPW is due to the simplicity of equation APF itself. Below shows the specification of the EPW that being used for the simulation purpose. The modelling of the EPW in MatLab and Simulink is based on the specification provided.

2.1. Electric Powered Wheelchair Specification

The list of authors should be indented 25 mm to match the abstract. The style for the names is initials then surname, with a comma after all but the last two names, which are separated by 'and'. Initials should not have full stops—for example A J Smith and *not* A. J. Smith. First names in full may be used if desired. If an author has additional information to appear as a footnote, such as a permanent address or to indicate that they are the corresponding author, the footnote should be entered after the surname.

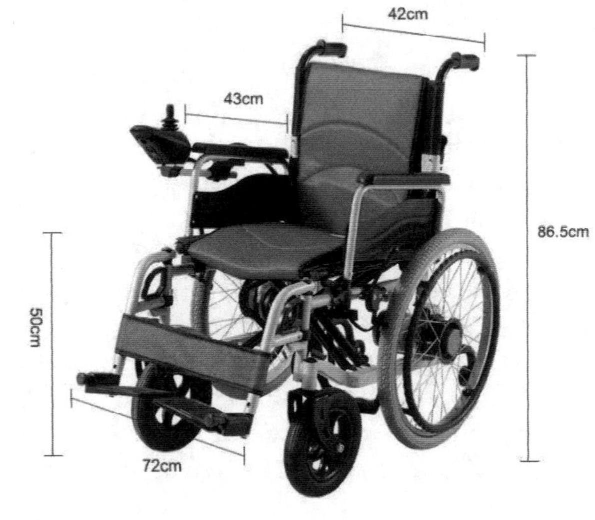

Figure 1. The commercial Electric Power Wheelchair used in the experiment with the measurements.

The EPW installed with the two free rolling wheel or castor wheels in front of the EPW. The wheel could rotate in 360 degrees to ease the mobility and maneuvering the EPW. The wheelchair is driven with the DC motors installed along with the wheels (in the middle).

Figure 2. The castor wheel equipped along with the Electric Power Wheelchair.

Figure 3. The joystick on the EPW where the user can justify the speed of the wheelchair.

Table 1. The specifications of the EPW [15]

Specifications	Capabilities
Motor Power	250W per motor
Climbing Ability	<13 Degree
Battery	24V,12A
Battery Endurance	10-15 Km
Speed Mode	6 Speed Adjustment
Brand	Jikang
Model	Tw100
Wheel Size	10 Inch (Front), 22Inch (Rear)
Weight	34 Kg (Without Battery)
User	Elder and Disables

All of the specifications from the tables and figures are directly from the manufacturer. For the experiment some modification has been made on the EPW in order to ease the installation of the sensors that will attached to the EPW.

2.2. DC Motor in EPW

This section will discuss on the equations involve for the EPW. These equations will be utilized in the Matlab and Simulink for modelling and simulation. Generally, the torque of the EPW generated from the DC motor. The DC motor equation can be seen as follows [16].

$$T_m = K_t i \tag{1}$$

The motor torque, T_m produced from the DC motor, constant factor, K_t and the armature current of the DC motor, i. The motor torque is directly proportional with the armature current by constant factor.

IOP Conf. Series: Materials Science and Engineering 1068 (2021) 012012 doi:10.1088/1757-899X/1068/1/012012

$$e_b = K_e \dot{\theta} \tag{2}$$

The back Electromotive Force (EMF), e_b is directly proportional to the angular velocity $\dot{\theta}$ of the shaft by the constant factor, K_e.

$$K_t = K_e = K \tag{3}$$

The back EMF constant, K_e and motor torque constant, K_t are considered equal thus it will denoted as, K.

From the equation (1), (2) and (3), the equations can be derived to formulate Newton's Second Law accordingly;

$$J\ddot{\theta} + b\dot{\theta} = Ki \tag{4}$$

The above equation follows the Newton's Second Law where moment of inertia is, J angular acceleration denoted as θ and motor bearing is denoted as b.

The Kirchhoff's Voltage Law is also applied in order to be implemented into the Matlab and Simulink. The Kirchhoff's Voltage Law can be seen as follows;

$$L\frac{di}{dt} + Ri = V - K\dot{\theta} \tag{5}$$

L indicates the inductance, R as the resistance in the motor and V represents the voltage of the motor. The above equations will be implemented for modelling the EPW model.

The parameters for the DC motor of the EPW that being applied in the simulation is based on the real model of the EPW itself.

Table 2. The parameters of DC motor

Parameters	Value
K	2.8842
J	0.175kg/m^2
b	0.1Nms
L	0.0002 H
R	3.116 Ohm
V	12V

2.3. Potential Field Method

As being aware, the Potential Field Method is another approach in path planning method. This research is mainly focused on the obstacle in a static condition and the path is pre-determined. The idea is to implement the artificial potential field in the environment enabling the EPW move toward the goal or destination and navigates away before reaching the obstacle. The EPW is represented as a particle under a scalar potential field

a. Attractive Field and Force

The attractive field is under one of Artificial Potential Field Method. The attractive field will be drawing the EPW towards the goal or the destination for the EPW.

$$U_{Att} = \frac{1}{2}\varphi x^2 \tag{6}$$

Above shows the attraction potential field equation. Where, φ is an adjustable constant. The attraction force be shown by the equation below.

$$F_{Att}(x) = -\nabla U_{Att} = -\varphi|x_{EPW} - x_{Att}| \tag{7}$$

Where, φ is an adjustable constant, x_{EPW} is the current position of the EPW. x_{ATT}, is the position of an attraction point.

b. Repulsive Field and Force

Repulsive field or force is another scalar potential field under Artificial Potential Field Method. The repulsive field or force is applied on the obstacle to ensure the EPW navigates away upon approaching the obstacle.

$$U_{Rep} = \begin{cases} \frac{1}{2}\eta\left(\frac{1}{x} - \frac{1}{x_o}\right)^2, if\ x \leq x_o \\ 0, \qquad if\ x > x_o \end{cases} \tag{8}$$

The above shows the equation for repulsive field. Where, η is the adjustable constant.

$$F_{Rep} = \begin{cases} \frac{1}{2}\eta\left(\frac{1}{x} - \frac{1}{x_o}\right)\frac{1}{x^2}.\frac{(x_{EPW}-x_o)}{x}, if\ x \leq x_o \\ 0, \qquad if\ x > x_o \end{cases} \tag{10}$$

Equation above shows the repulsive force, Where;

$$x = |x_{EPW} - x_o| \tag{11}$$

x_{EPW}, is the current position of the EPW. x_{Att}, is the position of an attraction point.

These equations will be applied into the simulation in order to map the trajectory of the Electric Powered Wheelchair (EPW). The repulsive force will act as the input for the EPW to change the direction for navigating away from the obstacle.

c. Repulsive Force as Input.

The next step is to take the repulsive force value as input for navigating control of the EPW. The idea is to take the highest value of the repulsive force.

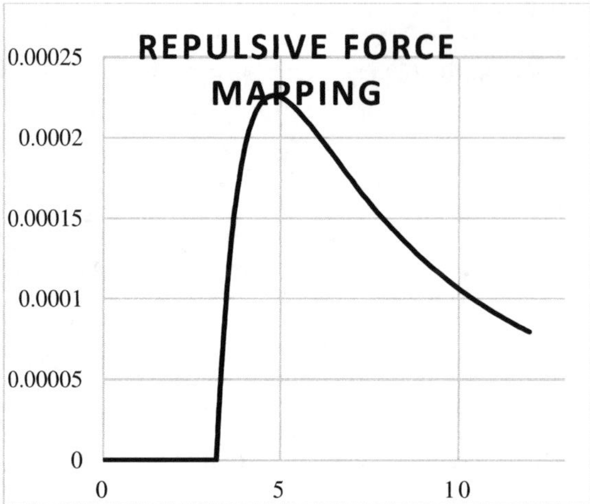

Figure 4: Repulsive Force Map on a Longitudinal Axis.

Figure: 4 shows the preliminary result gained from the repulsive force. The peak value of the repulsive force will be applied into the EPW control system for the EPW navigate away from the obstacle.

The idea is to assume the EPW equipped with distance sensors that would detect the obstacle. The obstacle will have the artificial repulsive force surrounding the obstacle. The sensors will detect the obstacle and assuming the force field around the obstacle also will be detected. The simulation will be conducted to study the effect of distance towards the repulsive force around the obstacle.

As in the theory, the repulsive force will be higher when approaching the obstacle or near the obstacle and become lower as move away from the obstacle. But the idea for this system is the distance that required to be activated is depends on the EPW sensors. The repulsive force value as the input will be varied but the output value will be kept constant. The output will be the differential drive motor of the EPW.

3. Results and Discussion

The results shown below are based on the simulation in the Matlab and Simulink.

3.1. Repulsive Force Response

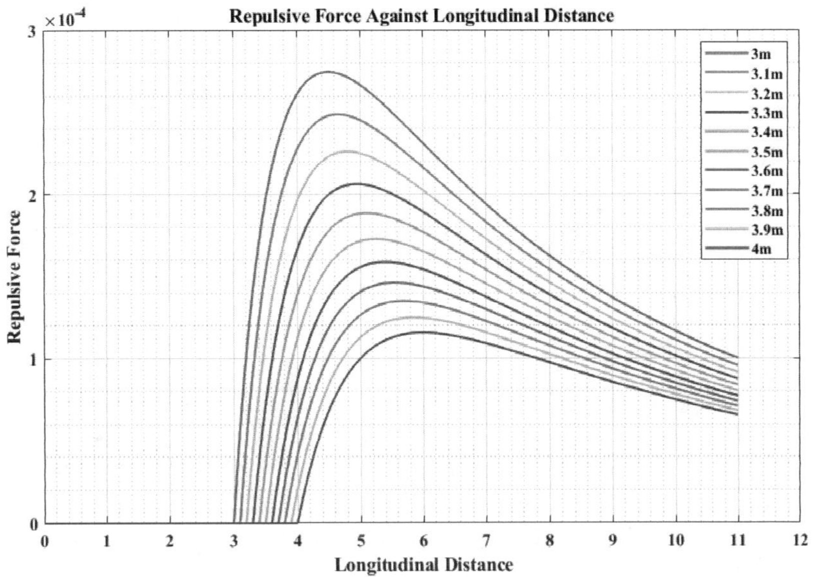

Figure 5. Repulsive Force Based on the Sensor Activation.

The above Figure: 5 shows the repulsive force response at different activation distance. The value of this response will be used as the control input for the EPW. This activation distances mainly to observe the trajectory of the EPW.

Table 3. The peak value of Repulsive Force based on the activation distance.

Distance, (m)	Repulsive Force (10^{-5}), (Kgm^2/s^2)
3	27.43
3.1	24.86
3.2	22.61
3.3	20.61
3.4	18.85
3.5	17.28
3.6	15.88
3.7	14.62
3.8	13.5
3.9	12.49
4	11.57

The Table 3 shows the results of the repulsive force when the EPW approaching the obstacle. The values that taken are the peak or the maximum value of the repulsive force. It is based on the distance activation that measured assumedly by the distance sensor of the EPW. The distance activation of the repulsive force is set by the user and not controlled by any algorithm or artificial intelligence.

3.2. EPW Trajectory

Below shows the trajectory of the EPW based on the repulsive force response.

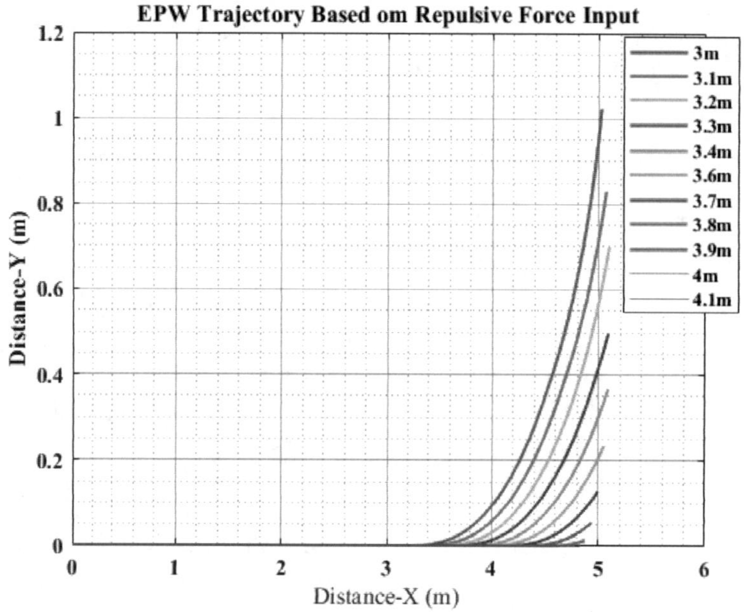

Figure 6. EPW Trajectory Based on Repulsive Force Response.

Figure: 5 shows the trajectory of the EPW based on the response of the repulsive force. The voltage ratio of each tire before the activation distance of repulsive force are 1:1 and that is 12V for each tire. Upon reaching the obstacle or the activation distance is activated, the ratio for the right and left tire is changed to 2:1.

From the figure, the EPW changes its direction before reaching 6m. It is due to the fact of the position of the obstacle is assumed to be at 6m but the activation distance for repulsive force is before reaching 6m. From the trajectory, it could assume that all the trajectories manage to avoid the obstacle but width of the EPW also must be considered in order the EPW to avoid the obstacle successfully.

This simulation conducted to study the most optimum activation distance of the repulsive force that can be take as the input for the EPW to navigate away from the obstacle. By considering the width of the EPW, the most optimum activation distance of the repulsive force is on 3.2m.

3.3. *EPW Trajectory Response with Repulsive Force for Obstacle Avoidance*

Result below shows the effect of full repulsive force towards EPW response.

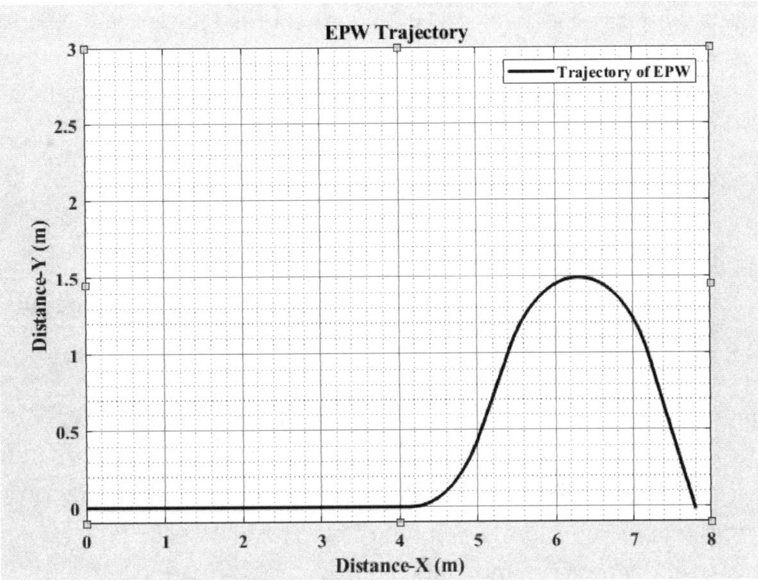

Figure 7. The Trajectory of EPW

Figure: 7 shows the trajectory of the EPW that affected by the repulsive force. The trajectory of the EPW is based on the repulsive force that artificially applied on the obstacle. The detail input voltage for the wheelchair as shown below;

Table 4. Conditions that affecting the right and left input voltage.

Conditions affecting the input voltage	Right Voltage Input, (V)	Left Voltage Input, (V)
Straight line	12	12
Obstacle Avoidance	10	3.6
Path Correction	3.6	10

Table 4 shows the voltage input for right and left for the EPW. Each condition gives the different trajectory response for the EPW. During straight line, the input voltage for right and left are same from starting point until the activation distance for repulsive force (approximately between 4 to 4.5 meter).

For the obstacle avoidance, the voltage input for left and right of the EPW as shown in the table above. The input voltage also influencing the trajectory of the EPW. If the input voltage is to high or low, the trajectory of the EPW will be affected either the EPW will be too closed with obstacle or too far from the obstacle. The path correction is another condition which is the EPW will correct the path towards its original path as shown on figure above.

The trajectory above gained from the optimum voltage of right and left tire of EPW. As being mentioned earlier, the obstacle is assumed to be on 6 meters from the starting point and the width and length of the obstacle is assumed to be 1 meter. From the simulations, the trajectory of the EPW shows that the it able to avoid the obstacle successfully with the repulsive force as the input response for the sensor on EPW.

4. CONCLUSIONS

From this research, the idea was to control the trajectory of the EPW based on the value of the repulsive force. From the simulation conducted, the highest value of the repulsive force is taken as the input for the EPW to navigate away from the obstacle.

This method has been proven could be applied to control the trajectory of the EPW with the value from the repulsive force by selecting the right value of activation distance of the repulsive force. It is important to select the precise sensor to detect the obstacle in order for the EPW to mitigate the obstacle. Even though the result is based on the simulation, the application of the system could be implemented in the real situation.

5. Acknowledgement

The authors would like to thank Universiti Malaysia Pahang (www.ump.edu.my) for the financial support through the UMP internal grant RDU190321.

REFERENCES

[1] J. Tang, Y. Liu, D. Hu, and Z. Zhou, "Towards BCI-actuated smart wheelchair system," *Biomed. Eng. Online*, vol. **17**, no. **1**, p. 111, 2018.

[2] H. Yashoda, A. M. S. Piumal, P. Polgahapitiya, M. M. M. Mubeen, M. Muthugala, and A. Jayasekara, "Design and development of a smart wheelchair with multiple control interfaces," in *2018 Moratuwa Engineering Research Conference (MERCon)*, 2018, pp. 324–329.

[3] P. Dey, M. M. Hasan, S. Mostofa, and A. I. Rana, "Smart wheelchair integrating head gesture navigation," in *2019 International Conference on Robotics, Electrical and Signal Processing Techniques (ICREST)*, 2019, pp. 329–334.

[4] M. Xiong *et al.*, "A Low-Cost, Semi-Autonomous Wheelchair Controlled by Motor Imagery and Jaw Muscle Activation," in *2019 IEEE International Conference on Systems, Man and Cybernetics (SMC)*, 2019, pp. 2180–2185.

[5] G. Bardaro, L. Bascetta, E. Ceravolo, M. Farina, M. Gabellone, and M. Matteucci, "MPC-based control architecture of an autonomous wheelchair for indoor environments," *Control Eng. Pract.*, vol. **78**, pp. 160–174, 2018.

[6] H. Zhang, W. Lin, and A. Chen, "Path planning for the mobile robot: A review," *Symmetry (Basel)*., vol. **10**, no. **10**, p. 450, 2018.

[7] C. Lamini, S. Benhlima, and A. Elbekri, "Genetic algorithm based approach for autonomous mobile robot path planning," *Procedia Comput. Sci.*, vol. **127**, pp. 180–189, 2018.

[8] G. Li and W. Chou, "Path planning for mobile robot using self-adaptive learning particle swarm optimization," *Sci. China Inf. Sci.*, vol. **61**, no. **5**, p. 52204, 2018.

[9] M. Dorigo and T. Stützle, "Ant colony optimization: overview and recent advances," in *Handbook of metaheuristics*, Springer, 2019, pp. 311–351.

[10] C.-T. Yen and M.-F. Cheng, "A study of fuzzy control with ant colony algorithm used in mobile robot for shortest path planning and obstacle avoidance," *Microsyst. Technol.*, vol. **24**, no. **1**, pp. 125–135, 2018.

[11] K. Akka and F. Khaber, "Mobile robot path planning using an improved ant colony optimization," *Int. J. Adv. Robot. Syst.*, vol. **15,** no. **3**, p. 1729881418774673, 2018.

[12] Y. Chen, G. Luo, Y. Mei, J. Yu, and X. Su, "UAV path planning using artificial potential field method updated by optimal control theory," *Int. J. Syst. Sci.*, vol. **47**, no. **6**, pp. 1407–1420, 2016.

[13] L. Lifen, S. Ruoxin, L. Shuandao, and W. Jiang, "Path planning for UAVS based on improved artificial potential field method through changing the repulsive potential function," in *2016 IEEE Chinese Guidance, Navigation and Control Conference (CGNCC)*, 2016, pp. 2011–2015.

[14] L. Zhiyang and J. Tao, "Route planning based on improved artificial potential field method," in *2017 2nd Asia-Pacific Conference on Intelligent Robot Systems (ACIRS)*, 2017, pp. 196–199.

[15] "Shenyang Richwinner Trade Co., Ltd. - Medical bed, Wheelchair, Walking aids, Commode chair, Body posture." https://richwinner.tradees.com/ (accessed Oct. 13, 2020).

[16] "Control Tutorials for MATLAB and Simulink - Motor Speed: System Modeling." http://ctms.engin.umich.edu/CTMS/index.php?example=MotorSpeed&section=SystemModeling (accessed Oct. 13, 2020).

Modeling and Simulation of an Energy Management System for Plug-in Hybrid Electric Recreational Boat (PHERB) using Penambang Boat Driving Cycle

A.R Salisa[1,2,3], E.A.E.S Shahiran[1] & J.S Norbakyah[1,2,3]
[1]Faculty of Ocean Engineering Technology and Informatics, Universiti Malaysia Terengganu, 21030 Kuala Nerus, Terengganu, Malaysia.
[2]Energy Storage Research Group (ESRG), Universiti Malaysia Terengganu, 21030 Kuala Nerus, Terengganu, Malaysia.
[3]Renewable Energy and Power Research Interest Group (REPRIG), Universiti Malaysia Terengganu, 21030 Kuala Nerus, Terengganu, Malaysia

E-mail: bakyahjabar@umt.edu.my

Abstract. In order to ensure that the vehicle achieve an improvement in energy efficiency, reduction in emissions and fuel use, the good energy management system needed. This paper presents an innovation of water transportation namely Plug-in Hybrid Electric Recreational Boat (PHERB) on modeling and simulation results of an energy management strategy (EMS) for PHERB. In this work, firstly, through a power flow analysis, the vehicle components are sized to meet the expected power and energy requirements. Then, the model is tested numerically in the MATLAB/SIMULINK environment using the Penambang Boat drive cycle with the proposed EMS. The simulation results show that the PHERB model can be used as a reference to build a hybrid electric boat in Malaysia environment.

1. INTRODUCTION

Many studies show that every conventional engine type emission rates of carbon dioxide and fuel consumption were high [1-4]. Concern about this issue, Plug in Hybrid Electric Recreational Boat (PHERB) is introduced to improve the performance of boat and reducing the emission and fuel consumption. PHERB consists of only one electrical machine (EM) which acts as either a motor or a generator at one time while the energy storage system (ESS) is equipped with an ultracapacitor (UC) for fast charging and unloading during regenerative braking and rapid acceleration. Figure 1 displayed the proposed PHERB powertrain.

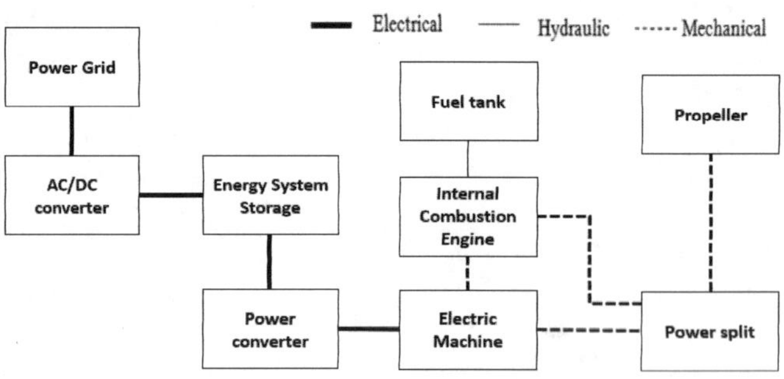

Figure 1. PHERB powertrain [5]

In this article, the suggested Energy Management Strategy (EMS) is introduced to the PHERB model to ensure that the vessel meets the driving efficiency goal without compromising the ideal working atmosphere to boost fuel economy and emissions. The PHERB model with special EMS has been mathematically developed numerically in MATLAB/SIMULINK. The Penambang Boat driving cycle was used in the PHERB model to evaluate PHERB efficiency.

2. METHODOLOGY

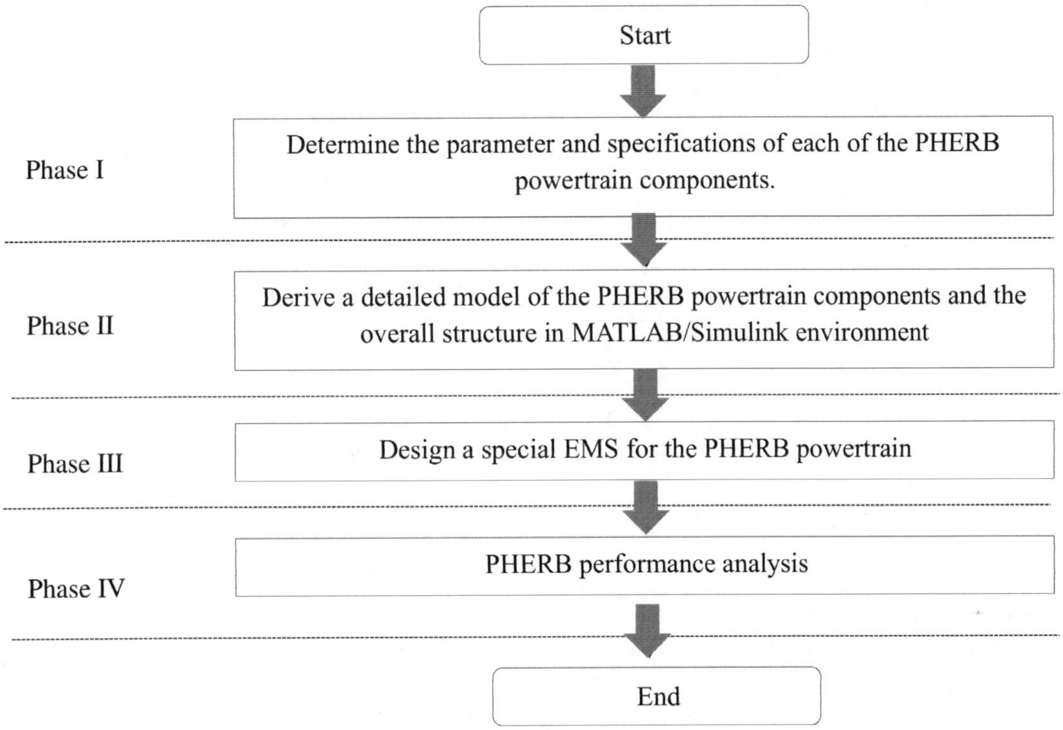

This research methodology is divided to four phases as illustrated in Figure 2.

Figure 2. Flowchart in this research.

2.1 PHERB parameter and specification

The design specification, parameter, and selection for the EM, ICE, and energy system storage (ESS) shall be carried out to define the key components of the PHERB powertrain. EM, ICE and ESS are sized according to the ship parameters, characteristics, and efficiency criteria (Table 1) based on the demand for steady-state boat capacity using the dynamic equation boat [6]. The type of boat chosen is a pleasure boat. In simulation, the length of the used boat is 12.4 m, and the density of the water is 1000 kgm-3. The enlargement of the boat model begins with the calculation of the energy and strength of the boat required for typical driving scenarios, based on the parameters and design characteristics of the boat, based on the PHERB specification, parameter, and requirement. By means of an analysis of the power flow, the size and the limit of each portion are calculated to satisfy the specifications accordingly. Table 2 displays the scale and parameters used for PHERB.

Table 1. Parameter, Specification and Configuration

Parameter and Specifications		
Type of boat	Conventional boat	PHERB
Configuration	Series	Series-Parallel
Length overall, L	12.4 m	12.4 m
Length at waterline, L_{WT}	11.0 m	11.0 m
Breath, B	1.8 m	1.8 m
Draught, T	0.64 m	0.64 m
Length between perpendicular, L_{PP}	10.67 m	10.67 m
Density of water, ρ	1000 kgm^{-3}	1000 kgm^{-3}
Total propulsive efficiencies, η_T	0.9	0.9

Table 2. PHERB Component Specification

Component	Specifications
ICE	20 kW @ 3000 rpm
EM	30 kW AC induction motor
ESS	5 kWh, 6 Ah

2.2 PHERB in MATLAB/SIMULINK environment

All parts in PHERB obtain a mathematical model is combined. The boat performance is simulated in the MATLAB/SIMULINK environment with a special EMS and different driving cycle. Overall structure of PHERB model is illustrated in Figure 3.

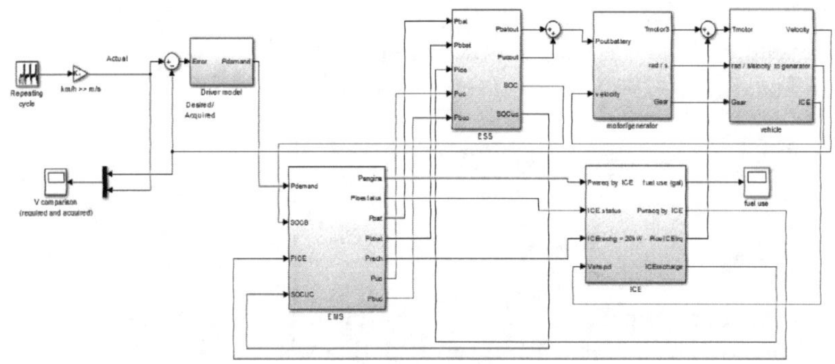

Figure 3. Overall structure of the PHERB model in MATLAB/SIMULINK environment

2.3 PHERB Energy Management Strategy

During this process, the special EMS was created and combined with the PHERB portion. EMS in vehicles is an important component because it can have a direct effect on vehicle efficiency [6]. Improving automotive energy conservation can deliver substantial benefits, such as lowering fuel usage, reducing emissions, lower maintenance costs, reducing noise pollution and improving driving efficiency and ease of use. [7]. EMS has two primary requirements that are reversed and forward-looking. There are six modes of operation that are mechanical reverse, regenerative reverse, EM only, EM and ICE mixed, ICE only and ICE recharged. All six modes of operation displayed in Figure 4 are based on the demand for forward or reverse vehicle power and the SOC level in the ESS. Three ESS SOC levels are medium, moderate and high. Another name for cruising in a water vehicle is forward mode and braking mode is also known as reverse mode.

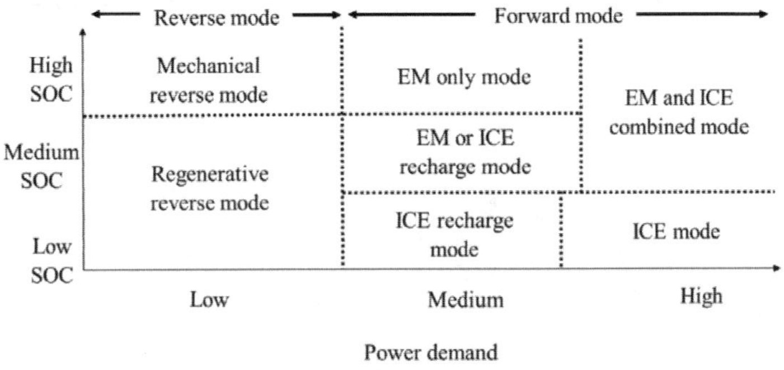

Figure 4. EMS proposed for PHERB

In PHERB EMS, there are two requirements for PHERB drives that are forward mode and reverse mode. The forward mode and the reverse mode are determined by the power demand flow. When the power demand is greater than zero, it is known to be forward mode, and vice versa, reverse mode. The flow chart is shown in Figure 5.

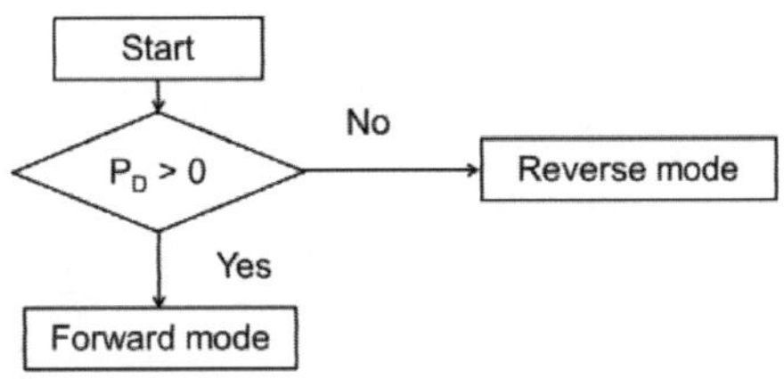

Figure 5. Driver condition for EMS

Figure 6 displays the number of six operations in forward and reverse mode. In forward mode, four operations involved are ICE mode, ICE recharged mode, ICE or EM mode, and EM mode only. Reverse mode, thus, requires two operations consisting of regenerative reverse mode and mechanical reverse mode, respectively.

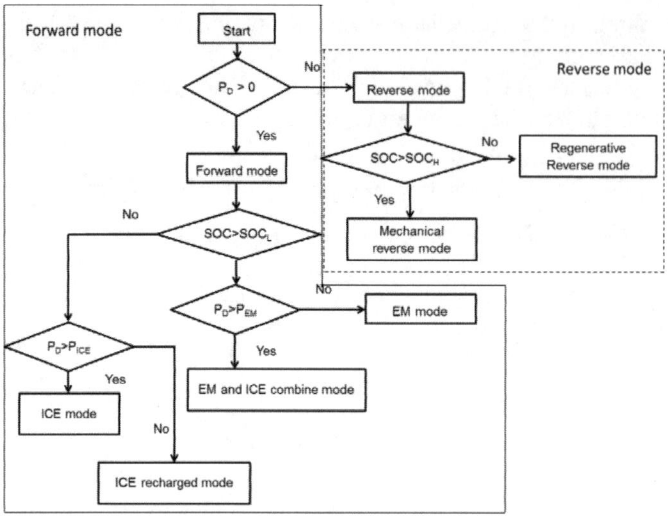

Figure 6. Flow chart of PHERB EMS

3. RESULTS AND DISCUSSION

Penambang Boat driving cycle is used to see the performance of PHERB performance in simulation. The PHERB model provided is used to simulate the driving efficiency and fuel consumption of the Penmabang Boat driving period. The Penambang Boat driving cycle is shown in Figure 7.

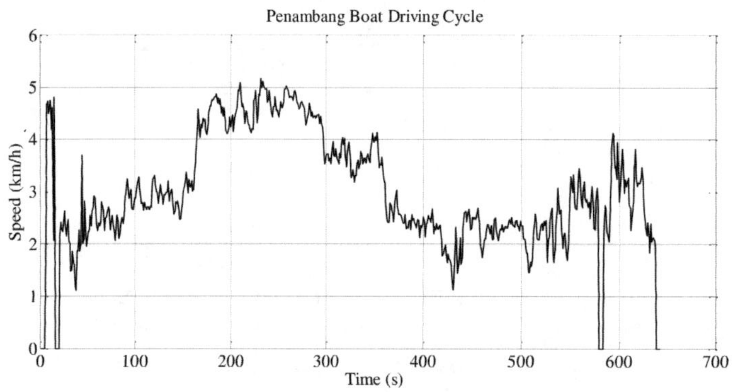

Figure 7. Penambang Boat Driving Cycle

3.1 PHERB Simulation

Figure 8 – Figure 10 shows the simulated ESS current, voltage and power usage of the Penambang Boat drive cycle. The peak currents are due to the strong demand for power to achieve accelerated movement of the boat at the respective periods. The negative values on the graph represent the regenerative braking events of the hard-breaking loop. In the ESS voltage table, the voltage increases during regenerative braking and decreases during high current discharge while the demand for electricity from the EM is at its highest.

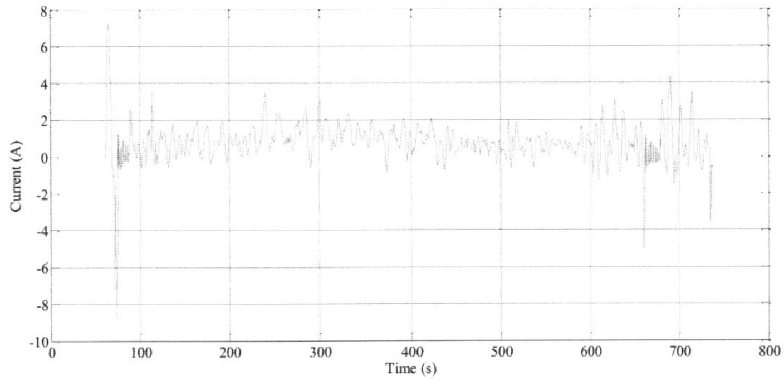

Figure 8. Current performance in PHERB ESS

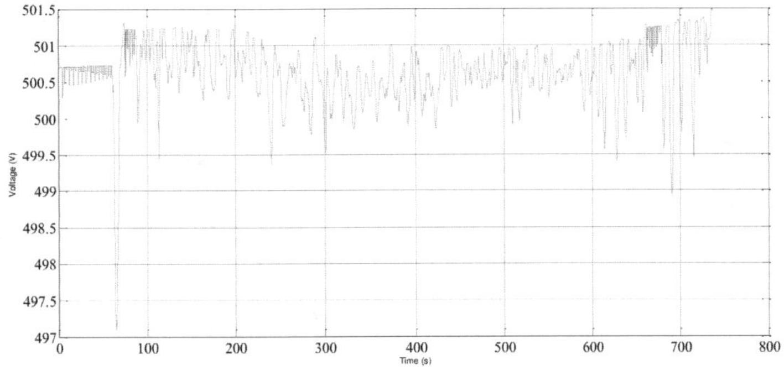

Figure 9. Voltage performance in PHERB ESS

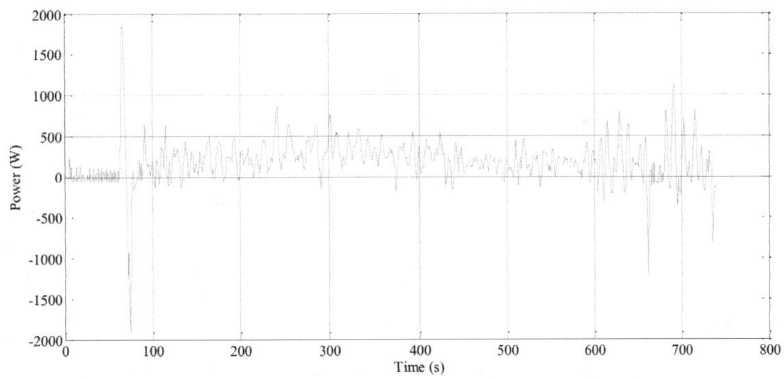

Figure 10. Power performance in PHERB ESS

Figure 11 – Figure 13 shows the simulated speed, torque and power of the EM. As shown in the figure, as the boat accelerates, the required EM torque increases rapidly, and when the speed is relatively steady, a much smaller torque is needed to overcome the resistance and drag the water to the boat. The average power demand from the EM is 5000 W at the speed level and the peak power demand is 1500 W during the acceleration at 39 rad/s.

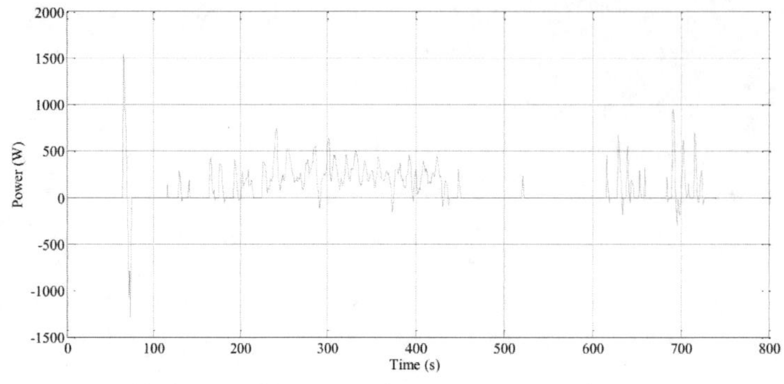

Figure 11. Power performance in PHERB EM

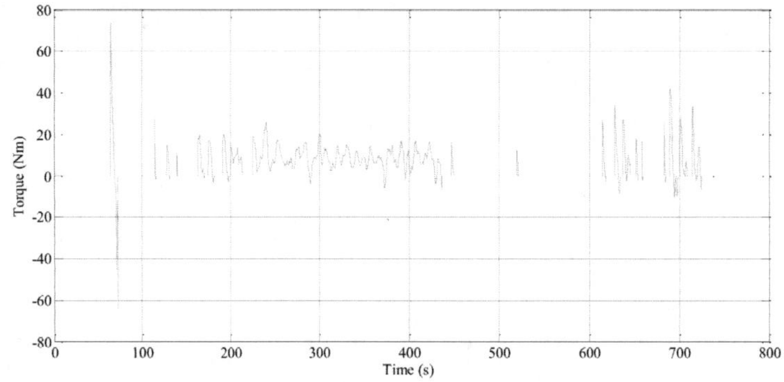

Figure 12. Torque performance in PHERB EM.

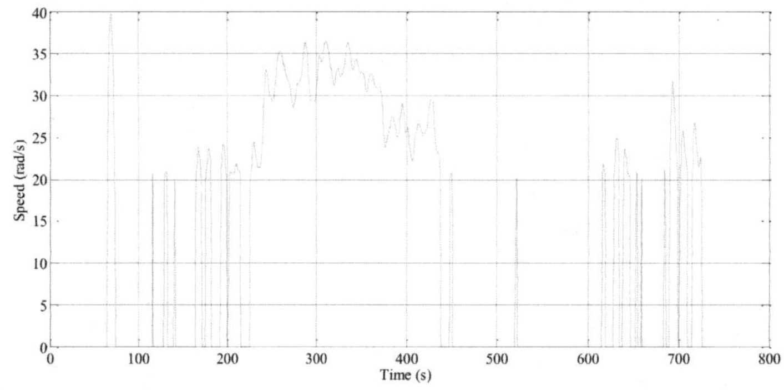

Figure 13. Speed performance in PHERB EM

Figure 14 – Figure 15 displays the simulated propulsion speed and torque specifications for the Penambang Boat drive cycle. The maximum torque of the propeller, 350 Nm, occurs when the boat accelerates from a standstill to a steady speed. The needed torque then decreases as the Penambang Boat drive cycle consists only of mild accelerations and decelerations.

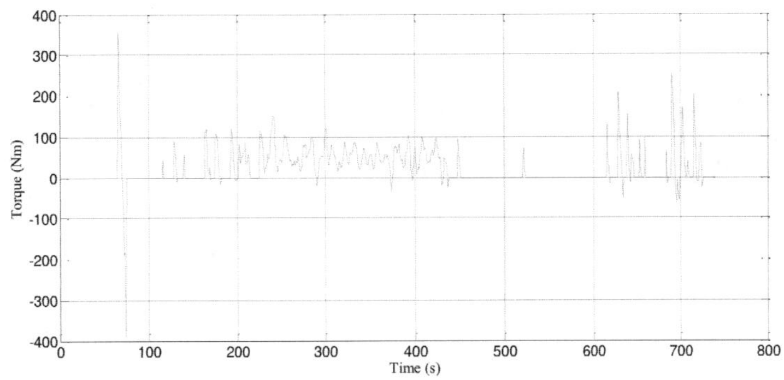

Figure 14. Torque performance in PHERB propeller

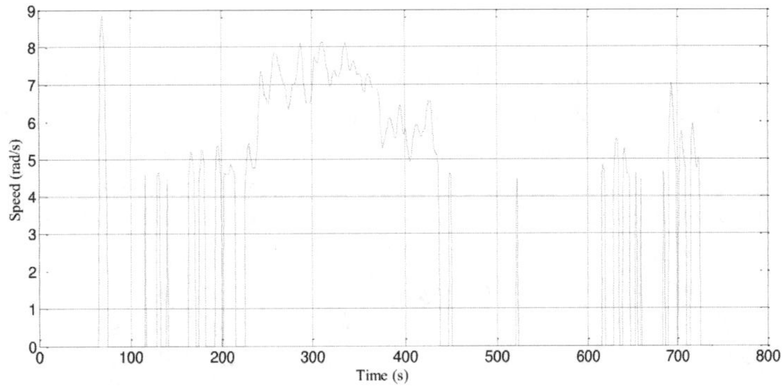

Figure 15. Speed performance in PHERB propeller

The simulated vehicle speed and the necessary speed for the Penambang Boat driving period is illustrated in Figure 16. As seen, the two curves are very well matched with a very small error of less than 0.05 m/s between goal and acquired speeds. Thus, it is expected that the PHERB will be able to provide driving efficiency as a traditional hybrid vessel under the actual driving conditions.

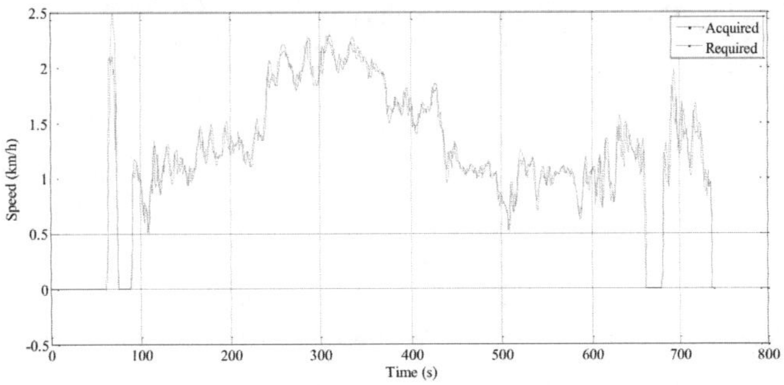

Figure 16. Acquired and Required speed performance in PHERB model.

4. CONCLUSIONS

The results of the PHERB subsystems in terms of ESS current, ESS voltage, ESS output power, EM speed, EM torque, propeller speed, and propeller torque show that the PHERB can be used as reference model to build plug in hybrid recreational boat in Malaysia environment.

ACKNOWLEDGMENTS

The authors would like to be obliged to Universiti Malaysia Terengganu (UMT) for UMT Scholarship, Ministry of Education Malaysia for providing financial assistance under FRGS (59353) grant and Faculty Ocean Engineering Technology and Informatics, UMT for all their technical and research support for this work to be successfully complete.

REFERENCES

[1] Zheng J., Yu R., Liu Y., Zou Y. and Zhao D. (2019). The Technological Progress of the Fuel Consumption Rate for Passenger Vehicles in China: 2009–2016. *Energies* 12:2384.
[2] Gao J., Chen H., Dave K., Chen J. and Jia D. (2020). Fuel economy and exhaust emissions of a diesel vehicle under real traffic conditions. *Energy Science Engineering*. **8**:1781–1792.
[3] Tan Tien Huynh, Minh Duc Le & Dinh Nghia Duong. (2019). Effects of butanol–gasoline blends on SI engine performance, fuel consumption, and emission characteristics at partial engine speeds. *International Journal of Energy and Environmental Engineering*. **10**:483–492.
[4] Speirs J., Balcombe P., Blomerus P., Stettler M., Achurra-Gonzalez P., Woo M., Ainalis D., Cooper J., Sharafian A., Merida W., Crow D., Giarola S., Shah N., Brandon N. and Hawkes A. (2020). Natural gas fuel and greenhouse gas emissions in trucks and ships. *Prog. Energy 2 012002.*
[5] Norbakyah J.S & Salisa A.R. (2019). Optimization of the fuel economy and emissions for plug in hybrid electric recreational boat energy management strategy using genetic algorithm. *International Journal of Power Electronics and Drive System*. **10(2)**:792-800.
[6] Hu, Y., Li, W., Xu, K., Zahid, T., Qin, F. and Li, C. (2018). Energy Management Strategy for a Hybrid Electric Vehicle Based on Deep Reinforcement Learning. *Applied Sciences*. **8**:187
[7] Khayyam, H., & Bab-Hadiashar, A. (2014). Adaptive intelligent energy management system of plug in hybrid electric vehicle. *Energy*. **69**:319-335

IOP Conf. Series: Materials Science and Engineering 1068 (2021) 012014 doi:10.1088/1757-899X/1068/1/012014

Glare prediction of solar PV system in airport environment: A scenario analysis of material, tilt and orientation

S. Sreenath[1], K. Sudhakar[2, 3,4, *], Yusop AF[2]

[1]Renewable Energy and Energy Efficiency Research Cluster, College of Engineering, Universiti Malaysia Pahang, 26600 Malaysia
[2]Faculty of Mechanical and Automotive Engineering Technology, Universiti Malaysia Pahang, 26600 Malaysia
[3]Energy Centre, Maulana Azad National Institute of Technology Bhopal, India
[4]Automotive Engineering Centre, Universiti Malaysia Pahang, 26600 Malaysia
*Email: sudhakar@ump.edu.my

Abstract The application of solar PV is suitable in airport areas for energy generation due to the availability of vast open spaces. At the same time, there is possibility of glare from solar PV array which may affect the visibility of pilots and air traffic controllers. This paper attempts to perform glare assessment for conceptual solar PV system located in an airport of India using ForgeSolar software under four different scenarios. These four scenarios are solar PV modules with smooth glass, smooth glass with anti-reflective coating (ARC), light textured glass with ARC and different tilt & orientation angles respectively. The location of the proposed site, details of solar PV array and characteristics of observers are provided. The occurrence and duration of glare for each scenario is obtained from the software. The impact of glare is assessed based on solar glare standard set by Federal Aviation Administration (FAA). Among the studied scenarios, safe value of glare is obtained in Scenario 4 only at specific values of tilt and orientation angle. A tilt angle of 15 degrees and orientation angle of 135 degrees is the suitable value with 99.1 % of theoretical energy output. It is observed that the yellow glare impact on ATC reduced in the descending order of Scenario 1 (2122 min), Scenario 2 (296 min) and Scenario 3 (0 min). It is concluded that proper selection of tilt and orientation angle of solar PV array aids in achieving zero minutes of glare for airport conditions.

Keywords: Airport, Aviation, Glare, Photovoltaic, Solar PV

1 Introduction

Solar PV technology is now well-known renewable energy source that is versatile, sustainable and environment friendly. The electricity consumption in airports usually comes from nearby grid which is owned by electric utility [1]. This electricity can be substituted with electricity generated from solar PV system. The vacant land areas in airport is suitable for solar PV installations. Airport based solar PV system helps to stabilise energy cost, mitigate carbon emissions, meet sustainability targets, create robust & reliable electricity [2].

However, the implementation of solar PV may affect the safe operation of airports. Anurag et al. (2017) reported three main roadblocks to solar PV installation in airports namely reflectivity & glare from PV array, interference to operation of radar and physical penetration into airspace [3]. In the technical guidance released by FAA, it is reported the possible glare & glint is a new and unforeseen issue from solar energy application in airports [4]. The reflections from solar PV surface may cause either glint or glare to pilots or airport staff which affects their visibility leading to incorrect decisions which is a concern to safe airport operations. Solar PV modules are integral part of any solar PV system. The top layer of PV module is made of glass material. A part of solar rays falling on PV surface reflects in different directions. These reflected rays may reach the cock pit of aircraft (landing & take-off) or window of ATC tower [5]. This may affect the visibility of pilots and air traffic controllers which is a serious aviation issue. In 2012, a portion of the solar PV arrays in Manchester-Boston airport was covered with wrap temporarily to prevent glare strike on staff of ATC tower. Later, the PV array was rotated 90 degrees to the east from the current position to mitigate glare problems [6]. Prior glare prediction helps in mitigating such problems, reduce unnecessary cost on reconstructions and avoiding delays in permissions and approvals. In this regard, a glare assessment tool called Solar Glare Hazard Assessment tool (SGHAT) was developed by FAA and Sandia National Laboratories. In the interim policy statement, FAA sets forth the standards for glare assessment and required prediction tool [7]. Now, this tool can be accessed by internal Sandia users only. However, ForgeSolar software, which is a licensed version of SGHAT tool, is available for public use with subscription fee. Sreenath et al. (2020) assessed annual glare from conceptual solar PV array in an airport in Malaysia and concluded that the selected area is not suitable for solar PV installations as per FAA's solar glare policy.

Some authors reported the energy performance of airport based solar PV systems [9]–[11]. However, the glare aspects of solar PV in airports are not mentioned in those literatures. In few literatures, the glare assessment is carried out for solar PV in airports using predication tool. None of them studies the effect of anti-reflective coating (ARC), glass texturing on glare occurrence and its duration. A comparative study of ARC, texturing and PV array orientations on glare reduction is not reported yet. This study is classified into four scenarios namely PV module with smooth glass & without ARC, PV module with smooth glass & with ARC, PV module with light texturing & with ARC, PV module in different tilt and orientation angles. This research work can be a reference material to staff considering solar projects in their airport. It is expected that this work will influence decision making and policies related to solar PV glare that are in revision or development phase. The objective of this paper is to assess the glare occurrence from solar PV array in airport under four scenarios and to compare the glare impact in these scenarios.

2. Methodology

2.1 Site selection

In this study, an international airport located in Kerala state, India is chosen. Using google earth pro tool, the location and details of airport are studied. A site lying close to terminal building is chosen for the proposed solar PV array. The selected site spans over an area of 9,129 m2. One kW solar PV system approximately requires 110 m2. So, 80 kWp solar PV system can be installed in the selected site. The ATC lies to the west of selected site and runway lies to the north of selected site. It is assumed that the energy generated from the proposed solar PV system is used to partially meet the energy demand of airport's terminal building. The location of site, position of ATC and pilot's approach path is shown in Figure 1. The movement of sun for the selected airport location is seen in Figure 2.

Figure 1. Bird eye's view of Kannur airport, India and details of observer positions (source: Google)

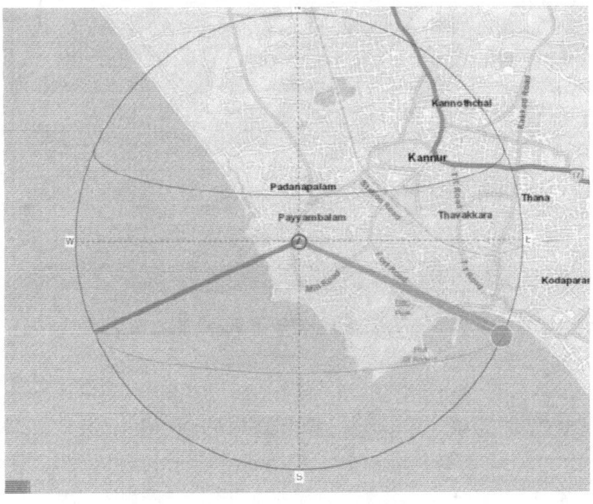

Figure 2. Sun's movement in the selected location of airport

2.2 Framing of scenarios for glare analysis

The site is chosen in such a way to study the effect of ARC, texturing and tilt & orientation of PV array. This study is classified into four scenarios namely PV module with smooth glass & without ARC, PV module with smooth glass & with ARC, PV module with light texturing & with ARC, PV module in different tilt and orientation angles. For each scenario, the duration and severity of glare is assessed using ForgeSolar software. The impact of glare on aviation safety is assessed based on solar glare policy of Federal Aviation Administration (FAA). The characteristics of four scenarios is shown in Table 1.

Table 1. Details of different scenarios considered for analysis

Name	Is ARC present	Is PV module textured	Tilt angle (degree)	Orientation angle (degree)
Scenario 1	×	×	10	180
Scenario 2	✓	×	10	180
Scenario 3	✓	✓	10	180
Scenario 4	✓	×	15 to 60	120 to 240

2.3 Glare analysis software: ForgeSolar

ForgeSolar is a glare prediction software that assess glare potential of solar PV array exclusively in airport areas. This software is a licensed version of SGHAT tool and it contains two assessment tools. The GlareGauge and GlareReduce tool of ForgeSolar software is used to predict the occurrence of glare and to optimize the PV design for glare mitigation. The glare assessment methodology and glare impact analysis follow the standards set by FAA (78 FR 63276). Initially, the user draws selected site using the interactive map feature provided by the software. The details of Solar PV array such as tilt angle, orientation angle, PV top layer covering are given in a separate tab. Then the location of observers who visual performance is to be assessed are provided in the map. The observer's considered in the present study are staff in ATC tower and pilots in approach path (flight path). The default values are considered for parameters such as height of ATC tower, viewing angle. Based on these input values, the Glaregauge tool provides the occurrence and duration of glare (monthly variation). As per FAA's solar glare policy, a proposed solar PV array must not possess potential for after-image on ATC tower (green or yellow glare) and low potential for after-image on final approach path of flight (yellow glare). This software allows to choose PV module with different top layer covering whose effects on the occurrence and duration of glare is studied. Using GlareReduce tool, the glare occurrence for different tilt and orientation of solar PV array is estimated. The inputs needed for the software such as flight path, Air Traffic Control (ATC) tower are accessed through publicly available sources.

3. Results and Discussions

3.1 Glare assessment for Scenario 1 (Smooth glass without ARC)

Glare assessment is carried out for solar PV array with smooth glass and without ARC. In this scenario, considerable duration of glare is predicted from the proposed solar PV array (Table 2). Hence, the visibility of pilots in Flight path and airport staff in ATC are affected by the glare. As per FAA guidelines, the glare from solar PV array is not safe for aviation (Figure 3). The annual duration of glare for Flight Path 2 is zero mins which can be attributed to its relative position from solar PV array. The predicted glare on flight path occurs at around 7 am and it last for less than 10 minutes only during five months (March, April, May, August, September). The glare impact on ATC tower is mainly yellow glare (more intense) which occurs in April, May, August, September (Figure 4).

Table 2. Duration of green and yellow glare at receptors for scenario 1

Receptor	Annual Green Glare (min)	Annual Yellow Glare (min)	Is it acceptable as per FAA's policy
Flight Path 1	437	0	✓
Flight Path 2	0	0	✓
ATC tower	43	2122	×

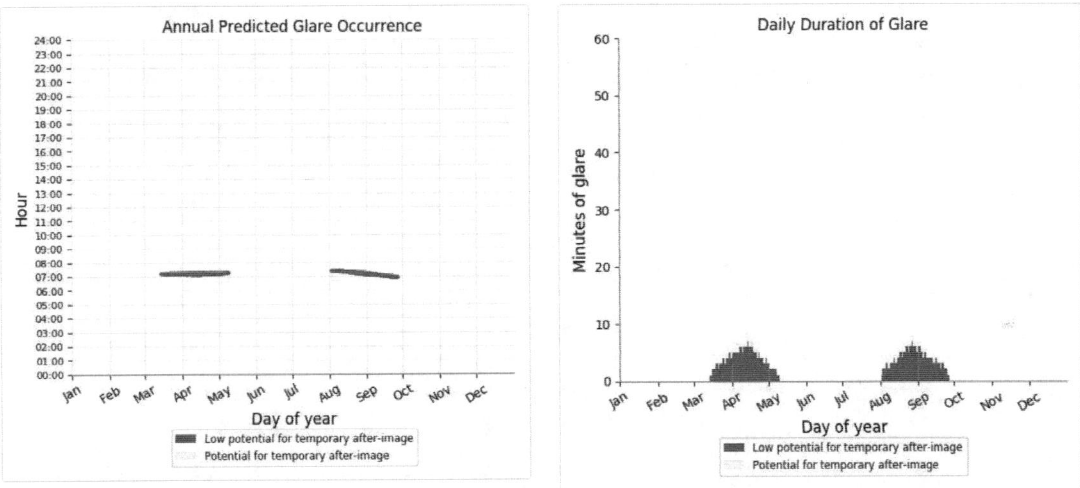

Figure 4. Variation of glare occurrence and duration at ATC for Scenario 1

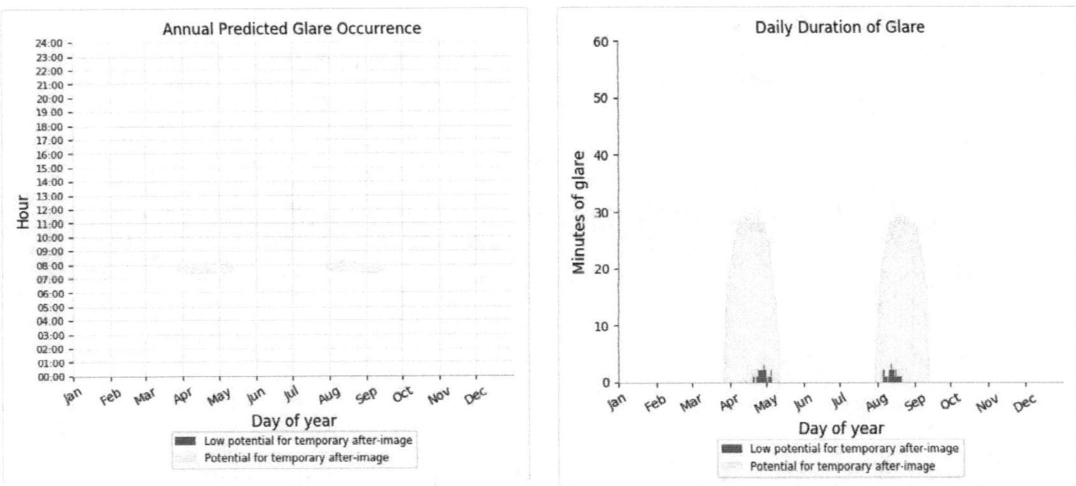

Figure 3. Variation of glare occurrence and duration at ATC for Scenario 1

IOP Conf. Series: Materials Science and Engineering 1068 (2021) 012014 doi:10.1088/1757-899X/1068/1/012014

3.2 Glare assessment for Scenario 2 (Smooth glass with ARC)

Glare assessment is carried out for solar PV array with smooth glass and with ARC. Since glare is predicted at ATC tower from the proposed solar PV array, the visibility of airport staff in ATC can be affected by the glare. The glare impact and duration does not adhere to FAA's policy (Table 3). The annual duration of glare for Flight Path 2 is zero mins as seen in Scenario 1. It is observed that the duration of yellow glare reduced by 86 % when smooth glass with AR coated PV module is considered (Figure 5). It means that the intensity of reflections reduced which caused the increase in the duration of green glare (less intense). The glare occurrence and duration for flight paths did not vary much from Scenario 1 (Figure 5). The glare impact on ATC tower changed from yellow glare in Scenario 1 to green glare in Scenario 2.

Table 3. Duration of yellow and green glare at receptors for Scenario 2

Receptor	Annual Green Glare (min)	Annual Yellow Glare (min)	Is it acceptable as per FAA's policy
Flight Path 1	529	0	×
Flight Path 2	0	0	✓
ATC tower	2493	296	×

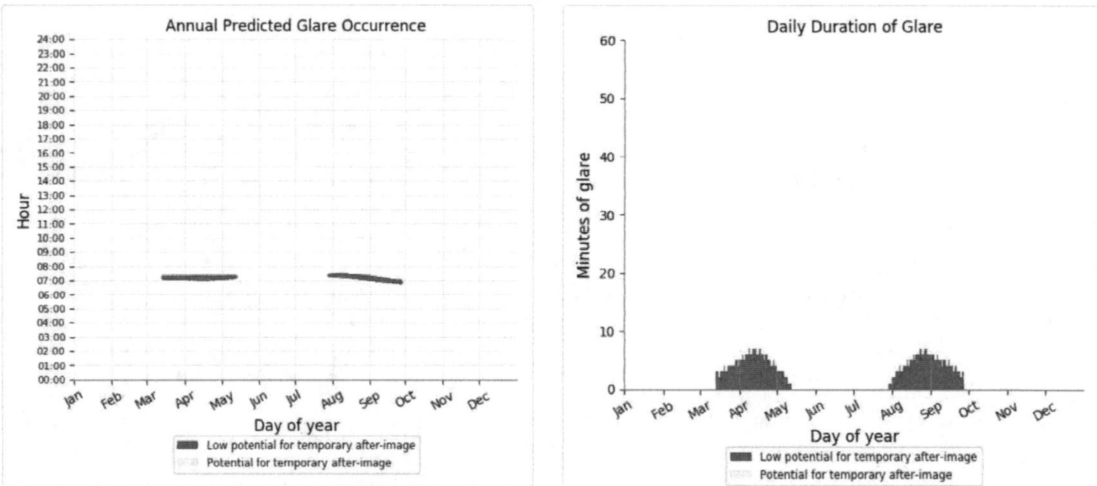

Figure 5. Variation of glare occurrence and duration at ATC for Scenario 2

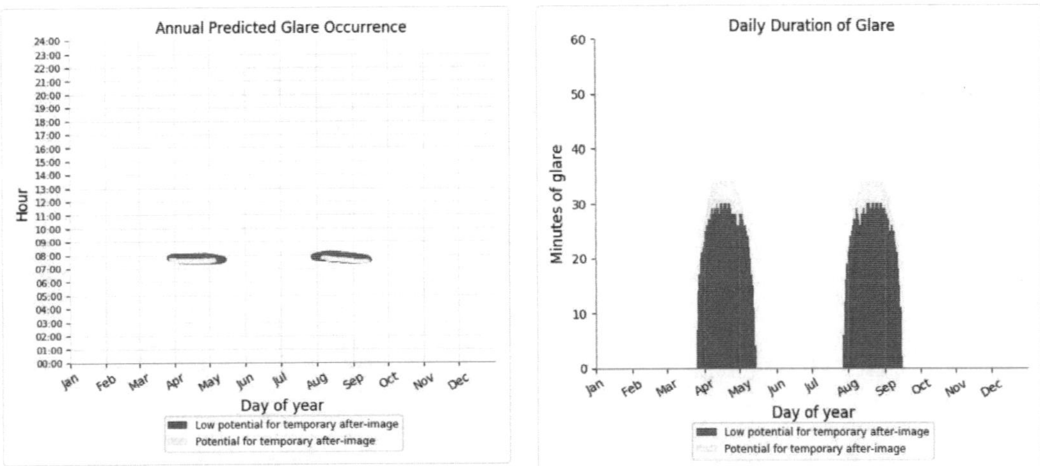

Figure 6. Variation of glare occurrence and duration at flight path 1 for Scenario 2

3.3 Glare assessment for Scenario 3 (Light textured glass with ARC)

Glare assessment is carried out for solar PV array with textured surface and ARC (less reflective). A considerable duration of glare is predicted from the proposed solar PV array in this scenario also (Table 4). The visibility of pilots in Flight path 1 and airport staff in ATC will be affected by the glare. The annual duration of glare for Flight Path 2 remains as zero mins (similar to Scenario 1 and Scenario 2). The duration of yellow glare on ATC tower is predicted as zero mins in this scenario. It can be attributed to reduced reflections from the chosen less reflective solar PV module. At the same time, the duration of green glare (ATC) increased from 2493 mins to 3038 mins which account to presence of low intense reflections (Figure 6). The predicted glare occurrence and duration for flight path is almost same as those observed for Scenario 1 and Scenario 2 (Figure 7).

Table 4. Duration of yellow and green glare at receptors for Scenario 3

Receptor	Annual Green Glare (min)	Annual Yellow Glare (min)	Is it acceptable as per FAA's policy
Flight Path 1	565	0	×
Flight Path 2	0	0	✓
ATC tower	3038	0	×

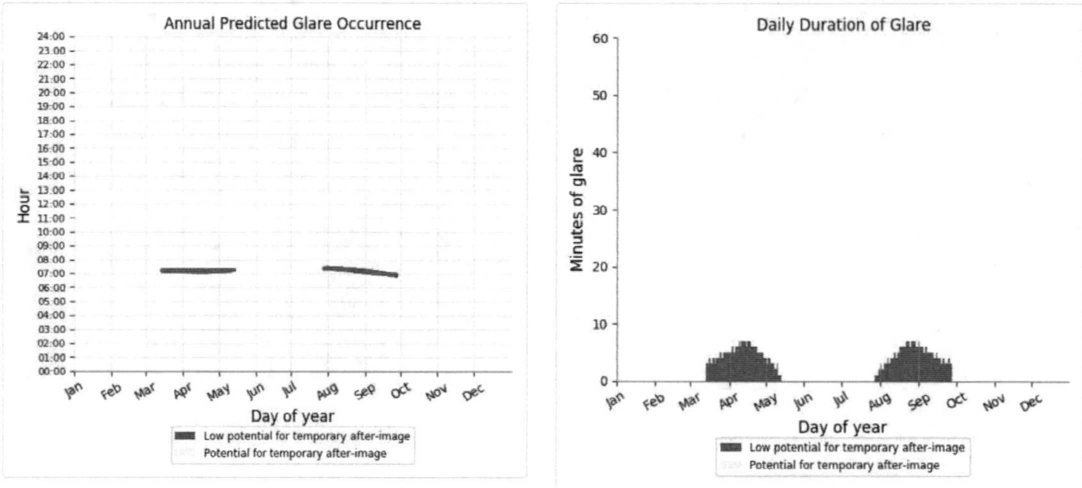

Figure 7. Variation of glare occurrence and duration at flight path 1 for Scenario 3

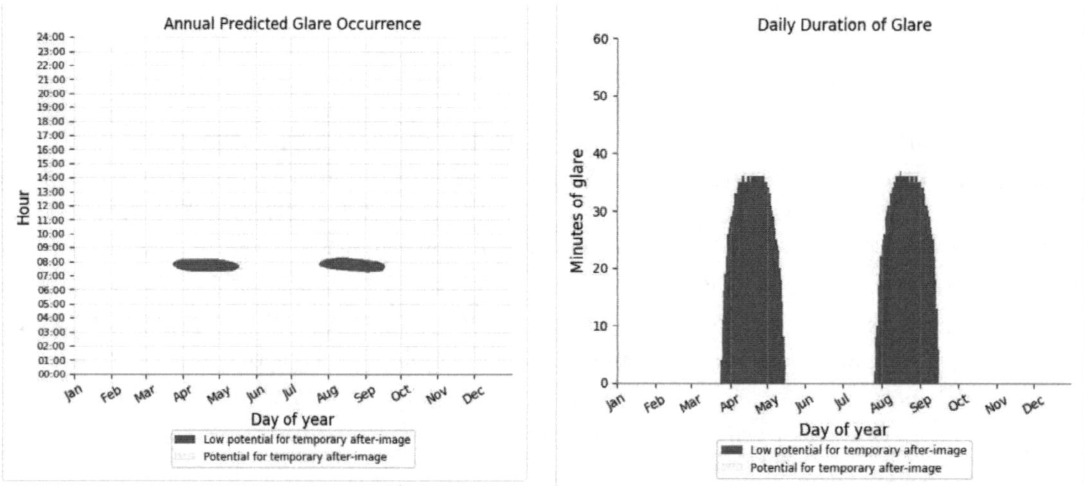

Figure 8. Variation of glare occurrence and duration at ATC for Scenario 3

3.4 Glare assessment for Scenario 4 (Variation of tilt and orientation)

In this scenario, the glare prediction for varying tilt and orientation of PV array is carried out keeping PV top layer covering as constant (smooth glass with ARC). The configurations with adherence to FAA's policy are (15°, 120°), (15°, 135°), (30°, 120°), (30°, 135°) , (30°, 150°) , (45°, 120°) , (45°, 135°), (45°, 150°), (60°, 120°), (60°, 135°) , (60°, 150°), (60°, 195°), (60°, 210°), (60°, 225°). Among these configurations, PV array tilted at 15 degrees and oriented at 135 degrees have highest percentage of theoretical maximum energy output (99.1 %). The duration of annual green and yellow glare for considered set of tilt and orientation angle is given in Table 5

Table 5. Variation of glare occurrence and energy output for different PV orientations

Tilt angle	Orientation angle	Annual Green Glare (min)	Annual Yellow Glare (min)	Expected energy generation (% max)	Is it acceptable as per FAA's policy
15	180	3,117	15	100	×
15	165	2,341	475	99.9	×
15	195	3,118	0	99.9	×
15	150	1,143	180	99.6	×
15	210	3,151	426	99.6	×
15	135	0	0	99.1	✓
15	225	3,340	0	99.1	×
15	120	0	0	98.3	✓
15	240	3,707	0	98.3	×
30	180	2,830	1	95.1	×
30	165	1,678	126	95.1	×
30	195	2,871	0	95.1	×
30	150	0	0	94.8	✓
30	210	0	0	94.8	×
30	135	0	0	94.4	✓
30	225	2,859	0	94.3	×
30	120	0	0	93.5	✓
30	240	2,941	0	93.4	×
45	120	0	0	84.4	✓
45	135	0	0	84.7	✓
45	225	5326	0	84.7	×
45	150	0	0	84.5	✓
45	210	5,555	0	84.5	×
45	240	3,114	0	84.3	×
45	165	785	21	84.2	×
45	195	4,329	0	84.1	×
45	180	2,915	0	84.0	×
60	120	0	0	72.1	✓
60	240	4,098	0	72.0	×
60	135	0	0	71.3	✓
60	225	0	0	71.2	✓
60	150	0	0	69.7	✓
60	210	0	0	69.7	✓
60	165	74	0	68.2	×
60	195	1,709	0	68.2	✓
60	180	3.063	0	67.5	×

3.5 Discussions

In the first three scenarios, the tilt and orientation angle are constant but top layer material of PV module surface is different. In Scenario 4, the top layer of PV module is made of smooth glass with ARC. The ARC helps in reduction of reflectivity of solar PV surface which in turn leads to better absorption of sunlight. A reduction in glare duration and its intensity is observed from Scenario 1 to Scenario 3. The duration of yellow glare is longer for smooth glass without ARC (Scenario1) than light textured PV module. In general, the PV modules that are available in solar market are coated with anti-reflective material. Light textured solar PV modules are not commonly manufactured for commercial applications. Nowadays, antiglare PV modules are specially manufactured for solar applications in airports. The acceptable value of safe glare obtained by variation of tilt and orientation angle is the cheapest method at first sight. The variation of tilt or orientation angle (Scenario 4) provide wide possibilities of glare mitigation as compared to glass texturing and AR coating in PV modules (Scenarios 1, 2, 3). However, the energy generation varies with the change in angles. Using glare prediction software, a trade-off between safe-glare and energy output can be estimated.

4 CONCLUSIONS

In this study, the glare assessment for different scenarios of PV array is carried out using glare prediction software. Among the studied scenarios, safe value of glare is obtained in Scenario 4 only for specific values of tilt and orientation angle. A tilt angle of 15 degrees and orientation angle of 135 degrees is the most suitable set with 99.1 % of theoretical energy output and 0 mins of yellow and green glare. The yellow glare duration on ATC reduced in the descending order of Scenario 1 (2122 min), Scenario 2 (296 min) and Scenario 3 (0 min). It can be attributed to the reduction in reflectivity from PV module surface. Zero minutes of glare is predicted for 14 sets of tilt and orientation angles with energy generation variation from 99.1 % to 68.2 %. It is concluded that the best way to mitigate glare is by varying the orientation and tilt angles. The best-case scenario is observed for solar PV array fixed at 15 degrees and oriented at 135 degrees (Scenario 4). The worst case of glare impact is obtained for solar PV array with smooth glass & without ARC (Scenario 1). The software used in this study is validated with real life situations. However, the developers report that the glare assessment methodology involves several assumptions and approximations. So, the actual results may differ from the predicted glare occurrence. Though glare impact standard set by FAA is considered in this study, the safe limit of glare may vary with different judications. The future work includes selection of different sites within airport, analysis of financial implication of each scenario, using different glare prediction software.

ACKNOWLEDGMENT
The authors are grateful to the financial support provided by the Universiti Malaysia Pahang (www.ump.edu.my) through the Doctoral Research Scheme (DRS) and PGRS1903172.

CONFLICT OF INTEREST STATEMENT
The authors declare that they have no known competing financial interests or personal relationships that could influence the work reported in the submitted manuscript. The findings, opinion, insights and discussions in this manuscript are solely of the authors and do not necessarily reflect the opinion of the any organization involved directly or indirectly. The assumptions and case studies reported within this article are only examples and are based on limited open-source information. The authors are not responsible for any consequences thereof with the use of information presented in the work.

REFERENCES

[1] S. O. Alba and M. Manana, "Energy research in airports: A review," Energies, vol. **9**, no. **5**, pp. 1–19, 2016, doi: 10.3390/en9050349.

[2] S. Sreenath, K. Sudhakar, and Y. Ahmad Fitri, "Airport-based photovoltaic applications," Prog. Photovoltaics Res. Appl., 2020, [Online]. Available: https://doi.org/10.1002/pip.3265.

[3] A. Anurag, J. Zhang, and J. Gwamuri, "General Design Procedures for Airport-Based Solar Photovoltaic Systems," pp. 1–19, 2017, doi: 10.3390/en10081194.

[4] J. Plante, F. A. A. Airport, and E. Division, "Technical Guidance for Evaluating Selected Solar Technologies on Airports," 2013.

[5] P. Devita and S. Barrett, "Energy technologies ' compatibility with airports and airspace : Guidance for aviation," vol. **8**, no. **4**, pp. 318–326, 2014.
[6] A. Kandt and R. Romero, "Siting Solar Photovoltaics at Airports Preprint," no. June, 2014.
[7] F. Register, "Interim Policy, FAA Review of Solar Energy System Projects on Federally Obligated Airports," 2013. [Online]. Available: www.federalregister.gov/articles/2013/10/23/2013-.
[8] S. Sreenath, K. Sudhakar, A. F. Yusop, E. Cuce, and E. Solomin, "Analysis of solar PV glare in airport environment: Potential solutions," Results Eng., vol. **5**, no. November 2019, p. 100079, 2020, doi: 10.1016/j.rineng.2019.100079.
[9] S. Sukumaran and K. Sudhakar, "Fully solar powered airport: A case study of Cochin International airport," J. Air Transp. Manag., vol. **62**, pp. 176–188, 2017, doi: 10.1016/j.jairtraman.2017.04.004.
[10] M. Mpholo, T. Nchaba, and M. Monese, "Yield and performance analysis of the fi rst grid-connected solar farm at Moshoeshoe I International Airport , Lesotho," Renew. Energy, vol. **81**, no. 2015, pp. 845–852, 2020, doi: 10.1016/j.renene.2015.04.001.
[11] M. H. Banda, K. Nyeinga, and D. Okello, "Performance evaluation of 830 kWp grid-connected photovoltaic power plant at Kamuzu International Airport-Malawi," Energy Sustain. Dev., vol. **51**, pp. 50–55, 2019, doi: 10.1016/j.esd.2019.05.005.

Study the Influence of Distance in Artificial Potential Field (APF) for Autonomous Emergency Braking System (AEB) on Longitudinal Motion

Z Abdullah[1], P M Heerwan[1,2]*I M Izhar[3] and M A Zakaria[2,3]

[1]Faculty of Mechanical and Automotive Engineering Technology, University Malaysia Pahang, 26600 Pekan, Pahang, Malaysia
[2]Automotive Engineering Centre, University Malaysia Pahang, 26600 Pekan, Pahang, Malaysia
[3]College of Engineering, University Malaysia Pahang, 26600 Pekan, Pahang, Malaysia
[4]Faculty of Manufacturing and Mechatronics Engineering Technology, University Malaysia Pahang, 26600 Pekan, Pahang, Malaysia
*Email: mheerwan@ump.edu.my

Abstract. Autonomous Emergency Braking (AEB) is one of the common control systems used by the vehicle in order to avoid the collision from the obstacle. AEB performance optimize mostly on dry road surface at low and medium speed. The slip tire model was derived in this paper in order to find the coefficient of road friction. The tire-road friction coefficient is used for the braking limit which is 450 N.m. In order to increase the ability of an AEB during presence of obstacle in front of the vehicle, combination of Time-to-Collision (TTC) and artificial potential field (APF) are proposed in this study. When the APF value was surpassed the threshold distance, the AEB will activate by itself. The limit for APF was designed based on assuming the dry road friction (0.9) condition and with present of static obstacle in front of the vehicle in longitudinal lane. Thus, the AEB system was designed considering on the dry road friction condition, time for Front Collision Warning (FCW)as well as Braking was included for the limit APF is developed. The combination of an additional distance with the maximum safety distance in the APF system will create the minimum safe distance from obstacle which is in the range of 2.0 meter after the vehicle stop entirely. The additional distance is influence by the product of the constant time setting which is t_s1 and t_s2 with current velocity. Then, the simulation results show that the proposed control strategy can adapt to the dry tire-road friction coefficient on the road.

1. INTRODUCTION

Nowadays, the capability to helps the human constraints during driving period in order to reduce road disaster is one of the stimulants of the autonomous vehicle's development recently. Almost 1.2 million deaths exist because of road accidents globally and it is assumed to boost in numbers by 2030. Advanced Driver Assistance Systems (ADAS) is, in this manner, created to ensure to protect the road user ADAS consists of Adaptive Cruise Control, Autonomous Emergency Braking, Anti-lock Braking System and many more as a merger of several complex system[1]. Advanced Driver Assistance Systems (ADAS) is an active safety system to prevent the collision by assisting the driver in certain conditions[1].

IOP Conf. Series: Materials Science and Engineering 1068 (2021) 012015 doi:10.1088/1757-899X/1068/1/012015

Autonomous Emergency Braking (AEB) is one of intermediate ADAS to maintain a distance or avoid from having collision[1].

The ability of the AEB system can be improve by integrate the system with artificial potential field (APF) risk assessment[2-4]. The fundamental idea is that the host vehicle gets attractive energy from the objective point and aversion from the obstacle and keeps away from crash to arrive at the objective position under the resultant force[5, 6]. Besides, the artificial potential field has the favour of limited amount of arithmetic and smoothness of the planned path[6]. The resultant force occurs around the obstacle when the spacing headway of the host vehicle cross the border of the maximum safety distance from the obstacle[7].

Even though the load of commercialized an AEB system, near miss accident still happens, where the distance of the host vehicle is too close to the obstacle[5]. So, to avoid this phenomenon, the enhancement of the AEB product which acknowledge and manage the safe distance towards the collision point is develop. The minimum safety of the vehicle after fully stop is around 2.08 to 3.3 m from the obstacle[8]. Subsequently, conventional Autonomous Emergency Braking System (AEB) have limitation in avoiding collision when the vehicle moves in medium and high speed[5]. This paper aim to do analysis the distance of an AEB activation at each point of the phase; front collision warning and full braking in order to make sure the host vehicle stop entirely 2.0 m from the obstacle. The combination of the additional distance; the production of constant time setting with current velocity, with maximum safety distance in the Artificial Potential Field System (APF) will directly increase or decrease the initial distance of each phase of an AEB activation and affect the minimum safety distance of the vehicle after entirely stop.

2. VEHICLE MODEL

A simplified vehicle dynamics model is developed in this paper to analyse the vehicle dynamics behaviour. Figure 1 shows a simplified model comprises of longitudinal model[9].

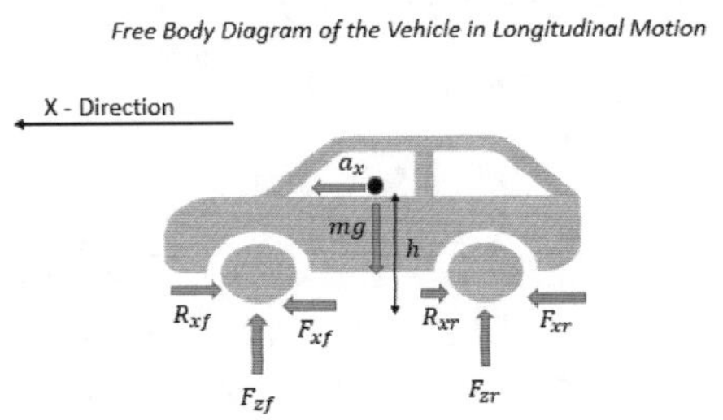

Figure 1. Analysis in Longitudinal Motion of Vehicle Model

Table 1. Specification of the Vehicle

Details	Symbol	Value	SI unit
Mass (Kerb Weight)	m	1330	kg
Center of gravity (c.g) length towards frontal part	Lf	1.107	m
Center of gravity (c.g.) length towards rear part	Lr	1.643	m
Height of center gravity	h	0.479	m
Effective radius of the tire	r_{eff}	0.393	m

3. LONGITUDINAL DYNAMIC MODEL

In the numerical analysis, the equalization of the forces applied on the vehicle is considered. Then, the vehicle is expecting driving on a level and straight street. By assuming the force at left and right tires are same, a bicycle model is used in the analysis. The linear longitudinal forces are expressed follows:

$$ma_x = F_{xf} + F_{xr} - R_x - D_a V_x^2 \tag{1}$$

After balancing the moment with respect to contact points between tires and surface, the normal forces exerted on front and rear tire could be composed as in equation (2) and equation (3)[9]:

$$F_{zf} = (mgL_r - ma_x h - D_a V^2 h_a) / (L) \tag{2}$$

$$F_{zr} = (mgL_f + ma_x h + D_a V^2 h_a) / (L) \tag{3}$$

Where m is the total mass of the vehicle, a_x is the longitudinal acceleration/deceleration, F_{xf} and F_{xr} are the front and rear wheel traction/braking forces, $R_x = R_{xf} + R_{xr} = C_{roll}mg$ is the rolling resistance force with C_{roll} being the rolling resistance coefficient, D_a is the aerodynamic drag force constant which is in small value and assume to be zero. V_x is the longitudinal velocity. Let L_f be the distance from $c.g.$ to the front axle; L_r the distance from $c.g.$ to the rear axle and $L = L_f + L_r$ be the wheelbase of the vehicle. The nonlinear longitudinal forces using Dugoff's tire model are expressed as in equation (4)[9]:

$$F_x = C_\sigma \frac{\sigma_x}{1+\sigma_x} f(\lambda) \tag{4}$$

$$f(\lambda) = \begin{cases} (2-\lambda)\lambda, & \lambda < 1 \\ 1, & \lambda \geq 1 \end{cases} \tag{5}$$

$$\lambda = \frac{\mu F_z (1+\sigma_x)}{2[(C_\sigma \sigma_x)^2 + (C_a \tan(a))^2]^{1/2}} \tag{6}$$

$$\mu_{brk} = -1.15k\{e^{-35\sigma_x} - e^{-0.35\sigma_x}\} \tag{7}$$

where σ_x be the longitudinal slip ratio of the tire under consideration and the longitudinal tire stiffness by C_σ. F_z is the vertical force on the tire while μ is the tire-road friction coefficient and k is the parameter of the road condition. The longitudinal slip ratio during braking is defined as in equation (8)[10] :

$$\sigma_x = \frac{V_x - r_{eff}\,\omega_w}{V_x} \tag{8}$$

Let r_{eff} be the effective radius of the tire and V_x as the longitudinal velocity while ω_w act as a rotational velocity of the tire. The equation for the decelleration, a of host vehicle with considering the tire-road friction coefficient is expressed as in equation (9) :

$$a = F_x / m \tag{9}$$

Where m is a mass of the vehicle and F_x is the nonlinear longitudinal force.

3.1 Autonomous Emergency Braking (AEB) System Algorithm

Autonomous emergency braking system is a system in the vehicle for executing the brake when the sensor detects the obstacle in front of the host vehicle[11, 12]. AEB is a system which helps the vehicle to avoid the collision by execute an emergency braking during appearance of the static obstacle[11]. The flow of the system is quite different with a conventional AEB. Assuming an obstacle occur in front of a vehicle within the width of the vehicle; then, the collision decision is simplified as longitudinal collision prediction using an Artificial Potential Field (APF) as shown in Figure 2. The system will be turned on when the value of APF violated the threshold distance between the vehicle and obstacle.

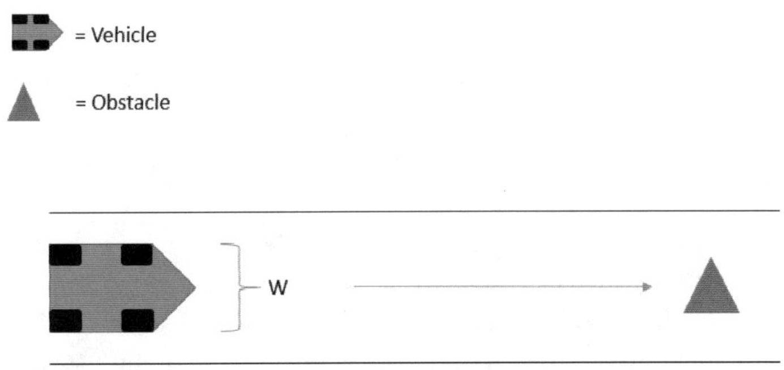

Figure 2 Driving scenario during presence of static obstacle

3.2 Time to Collision Risk Assessment

The time to collision (TTC) is defined as the time that a driver can use to reduce the speed of a vehicle by braking to avoid collision with the front target. The larger the TTC value is, the lower the risk of a collision will be, and vice versa. The evaluation of the TTC is basically from the kinematic model. Besides, the time to collision, TTC is calculated by deriving the equation of the kinematic model when the host vehicle approaches the obstacle as express in equation (10)[4, 13]:

$$|p| = \begin{cases} -vt & , \quad a = 0 \\ -vt + \frac{1}{2}at^2 , & a \neq 0 \end{cases} \tag{10}$$

The time to collision can be obtained by rearranged the equation of kinetic model as shown in equation (11):

$$|TTC| = \begin{cases} \frac{-v \pm \sqrt{v^2+2pa}}{a}, & v \geq 0 \; and \; a < 0 \end{cases} \tag{11}$$

3.3 Artificial Potential Field

The philosophy of the APF calculation can be schematically depicted in accompanying manners. The host vehicle moves in a power field. The location to be arrived is an attractive pole, $U_{goal}(x)$ for the end effector and obstacle, $U_{obs}(x)$ are repulsive range for the host vehicle are expressed as in equation (12)[6]:

$$U_{art}(x) = U_{goal}(x_d) + U_{obs}(x) \tag{12}$$

Let be ρ_r is the spacing headway of the vehicle and n represent the repulsive gain coefficient. The ρ_{or}, is a given point in the closeness to the obstacle. The potential field occur when the object and obstacle follow this condition ($\rho_r \leq \rho_{or}$) [14]. The repulsive force equation was derived and shown as in equation (14)[14].

$$F_{rep}(X) = \begin{cases} n\left(\frac{1}{\rho_r} - \frac{1}{\rho_{or}}\right)\frac{1}{p_r^2}, & \rho_r \leq \rho_{or} \\ 0, & \rho_r > \rho_{or} \end{cases} \tag{14}$$

3.4 Design of Vehicle Conditional Artificial Potential Field

The Vehicle Conditional Artificial Potential Field (VC-APF) is diverse from the conventional APF and will design only when condition is achieved, which is an AEB system will execute when the vehicle detects an obstacle in longitudinal lane. The smaller the space between the autonomous vehicle and the obstacle, the more repulsive force will be appeared centre on the obstacles. The manipulated range of the repulsion field is ρ_{or} and is related to the speed and the maximum deceleration of the autonomous vehicle, and the relation is explicit by equation (15)[7].

$$\rho_{or} = d_0 + (v_c \times TTC) + \frac{{v_c}^2}{2a_{max}} \tag{15}$$

Where ρ_{or} represents the safe distance while driving, and d_0 represents the minimum safe distance after the vehicle stop. TTC is the time headway of the vehicle to the obstacles, and v_c represents the current speed of the autonomous vehicle while a_{max} represents the maximum deceleration of the vehicle [7]. The value of repulsion force depends on each phase of threat level which is expressed by Equation (16) to equation (18).

Full braking;

$$F_{rep}(X) = \frac{1}{2}n\left(\frac{1}{\rho_r} - \frac{1}{d_0 + \left((v_c \times TTC) + \frac{{v_c}^2}{2a_{max}}\right)}\right)\frac{1}{{\rho_r}^2}$$

$$if, \; \rho_r \leq \rho_{or} - v_c \times t_{s1} \tag{16}$$

Warning Signal.

$$F_{rep}(X) = \frac{1}{2}n\left(\frac{1}{\rho_r} - \frac{1}{d_0 + \left((v_c \times TTC) + \frac{{v_c}^2}{2a_{max}}\right)}\right)\frac{1}{{\rho_r}^2}$$

$$if, \; \rho_{or} - (v_c \times t_{s1}) \leq \rho_r \leq \rho_{or} + (v_c \times t_{s2}) \tag{17}$$

No signal.

$$F_{rep}(X) = 0 \quad \text{if, } \rho_r > \rho_{or} + (v_c \times t_{s2}) \tag{18}$$

4. RESULTS AND DISCUSSION

The Potential Field profitably present threat assessment in relationship to the relative distance between host vehicle and obstacle data, provided by the frontal sensor. As shown, the threat assessment field of an AEB start to give warning signal at certain value depends on the velocity of the vehicle and repulsive force start to produce subsequently give braking after the vehicle enter the braking phase. The AEB system activate the desired braking torques, as it reaches maximum torque braking which is 450 N.m and the vehicle will stop entirely after execution of the braking is in the range of 2 m from the obstacle. The stopping distance of the vehicle directly influence by the combination of the additional distance with the maximum safety distance of the APF system. The additional distance data is consisting of the production of the vehicle current velocity, $\boldsymbol{v_c}$ with constant time setting which are $\boldsymbol{t_{s1}}$, and $\boldsymbol{t_{s2}}$. Table 2 below shows that all important result that have been recorded during the simulation. Next, the graph in Figure 3 and 4 for (a), (b), (c), (d), (e) shows that the condition of the vehicle which is in varies current velocity, $\boldsymbol{v_c}$ = 45 and 65 km/h and simulate at same initial relative distance of the vehicle from the obstacle which is at Pr = 250 m.

Table 2. Result of the Simulation of an AEB system

Velocity of vehicle (km/h)	45	65
Torque for braking (N.m)	450	450
Distance from obstacle when vehicle enter warning signal phase (m)	74.49	130.46
Distance from obstacle when vehicle enter braking phase (m)	36.91	76.10
Time setting, t_{s1} (s)	0.255	0.255
Time setting, t_{s2} (s)	1.245	1.245
Minimum safety distance after braking (m)	2.0	2.0

Velocity of the vehicle = 45 km/h

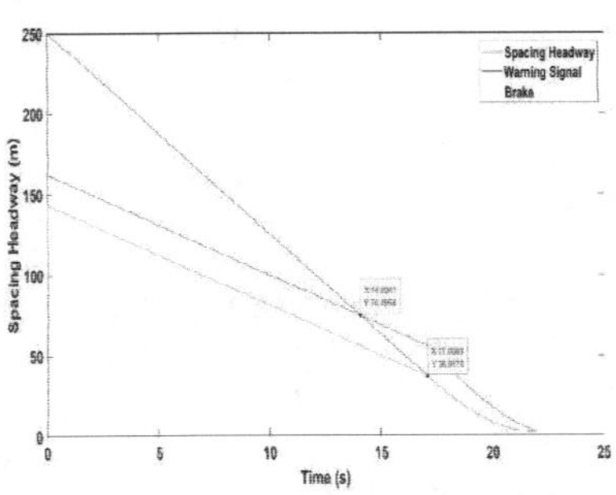

Figure 3 (a) Point of activation an AEB system

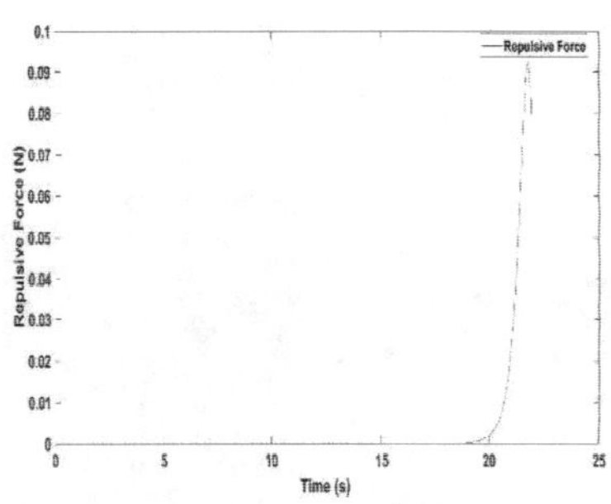

Figure 3 (b) Effect of APF during Braking

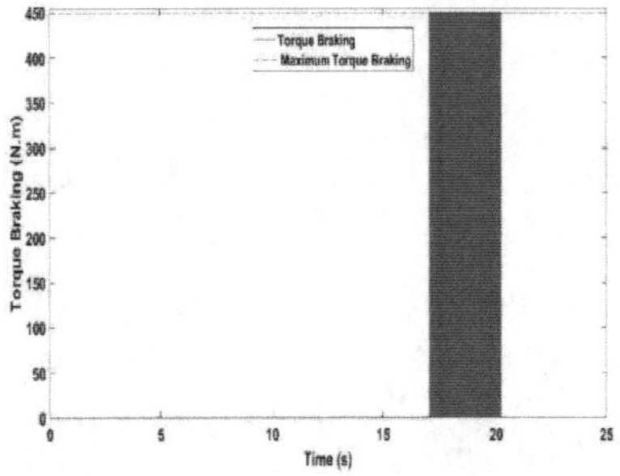

Figure 3 (c) Relationship between Torque Braking and Time

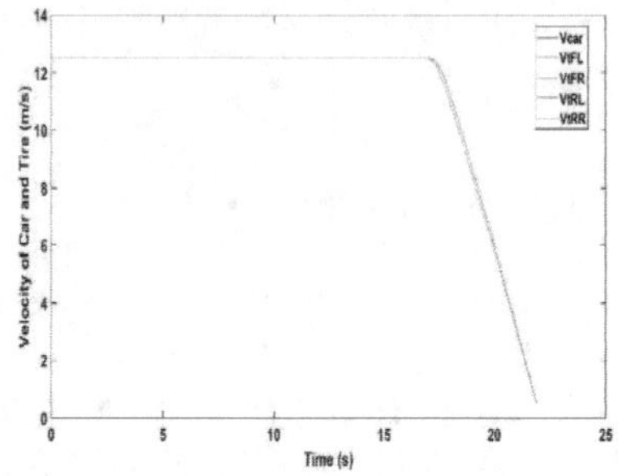

Figure 3 (d) Relationship of Velocity of the Vehicle with Velocity of each Tire

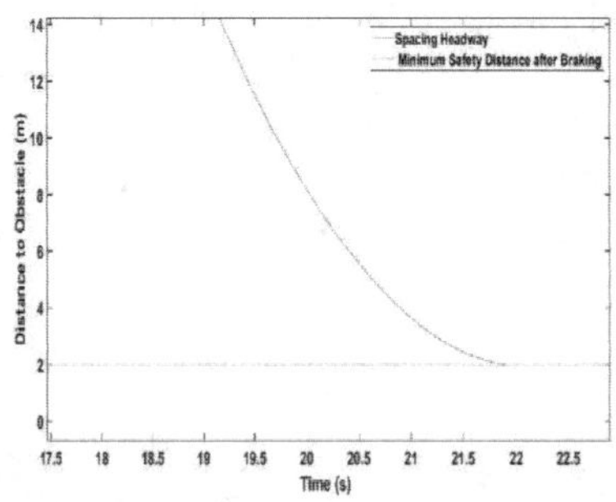

Figure 3 (e) Minimum of Safety Distance from Obstacle

Velocity of the vehicle = 65 km/h

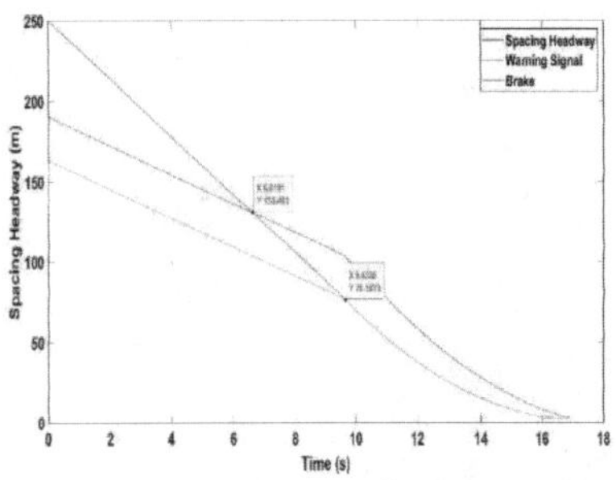

Figure 4 (a) Point of activation an AEB system

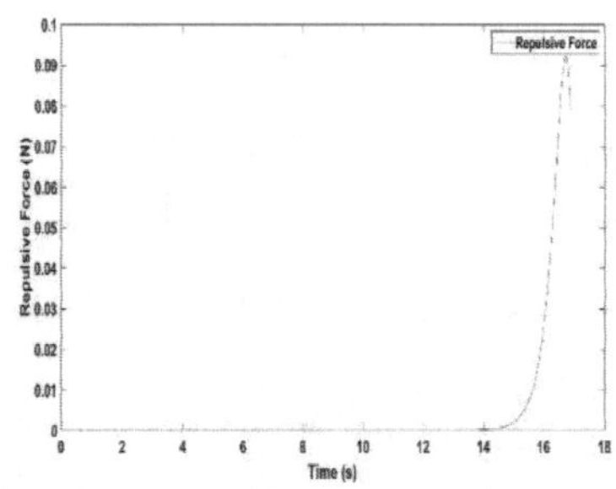

Figure 4 (b) Effect of APF during Braking

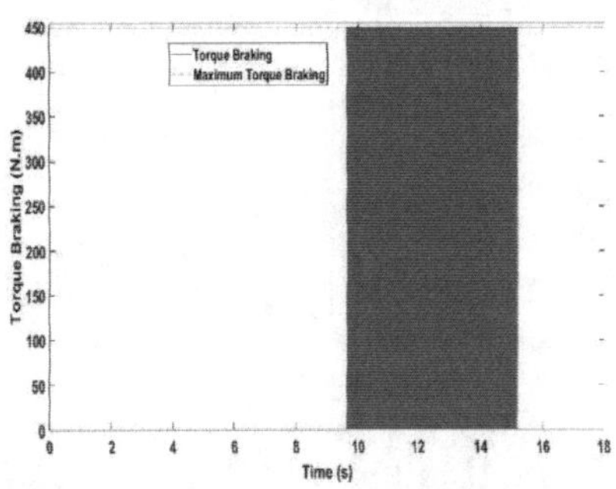

Figure 4 (c) Relationship between Torque Braking and Time

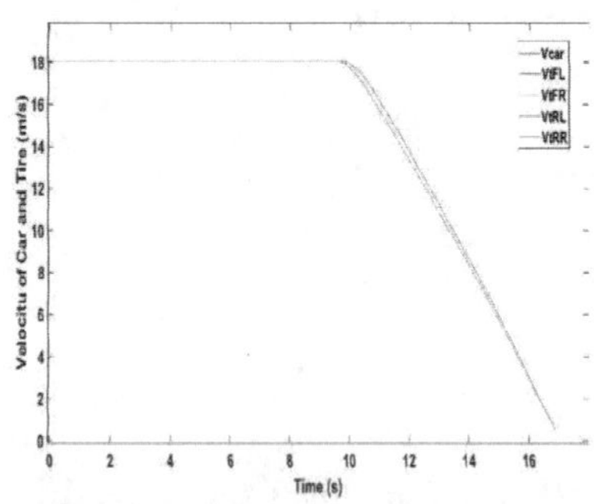

Figure 4 (d) Relationship of Velocity of Vehicle with Velocity of each Tire

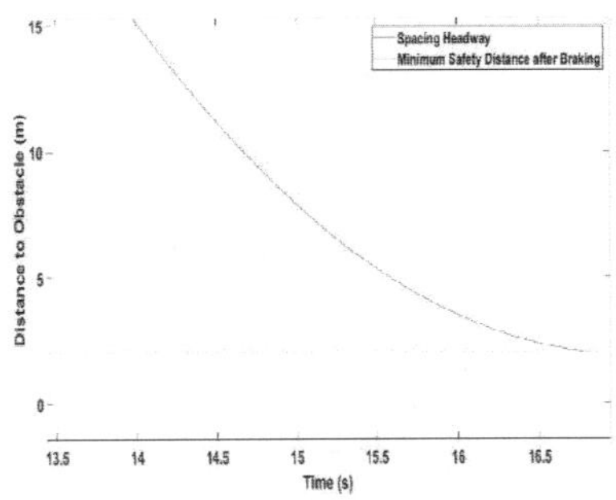

Figure 4 (e) Minimum of Safety Distance from Obstacle

5. CONCLUSIONS

In conclusion, as the development to the common Autonomous Emergency Braking system, the authors introduce an assimilation of AEB system with the Potential Field Risk threat strategy to avoid the close-miss incidents. APF profitably calculate the threat of the frontal static obstacle. Based on the threat measurement, the host vehicle successfully activates the desired braking actuations to grant it for a full vehicle stopping. The inclusion of PF into the AEB grant the vehicle to manage the safe distance of 2.0 m from the obstacle. For future works, more complicated scenarios and higher speed of the vehicle will be simulate for the improvement of the system. Varied driving style arrangement in the emergency braking scheme by varies drivers shall be investigated to strengthen the AEB achievement. This research hopefully will enhance the system of an AEB for the future work.

ACKNOWLEDGEMENT

The task granted in this researched is supported by University Malaysia Pahang for the financial backing via the UMP internal grant RDU192219. The authors would like to thank Mohamad Heerwan bin Peeie from Automotive Engineering Centre of UMP Pekan, Pahang.

REFERENCES

[1] K. Hojjati-Emami, B. Dhillon, and K. Jenab, "Reliability prediction for the vehicles equipped with advanced driver assistance systems (ADAS) and passive safety systems (PSS)," *International Journal of Industrial Engineering Computations,* vol. **3**, no. **5**, pp. 731-742, 2012.

[2] Y. Ruan, H. Chen, and J. Li, "Longitudinal Planning and Control Method for Autonomous Vehicles Based on A New Potential Field Model," SAE Technical Paper0148-7191, 2017.

[3] H. Wang, Y. Huang, A. Khajepour, Y. Zhang, Y. Rasekhipour, and D. Cao, "Crash Mitigation in Motion Planning for Autonomous Vehicles," (in English), *IEEE Transactions on Intelligent Transportation Systems,* Article vol. **20**, no. **9**, pp. 3313-3323, 2019, Art. no. 8617711.

[4] W. Yang, X. Zhang, Q. Lei, and X. Cheng, "Research on longitudinal active collision avoidance of autonomous emergency braking pedestrian system (AEB-P)," (in English), *Sensors (Switzerland),* Article vol. **19**, no. **21**, 2019, Art. no. 4671.

[5] U. Z. A. Hamid *et al.*, "Autonomous emergency braking system with potential field risk assessment for frontal collision mitigation," in *2017 IEEE Conference on Systems, Process and Control, ICSPC 2017*, 2018, vol. 2018-January, pp. 71-76: Institute of Electrical and Electronics Engineers Inc.

[6] O. Khatib, "Real-time obstacle avoidance for manipulators and mobile robots," in *Autonomous robot vehicles*: Springer, 1986, pp. 396-404.

[7] C. Li, X. Jiang, W. Wang, Q. Cheng, and Y. Shen, "A Simplified Car-following Model Based on the Artificial Potential Field," *Procedia Engineering,* vol. 137, pp. 13-20, 2016/01/01/ 2016.

[8] W. Yang, H. Zhao, and H. Shu, "Simulation and verification of the control strategies for aeb pedestrian collision avoidance system," (in Chinese), *Chongqing Daxue Xuebao/Journal of Chongqing University,* Article vol. **42**, no. **2**, pp. 1-10, 2019.

[9] R. Rajamani, *Vehicle dynamics and control*. Springer Science & Business Media, 2011.

[10] M. Abe, *Vehicle handling dynamics: theory and application*. Butterworth-Heinemann, 2015.

[11] I. A. Kulikov, I. A. Ulchenko, and A. V. Chaplygin, "Development and testing of a collision avoidance braking system for an autonomous vehicle," (in English), *International Journal of Innovative Technology and Exploring Engineering,* Article vol. **8**, no. **12**, pp. 703-709, 2019.

[12] H. K. Lee, S. G. Shin, and D. S. Kwon, "Design of emergency braking algorithm for pedestrian protection based on multi-sensor fusion," (in English), *International Journal of Automotive Technology,* Article vol. **18**, no. **6**, pp. 1067-1076, 2017.

[13] I. C. Han, B. C. Luan, and F. C. Hsieh, "Development of Autonomous Emergency Braking control system based on road friction," in *2014 IEEE International Conference on Automation Science and Engineering, CASE 2014*, 2014, vol. 2014-January, pp. 933-937: IEEE Computer Society.

[14] K. Gao *et al.*, "Conditional artificial potential field-based autonomous vehicle safety control with interference of lane changing in mixed traffic scenario," *Sensors,* vol. **19**, no. **19**, p. 4199, 2019.

Effect of algae fuel addition on fuel consumption and thermal efficiency of single cylinder diesel engine

Hazim Sharudin[1,a], N.A. Rahim[2], N.I. Ismail[2], Sharzali Che Mat[2], Nik Rosli Abdullah[3], N.W.M. Zulkifli[4], M.N.A.M. Yusoff[4]

[1]Faculty of Faculty of Mechanical Engineering, Universiti Teknologi MARA, Cawangan Johor, Kampus Pasir Gudang, 81750, Johor, Malaysia
[2] Faculty of Mechanical Engineering, Universiti Teknologi MARA, Cawangan Pulau Pinang, Kampus Permatang Pauh, 13500 Permatang Pauh, Pulau Pinang, Malaysia
[3]Faculty of Mechanical Engineering, Universiti Teknologi MARA, 40450, Shah Alam, Malaysia
[4]Centre for Energy Sciences, Faculty of Engineering, University of Malaya, Department of Mechanical Engineering, 50603 Kuala Lumpur, Malaysia

[a]Corresponding author: hazim@uitm.edu.my

Abstract. The objective of this paper is to evaluate the brake specific fuel consumption (BSFC) and brake thermal efficiency (BTE) of algae fuel blended in diesel fuel. Fuel blends of 1% of algae fuel (B1), 2% of algae fuel (B2), 3% of algae fuel (B3) and 4% of algae fuel (B4) to each 100 ml of fuel blend volume were prepared. Each algae-diesel fuel blend was tested on single-cylinder diesel engine by varying the engine loads (0, 2, 4 and 6 Nm) at a constant engine speed of 1500 rpm. Different BTE and BSFC engine performance levels were the measured parameters. Based on the result of the engine performance tests, it can be concluded that B4 blends in diesel fuel produced better BTE with 17% at 4 Nm engine load than diesel fuel. Meanwhile, the BSFC was lower for B4 with 16% compared to diesel fuel at lower engine load.

1. INTRODUCTION

Current trend of energy supply and consumption is flagrantly unsustainable from environmental, economic and sociological aspects. As the vital movers of economic growth, the continuing availability of energy is very important. A high energy consumption level contributes to a high standard of living in developed countries. Other than renewable energy and nuclear energy, petroleum products are one of the significant energy sources globally. Petroleum oil, coal, and natural gas are examples of fossil fuels. As the population growth has increased, the demands of fossil fuel have also been rising and will cause the depletion of fossil fuel. This urges researchers to find a solution to overcome the problem. New alternative fuel research has initiated in the late 1970s and early 1980s [1]. Alternative fuels, other than crude oil, can be applied with little or no alteration with contemporary petrol or diesel internal combustion engine. The alternative fuels become the interest of researchers because it is frequently generated from domestic resources that mitigate energy dependency. Alternative fuels also typically reduce vehicle exhaust emission that will reduce air pollution. Many studies have been conducted on alternative fuels, especially in transport vehicles, and they have revealed various possible alternative fuels, such as biofuels, that can be used, namely biodiesel, methanol, ethanol, straight vegetable oils (SVO) and hydrogen [1,2].

Based on the classification of biofuels, it is shown that algae are the substrate from the third generation of secondary biofuels that is linked to utilization of carbon dioxide as its feedstock [3,4]. Biodiesel produced by algae is called algae biodiesel. Many advantages of biodiesel have been discovered, such as higher combustion efficiency, higher biodegradability, domestic origin and reduce the dependency on imported petroleum [5,6]. As the liquid transportation fuel alternative, algae fuel has been recommended as a sustainable solution and economically viable alternative [7]. In comparison to other crops, algae are the most favorable oil source owing to their rapid growth rate, ability to grow in broad range conditions, potential for higher yield rates and biodiesel properties of algae fuel which is comparable to the standard biodiesel [7-9]. Hence, algae have become a great interest as a source of biodiesel because it is more effective. Different types of fuels can also be suited to different species of algae [8].

However, studies related to the properties and performance of algae fuel applied on diesel engine are scarce [10,11]. Most of the existing studies have focused on the inclusion of algae with alcohols and diesel fuel but not raw algae. Application of biodiesel on diesel engine has drawback owing to high emission of carbon dioxide (CO_2) [12]. The emission of oxides of nitrogen (NO_x) from the use of biodiesel is slightly higher than the use of diesel fuel [13,14]. By using biodiesel, the reduction in heating value increases the BSFC in percentage [15]. The increasing percentage of poon oil in the blend increases the exhaust gas temperature [16].

Therefore, the focus of this research is to investigate the properties of different volume percentage of blended algae fuel with diesel fuel as an alternative fuel. The properties of optimum blends of algae fuel and diesel fuel need to be clarified to facilitate further studies on biodiesel. In this research, the properties measured were calorific value, density and viscosity. Then, the effect of algae-diesel blends on engine performance was investigated, specifically for BSFC and BTE.

2. Materials and Methods

2.1. Algae Fuel Preparation

2.1.1. Algae oil separation from water

The algae used in this study were collected from the ponds of Sekolah Menengah Sains Tun Syed Sheh Shahabudin, Bukit Mertajam. Filtration and drying methods were used for separating the algae from water. Figure 1 shows the vacuum filtration machine used to obtain the sediment of algae due to pressure exerted from the vacuum. The sediment obtained was dried to remove moisture for oil extraction process. In the drying process, the industrial oven was used to dry the sediment of algae at 40^0C for 18 hours.

2.1.2. Algae fuel extraction

The expeller/press method, solvent extraction process by hexane, and supercritical fluid extraction (SFE) method are popular methods available to extract oil from algae. The SFE method was selected as it is more efficient in extracting nearly 100% of oils than the conventional solvent separation methods [8]. The SFE method, shown in Figure 2, involved heating and liquefying the carbon dioxide below pressure so as to have the properties of gas and liquid. Algae fuel was obtained through extraction for about 1 hour for each process at temperature of 50^0C and at 300 bar pressure by using CO_2 liquefied fluid as solvent (90%) and ethanol as co-solvent (10%). Finally, the BUCHI rotary evaporator shown in Figure 3 was utilized for solvent removal from the oil obtained, leaving only pure algae fuel. The evaporation process occurred about 5 minutes with a speed of 135 rpm and the bath temperature was set at 60^0C.

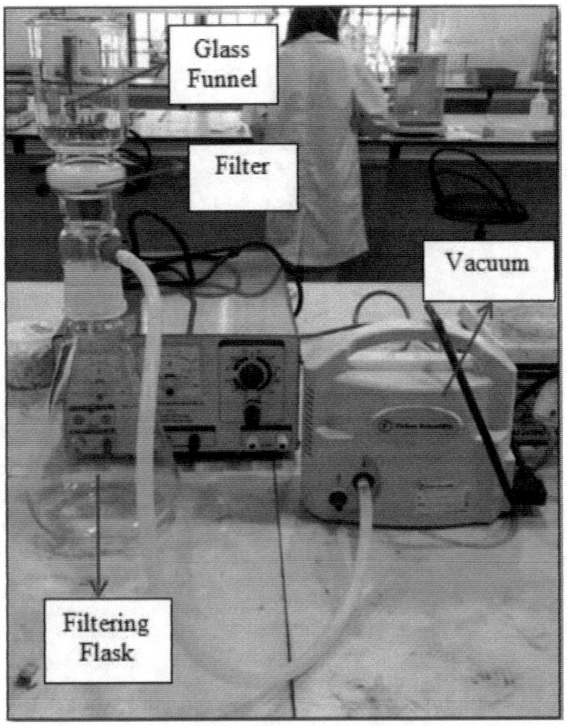

Figure 1. Vacuum filtration machine

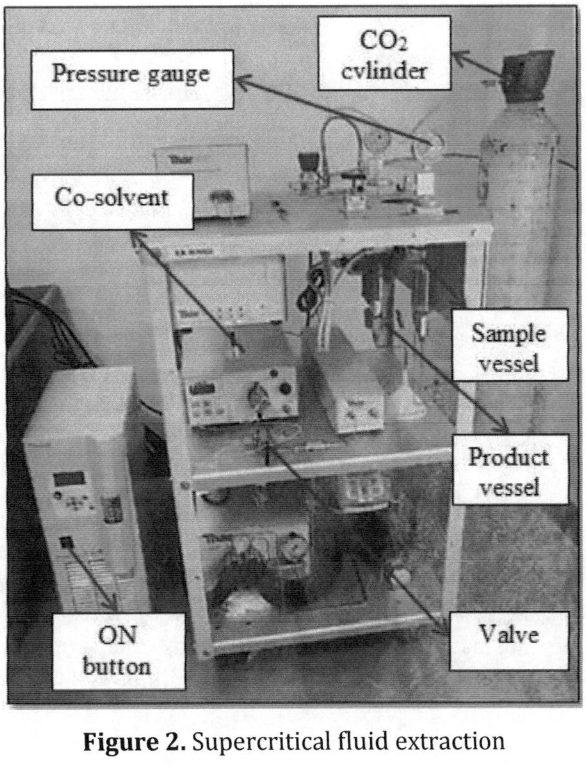

Figure 2. Supercritical fluid extraction

Figure 3. The BUCHI rotary evaporator

2.2. *Fuel Blending Preparation*

The blends of algae biodiesel fuel were prepared based on volume for engine testing, and essential properties were remained within the acceptable limit in accordance to the American Society for Testing Material (ASTM) standard. The volume percentages of algae fuel were selected from 1% to 4% out of 100 ml blended with diesel fuel. Table 1 summarizes the volume percentage of algae and respective nomenclature to differentiate each blend. All the diesel-algae mixtures were continuously stirred using the magnetic stirrer and hot plate for 15 minutes at room temperature and 500 rpm speed.

Table 1. Fuel blends and their nomenclature.

Fuel Blend	Percentage of Algae	Nomenclature
100 ml Diesel	0%	D100
99 ml Diesel + 1 ml Algae fuel	1%	B1
98 ml Diesel + 2 ml Algae fuel	2%	B2
97 ml Diesel + 3 ml Algae fuel	3%	B3
96 ml Diesel + 4 ml Algae fuel	4%	B4

2.3. *Fuel Blending Preparation*

Each developed sample was subjected to density, viscosity and calorific value test following the international standard procedures. In engine testing, a measurement of fuel properties is required to recognise the characteristics of fuel blends that may affect the performance and emissions of the engine. Additionally, the properties of the blended fuel were also analysed as they are critical factors for engine requirement, consumer satisfaction and industry standards. The properties of fuel blends obtained are shown in Table 2. For comparison, the properties of diesel fuel (D100) were also provided.

IOP Conf. Series: Materials Science and Engineering 1068 (2021) 012016 doi:10.1088/1757-899X/1068/1/012016

Table 2. Properties of diesel and fuel blends

Fuel property	Unit	Diesel fuel (D100)	1% of algae fuel (B1)	2% of algae fuel (B2)	3% of algae fuel (B3)	4% of algae fuel (B4)	ASTM Test Method
Density (at 15 ^{0}C)	kg/m^3	850	845.8	845.2	844.8	844.6	ASTM-D 4052
Viscosity (at 40 ^{0}C)	mm^2/s	2.6	3.069	3.003	2.943	2.920	ASTM-D 445
Calorific Value	kJ/kg	45500	44546	44529	44382	44190	ASTM-D 4809

2.4. Engine Setup

Engine performance experiment was conducted using a vertical single-cylinder, four-stroke, and air-cooled diesel engine by the Yanmar TF-M by Yanmar Co. Ltd. This engine was combined with eddy current dynamometer with maximum power input of 20 kW at 2450 to 10000 rpm. Next, the control panel was used to vary the load, and the speed of the engine was adjusted. The full details of engine specifications are given in Table 3, and the schematic diagram of the engine setup is shown in Figure 4. The experiment was carried out using algae fuel mixed with different volume percentage of diesel fuel (D100): 1% of algae fuel (B1), 2% of algae fuel (B2), 3% of algae fuel (B3) and 4% of algae fuel (B4) to 100 ml of fuel blend volume.

Table 3. Engine specification for experiment.

Parameters	**Values**
Type	Vertical cylinder, 4-cycle air cooled diesel engine
No of cylinders	1
Bore x Stroke	70mm x 57mm
Combustion	Direct injection
Maximum Output	3.5kW
Engine Rated Speed	3600 rpm

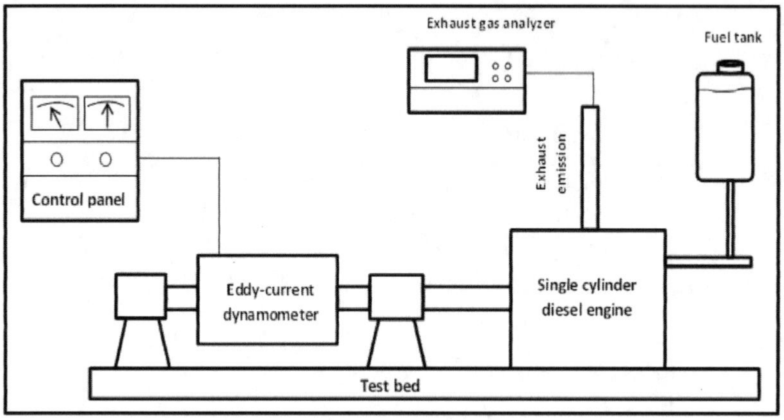

Figure 4. Schematic diagram of the engine test set up

2.4.1. Engine test condition

First, the single-cylinder diesel engine was warmed-up by running it for 15 minutes using D100 to ensure that the engine operation was perfectly stable. For the engine test condition, the engine speed was set at a constant 1500 rpm, while engine load was varied (2, 4, 6 Nm). Then, the D100 was changed to the fuel blends for running the engine to ensure the removal of residual diesel from the fuel lines. Each load was left to run until the volume of the fuel dropped for every 10 ml, and the time taken for each 10 ml drop was recorded. Finally, the fuel blends were changed with the next fuel samples, and the engine was left running until it reached a stable condition before the required data were recorded. The engine performance test was performed and repeated for three times to get an average data of the experiment. Based on the data for fuel consumption and calorific values obtained from the experiment, the engine performance of the blends was obtained.

BSFC signifies the level of efficiency of engine fuel consumption in its work performance and fuel consumption is determined based on mass flow rate per unit of time($\dot{m}_f$). It is one of the crucial parameters because the engine was operated with various fuel types, indicating different levels of fuel performance either economic or otherwise. The BSFC parameter is defined as the ratio of mass flow rate per unit time divided by power output as expressed by equation (1):

$$BSFC = \frac{\dot{m}_f}{Engine\ power} \quad (1)$$

Where, $\dot{m}_f$ is fuel mass flow rate (g/hr), engine power; $\frac{2\pi NT}{60000}$ and N is rotational engine speed (rpm) and T is torque.

BTE is described as the energy ratio in brake power (*bp*) divided by fuel energy input, which is the product of mass of fuels ($\dot{m}_f$). It is then multiplied by lower heating value of fuel blend, or known as fuel calorific value. This efficiency demonstrates the generation of output by the engine with the heat from the supplied fuel. Equation (2) denotes the BTE calculation.

BTE (%):

$$\eta_{bth} = \frac{3600}{calorific\ value\ x\ BSCF}\ x\ 100 \quad (2)$$

Where, η_{bth} is t brake thermal efficiency (%), and calorific value is the blended fuel heating value (MJ/kg)

3. RESULTS AND DISCUSSION

The results obtained from the engine performance tests are discussed and the graphs are presented in this section. The engine performances, namely BSFC and BTE, are discussed based on the research objectives.

3.1. *Brake Specific Fuel Consumption*

Figure 5 shows the BSFC with engine load at constant 1500 rpm speed. Generally, higher consumption of fuel is observed for power generation owing to higher loss of heat in the combustion chamber walls at low engine speed (1500 rpm), which results in poor combustion efficiency. At 2 Nm engine load, D100 with 1390.537 g/kwh has the highest BSFC compared to other fuel blends. It could be due to the combined effects of viscosity, fuel density and low calorific value of the blends. For 2 Nm engine load, it was shown that the algae-diesel blends of B1, B2, B3 and B4 were lower with an average reduction of 23, 14, 19 and 16%, respectively, relative to D100. A research stated that the BSFC decreases when the engine speed is lower than 1300 rpm, and shows ascendant trend when the engine speed increases [10]. Besides, the decrease in BSFC in relation to the load is due to the increase in heat energy conversion into mechanical work with increasing combustion temperature [11].

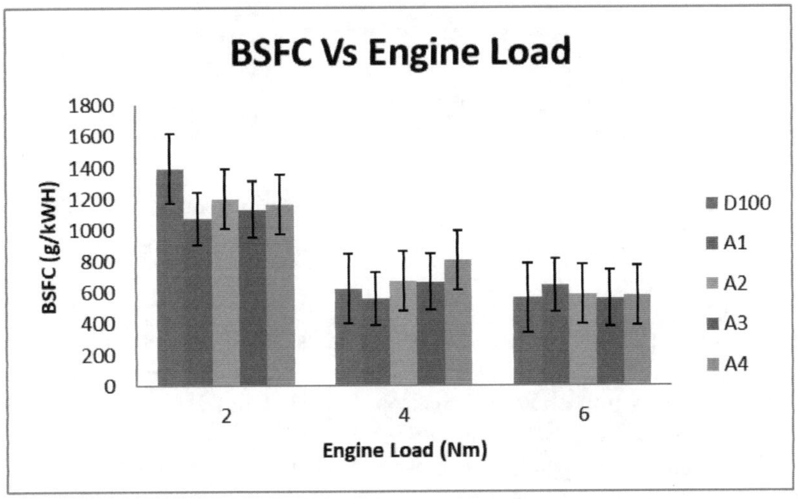

Figure 5. BSFC against engine load at constant 1500 rpm speed.

Besides that, 11% decrease of B1 was recorded with 554.354 g/kwh at the engine load of 4 Nm as compared with D100. This can be related to two major factors, namely the decrease in calorific value in fuel and effect of fuel density. This result also showed the decrease of BSFC value due to the increment of mass fuel consumption at lower torque and lower thermal efficiency, which has also been described previously in past research [12]. Moreover, BSFC of the algae biodiesel decreased when the engine load increased. However, it can be observed that the BSFC for D100 in Figure 5 was the lowest among the algae-diesel fuel blends at higher engine torque (6 Nm). This is because BSFC is proportional to the calorific value and the amount of fuel injected; hence the BSFC for biodiesel is higher than the BSFC for pure diesel fuel for a constant engine output due to lower calorific value of biodiesel [13,14]. Comparable results were also reported by other studies as higher BSFC produced by the use of biodiesel compared to diesel [15,16]. This is also due to the higher density in the volumetric injection system, which causes higher BSFC [16,17].

3.2. Brake Thermal Efficiency

Variation in the BTE of D100, B1, B2, B3 and B4 at various engine loads with constant 1500 rpm speed is presented in Figure 6. It can be observed in Figure 6 that the BTE increased for all fuel blends when the engine load increased. Higher concentration of algae biodiesel improved the BTE of the engine. BTE for B2, B3 and B4 blends increased gradually along with the engine load but B1 showed 14% decrement, unlike D100 at load 6 Nm. This could be due to insufficient air during the experiment, resulting in lower oxygen concentration that causes incomplete combustion of the blends [15,18]. Besides that, the BTEs of B1, B2, B3 and B4 blend were higher, with an average increase of 33, 17, 17 and 17%, respectively, relative to D100.

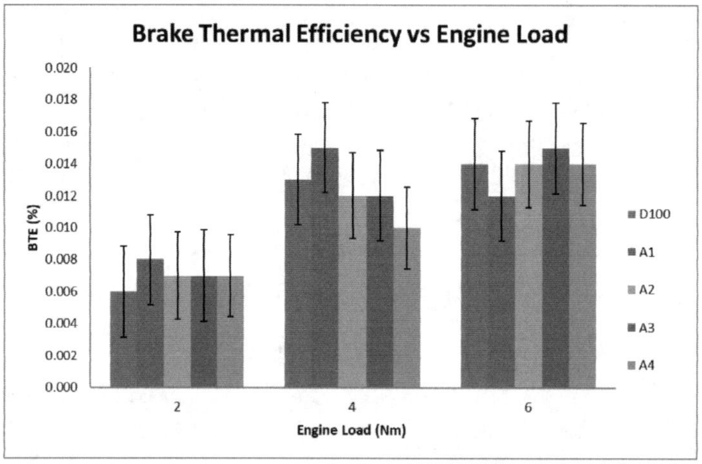

Figure 6. BTE against engine load at constant 1500 rpm speed.

Generally, the BTE increases in two circumstances: when the concentration of the biodiesel in the blend increases, and when the speed of the engine increases at full load. This is also supported by the findings of previous researchers when BTE for the blends somewhat increases not similar to diesel fuel at high loads of both speeds and all injection timings [11]. High BTE shows that few amounts of unburned hydrocarbons are present in the engine, which leads to better and complete fuel combustion [15]. However, for certain engine speed, a decrease in BTE for all fuel blends indicates insufficient air that causes incomplete combustion of the fuel. Previous research has shown that at lower engine loads, BTE decreases for blends, leading to delay of combustion process and heat release in the expansion stroke which affects cylinder temperature [11]. Moreover, the amount of oxygen content in the blend affects the BTE [19].

4. CONCLUSION

Based on the result of this experiment, it can be concluded that the B4 blends in diesel fuel produced better BTE with 17% at 4 Nm engine load than diesel fuel. This could be affected by higher viscosity and oxygen content in the blends. Meanwhile, the BSFC was lower for B4 with 16% compared to diesel fuel at lower engine load. This was due to higher density and lower calorific value of algae biodiesel compared to diesel fuel. In conclusion, algae are suitable to reduce the usage of diesel fuel and replace diesel fuel in diesel engines.

It is suggested that the use of algae fuel to improve the strategies in reducing the NO_x emission for biodiesel combustion is explored in future studies. The utilization of various algae fuel ratios can also be investigated. Type of engine used in future studies can also be changed to other engine types such as those which are not used for transport vehicles. Besides, the variables and constant of engine testing parameter can also be changed to determine the effects on the emission characteristics, exhaust gas temperature and BSFC.

ACKNOWLEDGMENTS

The authors would like to express sincere gratitude to the Faculty of Mechanical Engineering (Universiti Teknologi MARA, Penang), Faculty of Mechanical Engineering (Universiti Teknologi MARA, Shah Alam) , Department of Mechanical Engineering (Universiti Malaya) and Sekolah Menengah Tun Syed Sheh Shahabudin for their benevolent assistance in executing this research. The authors also would like to express sincere gratitude to Universiti Teknologi MARA (UiTM) for the opportunity to trigger this research and Fundamental Research Grant Scheme (FRGS) from the Ministry of Higher Education Malaysia (Grant No FRGS/1/2019/TK10/UiTM/02/16) for the financial support.

REFERENCES

[1] Tüccar G, Tosun E, Özgür T, Aydın K 2014 *Fuel* **132** 7–11
[2] Murugesan A, Umarani C, Subramanian R, Nedunchezhian N 2009 *Renew Sustain energy Rev. 13(3)* 653–62
[3] Salvi BL, Subramanian KA, Panwar NL 2013 *Renew Sustain Energy Rev.* **25** 404–19.
[4] Fatima N, Mahmood MS, Hussain I, Siddique F, Hafeez S 2016 *Energy Sources, Part A Recover Util Environ Eff* **38(15)** 2306–11
[5] Demirbas A 2009 *Appl Energy.* **86** S108–17
[6] Sharudin H, Ismail NI, Pahmi A, Mahadzir MM, Rahim NA 2019 *In: Journal of Physics: Conference Series* **1349** 012017
[7] Lee S, Speight JG, Loyalka SK 2014 *Handbook of alternative fuel technologies.* crc Press
[8] Demirbas A, Demirbas MF 2011 *Energy Convers Manag* **52(1)** 163–70
[9] Demirbas A, Edris G 2017 Energy Sources, *Part A Recover Util Environ Eff.* **39(8)** 827–34
[10] Chen X, Wang Z, Pan S, Pan H 2019 *Fuel* **235** 603–9
[11] Hulwan DB, Joshi S V 2011 *Appl Energy* **88(12)** 5042–55
[12] Bayındır H, Işık MZ, Argunhan Z, Yücel HL, Aydın H 2017 *Energy* **123** 241–51
[13] Asokan MA, Kamesh S, Khan W 2018 Energy **145** 238–45
[14] Dhar A, Agarwal AK 2014 *Fuel* **119** 70–80.
[15] Buyukkaya E 2010. *Fuel* **89(10)** 3099–105
[16] Lin Y-C, Hsu K-H, Chen C-B 2011 *Energy* **36(1)** 241–8
[17] Lapuerta M, Armas O, Rodriguez-Fernandez J 2008 P*rog energy Combust Sci.* **34(2)** 198–223.
[18] Abdullah NR, Ismail H, Michael Z, Rahim AA, Sharudin H 2015 *J Teknol.***76** (9)
[19] Varuvel EG, Mrad N, Tazerout M, Aloui F 2012 *Appl Energy.* **94** 224–31

IOP Conf. Series: Materials Science and Engineering 1068 (2021) 012017 doi:10.1088/1757-899X/1068/1/012017

Development of CNC Milling Machine for Small Scale Industry

Wasis Nugroho[1], Damhuji Rifai[1,*], Aminul Hakim Embong[1], Kamarul Adnan Abd Aziz[1], Ahmad Siraji Embong[1], Haswandi Awang[1], Mohd Faizal Ibrahim[1]

Faculty of Engineering Technology, University College TATI, 24000 Kemaman, Terengganu, Malaysia

*Email: damhuji@tatiuc.edu.my

Abstract. The development of small-scale industry (SME) has increased the demand of small-scale CNC machines since the implementation of machine in production of small-scale product shown a profitable economic value. The use of machines in the fabrication of small part has been proven to increase product production efficiency as well as reduce product costs. This is able to increase the business profits of small companies and push the business development to a higher level. The main purpose of this paper is to discuss the development of 3-axis CNC machines using integration between microcontroller as main controller and spindle drill. The main application of this built machine is to cut and carve low-hard materials such as wood and acrylic material. The design product to be fabricate is sent to the microcontroller through a serial communication system. The microcontroller will control the movement of the spindle drill according to the point coordinates to perform the desired engraving or cutting according to the desired design. Three tests to determine the quality of product production are carry out. The tests are surface section, circular test and axial perpendicular test which can determine the accuracy of cutting geometry such as position, straight, roll, tone, evaporate and perpendicular to the work section under certain tool grooves and feed rate. The results of tests carried out on the products produced by the 3-axis CNC machine developed showed that this machine has a cutting and engraving accuracy of 98.5% and a depth accuracy of 100% for wood and archival materials. This machine is suitable for carving 2D and 3D objects with a maximum size of 30 cm x 30 cm.

1. Introduction

The manufacturing process has undergone dramatic changes since the last 30 years. This change is boost by the introduction of a manufacturing machine that was fully control by a numeric control system (NC). The introduction of numerical control technology has allowed machine work direction to be monitor via computerized control (CNC). In this system, a microcontroller is use to carry out all numerical controls of CNC machines.

CNC machines have been prove to play an important role especially in the manufacturing, automotive and oil and gas industries. Many parts used in this industry require high accuracy and fidelity to ensure the quality of the final product produced and the safety of the workers involved. Competency in controlling and operating CNC machines has been prove to provide high income for workers. Competency in operate and maintenance of CNC machining provides high employment opportunities for graduate students in engineering and technology field. Machines that use CNC working principles are widely developed such as lathes, laser cutters and 3D printing machines. All of these machines have the same basic components, which consist of the integration of motor equipment, motor guides, main control components and computer software as a platform to control the motor spindle movement. The main differences are in the mechanical design and use of the means for the final operation.

With the advancement of numerical control systems, many CNC machines are develop for shaping objects especially for soft materials such as acrylic and wood. Most of these machines use a knife blade and laser as a cutter to form the desired product. The introduction of a control system driven by a microcontroller has allowed CNC machines to be produce at a low cost compared to CNC machines that use computer as a control system. The low cost of building CNC machines has allowed the SME industry to use these machines to produce their products. There are many investigators and researcher try to build a CNC machine at a minimum cost. Jayachandraiah [1] has made a comprehensive review regarding the construction of a low cost 3- Axis CNC machine. The results of his study found that using a microcontroller as a numerical guard can reduce the cost of development a CNC machine. He provided a mechanical built-in frame using easily available materials such as aluminium and carbon steel main frames.

In another study, Khanna et al. [2] have used LabVIEW as the main control system in fabricate 3-Axis CNC machine. The main purpose of the construction of this machine is to engrave objects for acrylic materials. There are other CNC machine study, [3] develop CNC machine specially to do the work of grinding, engraving and cutting PCB boards [4] [5] [6]. In addition, there are also studies that focus on algorithm development for CNC machines. [7] in his study have made improvements in the CNC machine control algorithm for linear and spherical interpolation. In another investigation, [8] has tried to optimize the selection of 3 - Axis CNC machine frame materials to ensure the low cost in order to ensure the SME industry can afford to buy the CNC machine. There is also researcher develop CNC machine whose main function is to do the work of draw objects [9]. The main objective of this machine development is to increase speed by minimizing errors and increasing the output of the resulting product.

In Malaysia, the development of CNC machines – 3 Axis particularly is still new. The increase in the number of machine fabrication companies capable of forming and building CNC machines is still low. In fact, many CNC machines used by SME are import from other countries. This is not good from the economic point of view since high price it also cause the outflow of money. In this regard, this study is to construct a low cost 3 –Axis CNC machine for use by SME manufacturing companies. The long-term implication of this study will give an impression of modernizing PKS in Malaysia with execution engines in their expenditure.

2.Structure Design

The structure of machine is an important component in the construction of a CNC machine. Machine structure must have high static and dynamic rigidity and have good machine tool damping response. This is because all the equipment involved in producing a CNC machine such as the motor and spindle motor will be integrate into the machine structure to form a complete CNC system. Therefore, a strong machine structure will ensure the machine can make precise movements and produce high quality products.

The structure designed in the development of this 3- Axis CNC machine is to ensure the structure has high rigidity. This is to ensure the CNC machine can produce a better lane width and can operate at higher fidelity. This high stiffness can also ensure the CNC machine operate at high speeds and so can produce better cutting and engraving quality. The CNC machine structure to fabricate is ensure to have the right joints and symmetry to reduce machine vibration during the cutting and engraving process.

Design of the 3-Axis CNC machine will have a working space dimension of 600 * 800 * 500 mm3. The target precision of the product to be produce is 50μm and the reproducibility in the environment is 10μm. The operating width path for the CNC machine is between (0-10Hz). This can ensure that the CNC machine has high efficiency.

The height of the CNC machine structure is 550mm without a spindle motor. 600mm structure width. The main frame of this CNC machine is fabricate using 6061 aluminium material. The built structure of this machine is connect using an M6 crew. The height of this machine is design to allow the machine have as much as 150mm of free movement when moving tin the Z-axis. The design of the 3-Axis CNC machine to be developed is as in Figure 1. The structural of the CNC machine design is purposely to using only one spindle mill to ensure the cost of producing this machine is at a minimum.

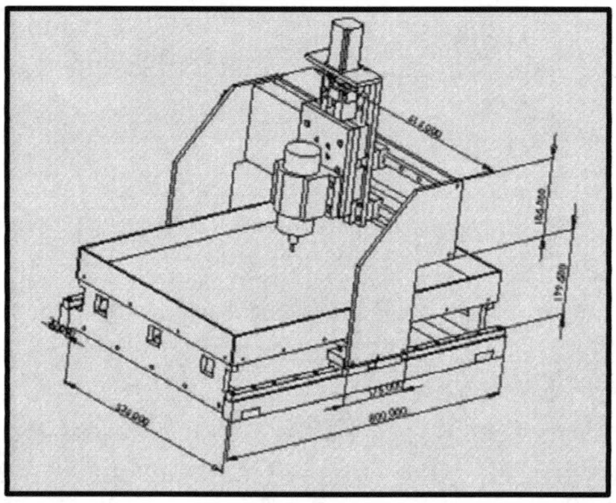

Figure 1. Propose design of closed-frame support structure for CNC milling machine prototype

In this structure, the contact surfaces with axes are purposely designed with allowances so a grinding process can be used on these surfaces to ensure good contact between axes module. All the angles are curved for better casting performance.

2.1 Machine assembling

The procedure of making the machine is listed below step by step. Each step has a corresponding figure listed in Figure 2.

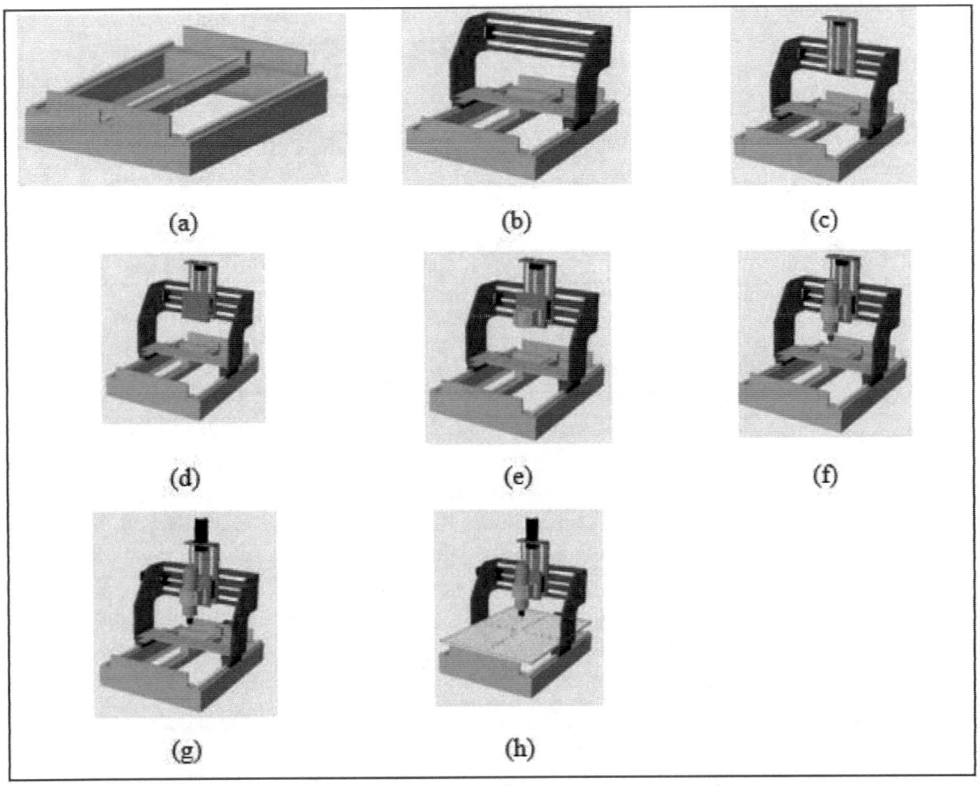

Figure 2. Assembly stage sequence

After the machine structure is fabricate, the alignment of the movements of the X, Y and Z axis is carefully carry out. This is to ensure the cutting and engraving of the product can be carry out at high precision to produce a quality product. The X and Y-axis is install sequentially from top to bottom. If the screw installation is not in the correct position, it is difficult to make the alignment again when using the top to down method of mounting the frame. Therefore, it is important to ensure that the design is properly plan at the initial stage of the development of machines that is during earlier stages process of design CNC machine.

3. Controller Design and Setup

The microcontroller used to control the CNC machine is DDCSV2.1 500KHz CNC 4-Axis Engraving Machine Controller Motion Control System. This controller use for the purpose of to control the stepper motor and it is come with electronic hand wheel which makes it easy to control the movement of the movement of CNC machine. This controller has been develop by Faster CNC and has control speed for each position at 4 milliseconds. High accuracy of position is the main advantage of this controller. The highest uniaxial output pulse is 500 KHz and this controller can be used to control both stepper and servomotor. Figure 3 shown the DDCSDV2.1 controller that come with hand wheel.

IOP Conf. Series: Materials Science and Engineering 1068 (2021) 012017 doi:10.1088/1757-899X/1068/1/012017

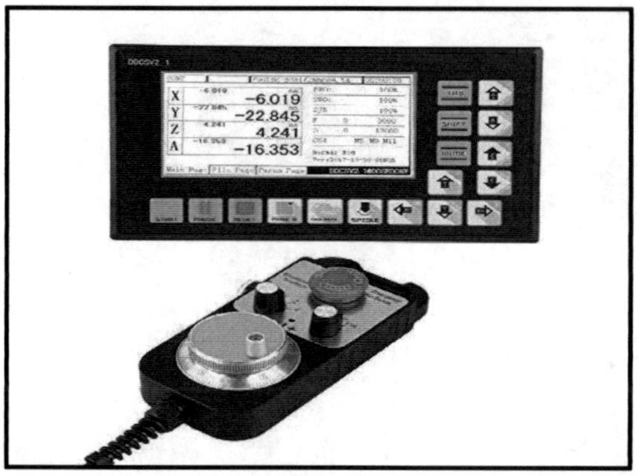

Figure 3. Motion controller DDCSDV2.1 with hand wheel

The main advantage of the DDCSV2.1 controller is that it can be configure using a variety of software interfaces. DDCSV2.1 is compatible with the MASTERCAM software on Microsoft Windows environments. MASTERCAM provides a terminal for programming of G codes.

DDCSV2.1 provides settings for configuring and tuning the movement of the motor. It provides basic tools for tuning and set PID parameters, program servo algorithm, DAC selection and configuration open loop testing, filters, and real-time motor monitoring. A user interface of the work space is shown in Figure 4.

Figure 4. Interface of DDCSV2.1 setting parameter page

The DDCSV2.1 next page interface is distributed as a CNC human machine interface (HMI) with built-in customizable standard features. The DDCSV2.1 also can be customized with respect to number of axes, type of machine, tool offset display, custom messaging, etc. A screenshot of the workspace is shown in Figure 5

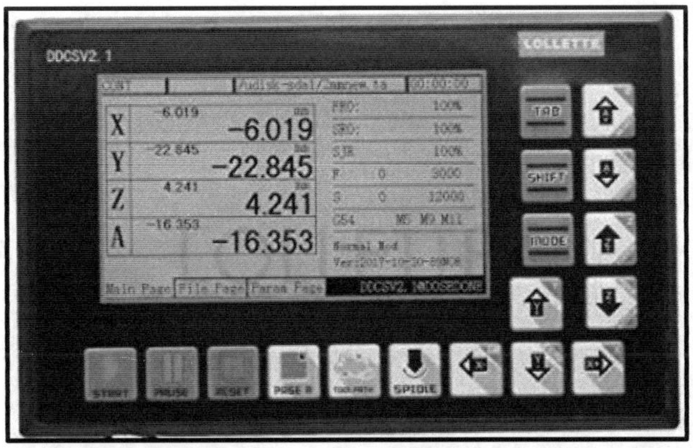

Figure 5. Interface of DDCSV2.1 NC

While a complete CNC solution has been provided by the controller to control and operate our mini CNC machine, the cost of the DDCSV2.1 motion controller alone is almost one over third our total budget. The controller provides a set of powerful software tools. We can use these tools to test our machine's precision and ensure everything is running properly.

3.1 Hardware connection and setup

Installation of the CNC machine is start by connecting all components with the DDCSV2.1 controller. The stepper motor is use to drive the CNC machine on the X, Y and Z-axes. The data from the decoder (pulse/rev) is use to determine the position of the motor spindle. For safety purposes, limit switch use as a indicator for the maximum motor movement of each axis. This switch serves as a sensor that will limit the motor spindle movement during the cutting and engraving process to be carriy out by a CNC machine. All of these components connect to a 24V power source. The negative terminal is connect to ground on the same source.

After the component wiring connection complete, the motors that drive the machine's CNC system are connected to the motor driver. The motor driver serves to move the stepper motor. It receives a signal from the controller and generates a signal in the form of a TTL quadrature that will drive the stepper motor. The encoder used in this machine build is an auxiliary type single-phase TTL signal encoder. It only has four channels: A +, B +, + 5V, and GND.

The encoder signal is use to locate the position stepper motor. X1-X4 is the encoding and quadrature mod. TTL cues are usually encode in quadrature mode and X4 mod can provide maximum resolution. In this machine, all coders are mount in their CW orientation across all axes. After properly configuring the encoding mode, we can roughly determine how many encoders the estimate corresponds to a 1cm step move.

4. Result and Discussions

The fabricated prototype mini-CNC machine is subjected to several different tests to determine its accuracy and repeatability. Three major tests are conducted in this study, that is surface flatness test and axes perpendicularity test.

4.1 Flatness test

Flatness test is done by machining a flat surface on a work piece and then measuring the flatness of the created surface. The flatness test can tell whether our XY axes are horizontal and moving in a plane as well as whether our working table is tilting or not.

Two different surfaces are created in two tool paths, one along X-axis and one along Y-axis, as shown in Figure 6 from the Mastercam simulation software. Theoretically machining a surface in two directions will not affect the surface feature. Therefore, any differences between two surfaces can tell the existence of error in Z-axis.

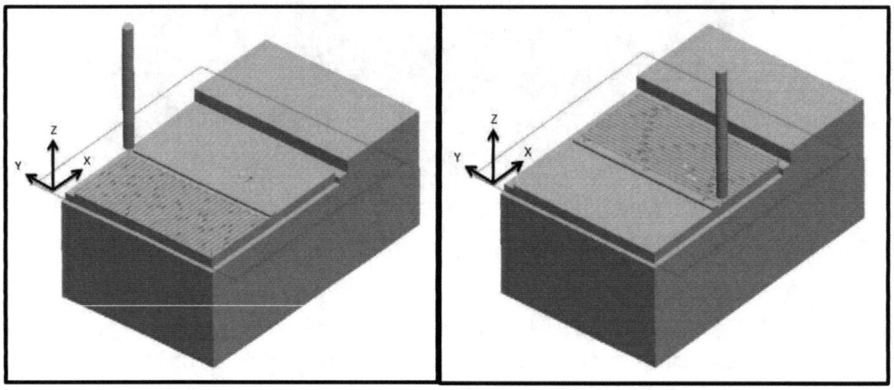

Figure 6.Flatness test X and Y direction tool path

For each surface created, six data paths are measured on the surface, three along X axis and three along Y axis, as shown in Figure 7.

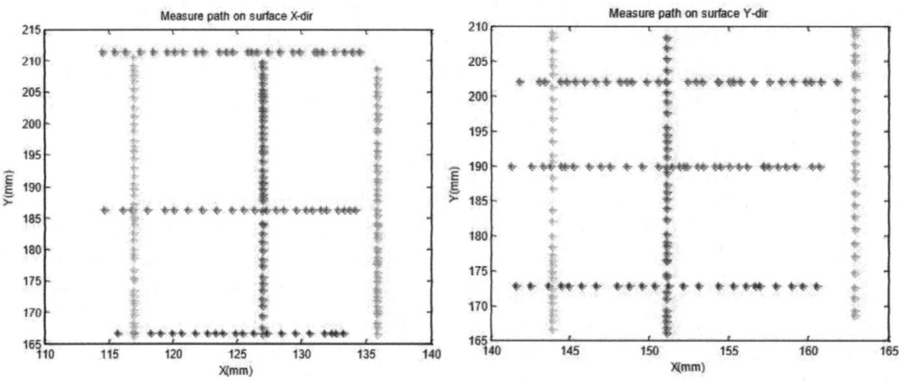

Figure 7. Measure path on surface of X and Y direction tool path

Ideally, measurement paths in all directions should all have constant height in Z direction. However, due to errors in XY axes, the surface is slightly tilt. Figure 8 a) shows the same three paths in X direction but measured on surface created by Y direction tool path. Figure 8 b) shows the profiles of three paths in Y direction measured on surface created by X direction tool path. Figure 8 c) shows the same three paths in Y direction but measured on surface created by Y direction tool path.

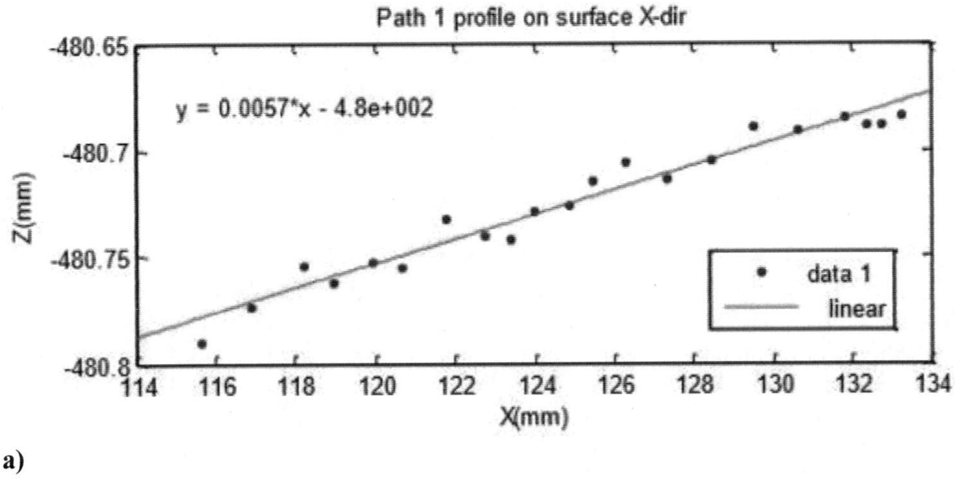

a)

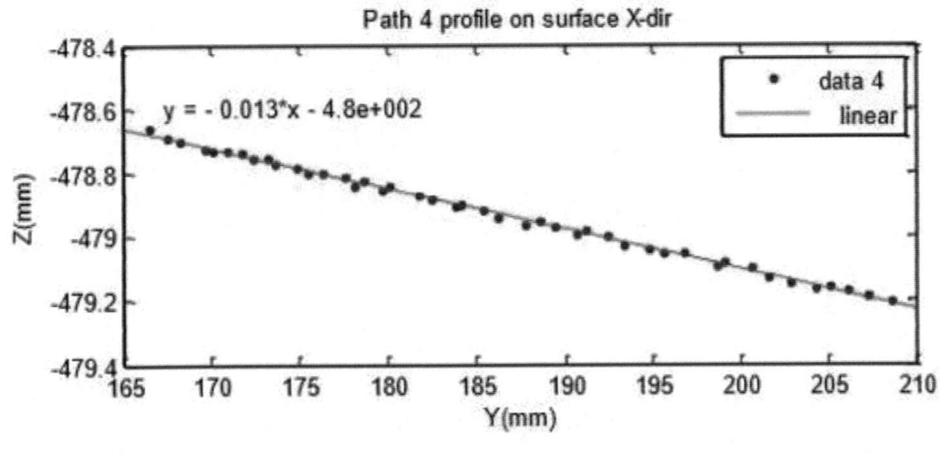

b)

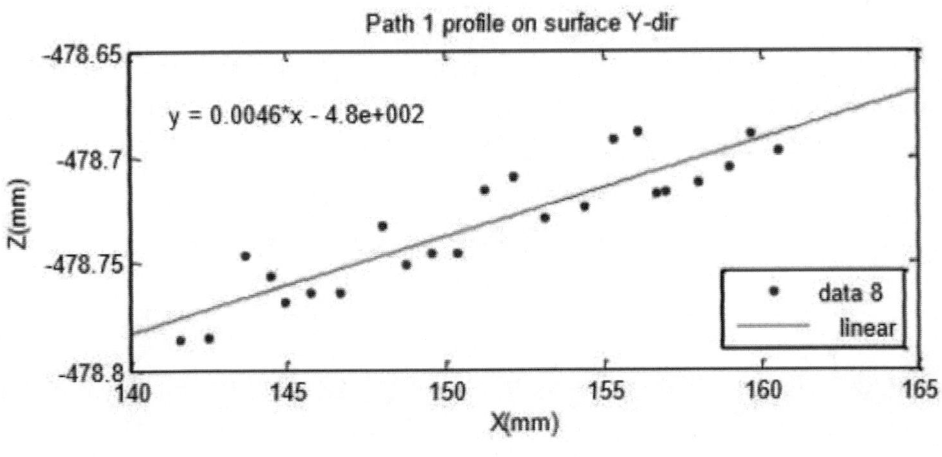

c)

Figure 8.Profiles of three paths: a) in X direction on X surface b) Y direction on X surface c) X direction on Y surface

Figure 8 shows the profile of surface measurements on the X and Y-axis. From this profile figure, it is clearly that the tool path on the X and Y-axis is at the tilt stage. The three data sets on the X, Y and Z-axis show their consistent surface profiles. The surface profile of the X-axis has a slope value of 0.005. This indicates that the surface profile is inclined towards the –X axis. Meanwhile, the surface profile in the Y direction has a slope value of -0.013. This represents -0.74 degrees of difference between the Y-axis direction and the horizon stage. The comparison of the surface profile show the Y-axis surface flatness are more consistent. Table 1 shows the residual norm of the linear fit model on both the X direction profiles and the Y direction profiles on the two surfaces. These data indicate that the surface profile in the Y-axis direction is smooth and flat. This comparison shows that the surface in the X direction axis is coarser than the surface in the Y-axis direction. For the path surface of the X tool, the cutting direction is less impressive, but the two profiles are still rougher than the Y direction profile on the path surface of the Y tool. On the other hand, the profile of this cutting direction is actually due to the Perspex fibre orientation of the material. The material used to carry out this surface profile analysis is plastic. Cutting plastics that do not follow the direction of alignment of the Perspex fibber material will result in rough cuts. The cutting process should align with plastic fibre orientation. Due to the inconsistent structure and density of the material, chip formation will vary along the cutting path, resulting in a non-uniform and uneven surface. This show in the Y-axis direction profile, which is parallel to the direction of the plastic fibre it produces a smoother and better profile of the cutting surface.

Table 1 Norm of residuals of linear fit

Tool path	X direction	Y direction
X tool path	0.030,0.054,0.041	0.05,0.026,0.051
Y tool path	0.033,0.080,0.084	0.0084,0.012,0.013

4.2 Perpendicularity test

The next test is perpendicularity test of X Y-axes. A tool path is created to cut the edge of a work piece purely in two directions: X and Y, as shown in Figure 2. After two flat vertical surfaces is created, the straightness of two surfaces can be measured using dial test indicator (DTI). The slope of two lines in XY plane can be measured. The angle between these two lines can be calculated using these two slope values.

Then, a linear fit was applied to these two sets of data individually. The slope of two straight lines can be measured. The norm of residuals of this linear fit also indicates the straightness of these two surfaces, which equivalently determines the straightness of XY axes movement. Figure 9 shows the linear fit of two measurements taken from two surfaces.

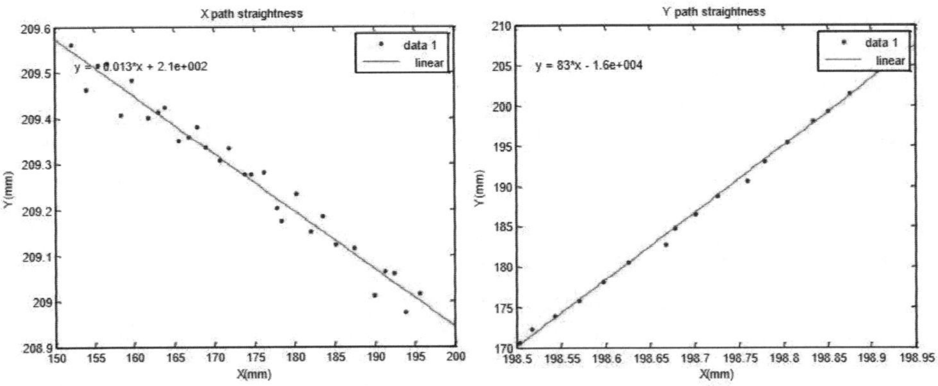

Figure 9.X and Y axis path straightness

The path in X direction has a slope value of -0.0126. The path Y direction has a slope value of 83.199. The angle between these two lines can be calculated using:

angle = tan-1(p11)-tan-1(p12)

= tan-1(-0.012632)-tan-1(83.199)

= 90.0351°

Therefore, there are 0.0351° differences between the X-axis alignments and Y-axis alignments. This is a relatively small value. The perpendicularity of XY axes is within the acceptable tolerance. The straightness in Y direction is better than that in X direction. The linear fit in figure 6.13 gives a R2 value of 0.9657 and the linear fit in gives a R2 value of 0.9981. Both still move straight, and no obvious curvature is observe.

Conclusion

This paper discusses the fabrication of a small-scale three-axis CNC milling machine that is suitable for use in small industries (SME). In order to determine the accuracy of the CNC machine, flatness test and perpendicularity test are perform on the products produce by this CNC machine. This test is perform to analyze the precision of cutting and engraving. Perspex material is use as a test material. The test results on this material shows that a margin of 0.0351° is not perpendicular to the XY axis. The deviation of 0.28° tilt in the X-axis and -0.74° tilt in the Y-axis. This deviation is due to improper alignment of the CNC machined structure. Where the main cause of this alignment problem is the incorrect design of the CNC machine and the installation method that uses the top to down method. The circumference tests showed that increasing the feed rate decreased machining precision and surface quality. Cutting tests using other materials is need to be conduct for further performance analysis.

References

[1] B. Jayachandraiah, O. V. Krishna, P. A. Khan, and R. A. Reddy, “Fabrication of Low Cost 3-Axis Cnc Router,” *Int. J. Eng. Sci. Invent.*, 2014.

[2] A. Khanna, A. Kumar, A. Bhatnagar, R. Tyagi, and S. Srivastava, “Low-cost production CNC system,” in *7th International Conference on Intelligent Systems and Control, ISCO 2013*, 2013.

[3] J. M. Kajal, R. N. Kranti, R. P. Pooja, and S. M. Vikas, “Automatic mini CNC machine for PCB drawing and drilling,” *Int. J. Recent Innov. Trends Comput. Commun.*, 2016.

[4] F. Calabrese and G. Celentano, "Design and realization of a STEP-NC compliant CNC embedded controller," in *IEEE International Conference on Emerging Technologies and Factory Automation, ETFA*, 2007.
[5] N. Muthupalaniappan, A. Maruthupandi, S. Pandian, A. J. P. Antony, and S. R. Pandian, "A low-cost web-based learning platform for CNC machining education," *6th Int. Conf. Adv. Comput. ICoAC 2014*, pp. 91–96, 2015.
[6] S. Gordon and M. T. Hillery, "Development of a high-speed CNC cutting machine using linear motors," *J. Mater. Process. Technol.*, 2005.
[7] M. Del Guerra and R. T. Coelho, "Development of a low cost Touch Trigger Probe for CNC Lathes," *J. Mater. Process. Technol.*, 2006.
[8] P. Girhe, S. Yenkar, and A. Chirde, "Arduino Based Cost Effective CNC Plotter Machine," *Int. J. Emergin Technol. Enginieering Res.*, 2018.
[9] A. N. M. Amanullah, Murshiduzzaman, T. Saleh, and R. Khan, "Design and Development of a Hybrid Machine Combining Rapid Prototyping and CNC Milling Operation," in *Procedia Engineering*, 2017.

Investigation of magnetic field effect on MRR, EWR and surface roughness during EDM of AISI420 tool Steel

A M Efendee[1,2], A Azhari[1], A R Hasnain[2], S Zainal Ariffin[2], M Mukhtar[2]
[1]Faculty of Manufacturing Engineering, Universiti Malaysia Pahang, 26600 Pekan, Malaysia
[2]Faculty of Engineering Technology, University College TATI, 24000 Kemaman, Malaysia

*Email: azmir@ump.edu.my

Abstract. Electrical discharge machine (EDM) is a machining process that is not affected by the toughness or hardness of the sample material but the electrical and thermal conductivity. In EDM process, good surface quality can only be produced if low peak current is used but the machining process will take a long time causing material removal rate (MRR) to be too low. While the accumulated machine debris in the machining zone can cause abnormal discharge and disrupt the material removal process. The present research aims to study the magnetic field effect on MRR, EWR and SR for EDM process improvement. In addition to MRR, electrode wear rate (EWR) and surface roughness illustrate the effectiveness of the EDM process. The installation of magnetic devices in the EDM machining area were implemented and the experiments were conducted using graphite electrode and AISI420 as the workpiece. Permanent magnets having 0.54 Tesla were applied to produce magnetic fields during EDM operations. The presence of this magnetic field also contributes to the effectiveness of the flushing process because the evaporated debris will be attracted and attached to the magnet, then purify the spark gap medium for the next discharge process. Dielectric that stays clean and sparks under magnetic field influence increases the effectiveness of the material removal process. Surface roughness from Ra measurement has recorded 12.6% to 28.1% improvement when magnetic devices were applied on EDM. The spark ignition delays and the refinement of EDM spark enhanced the surface quality compared to conventional EDM. Comparison of images through optical microscopy and SEM also proves that the Magnetic Field Assisted EDM (MFAEDM) method is capable of producing better surface quality. This MFAEDM shows that action to hybridize EDM is necessary to increase EDM competency by attaining both of machining efficiency and high quality of surface integrity.

1. Introduction

Several EDM hybrid techniques have been introduced to improve EDM machining performance. Those include dielectric-powder based EDM, dry EDM, ultrasonic vibration-assisted EDM and magnetic field-assisted EDM (MFAEDM) where is the main purpose is to solve EDM limitations. EDM material removal performance depends on the combination of electrode and machining parameters as these factors will determine the material removal rate, the electrode wear rate and the surface roughness. This study to present the performance of MFAEDM compared to conventional EDM of AISI420 tool steel using graphite electrode. Report by Ming, et al. [1] and Khan, et al. [2] suggest that MFAEDM is suitable for machining of ferromagnetic and non-ferromagnetic materials because of its advantages such as easy application, facilitating flushing system, economically competence and environmentally friendly. In this investigation, constant magnetic field was proposed and applied perpendicularly to the discharge channel. The use of magnetic fields in EDM machining was introduced to facilitate the removal of machined iron dust from machining areas. Perfect flushing circulation provides an excellent medium for the spark process and there is no delay in spark processing. The findings from Shabgard, et al. [3], Gholipoor, et al. [4], Singh Bains, et al. [5] show the use of magnetic fields can improve spark gaps cleaning because ferromagnetic material is extracted and sticks to the magnetic material rapidly.

Material removal process occurs when there is a spark which is caused by different potential of voltage on the electrode and sample material. Spark discharge begins with the formation of an ionization channel on the spark gap and it increases until there is a spark eruption [6]. The experiment used peak current (I_p) and pulse time (t_{on}) as the main machining parameters to study its effect on MRR, EWR and surface roughness (SR). Apart from the selected parameters, the MRR, EWR and SR are also subjected to the type of electrode used. The results from Haron, et al. [7], Lin, et al. [8] and Pavan and Sateesh [9] show that the MRR is higher and the EWR is lower with copper electrode than graphite electrode. However, graphite is increasingly becoming the material of choice as an electrode especially for roughing process [10]. The EWR is enhanced in the case of magnetic MFAEDM because effective ionisation improves discharge progression so the machining process can be completed in shorter time. The relationship of peak current and the surface roughness is evident as they are directly proportional to each other until it reaches the machining limit where the specimen surface burns out. Surface roughness increases with rising pulse time but once it reaches its maximum value, surface roughness decreases with increasing pulse duration. When pulse duration is too long, energy density in the discharge area decreases. Therefore, surface roughness tends to decrease with longer pulse duration [11]. Another important topic to review on MFAEDM output is micro-crack and white layer. The layers from EDM machining are composed of recast layer, heat-affected zone and base metal. The severely affected layers are formed on recast layer surface from rapid heating and cooling effects caused by the discharge process based on peak current values and pulse time duration [12].

2. Method and materials

The experimental arrangement was developed using electrode graphite as EDM tool, too steel AISI 420 as material specimen and kerosene as dielectric which will also act as insulation to avoid shortening between tool and workpiece. The electrodes used are cylindrical in size Ø25 × 40 mm mounted on the tool holder for EDM machining depth of 2mm cut. Table 1 shows the properties of graphite electrodes where electrical conductivity and thermal conductivity play an important role for the machining and value rate of the tool. The workpiece sample used is from the AISI 420 tool steel family, commercially it is better known as Stavax ESR which is widely used in mold making industry. This raw material has been milled into cube-shaped specimens in dimensions of 30 × 30 × 30 mm. Specimens have the composition of the mass elements Fe - 84.32%, C - 0.38%, Si - 0.9%, V - 0.3%, Cr - 13.6% and Mn - 0.5% [13][Pavan, 2020 #221]. The first part of the experiment was performed under the normal range labeled as conventional EDM. As for MFAEDM a set of permanent magnets with a combined magnetic flux density of 0.54 T was applied on the Charmilles Roboform EDM machine. The magnetic bar is clamped side by side with the AISI 420 sample interspersed by a 5 mm aluminum spacer. This is to avoid the sample being directly magnetized by the magnetic bar mounted close to it. An illustration of the schematic diagram of the study is shown in Figure 1. While the details of the experiment parameters were summarized in Table 2.

Table 1. Properties of graphite electrode.

Physical properties	Graphite
Electrical conductivity compared with silver (%)	0.11
Thermal conductivity (W/mK)	160
Melting point (^{0}C)	455

Specific heat (J/kgK)	0.17-0.2
Specific gravity at 20 °C (g/cm3)	1.75
Coefficient of thermal expansion (1/°C)	7.8 x10^{-6}

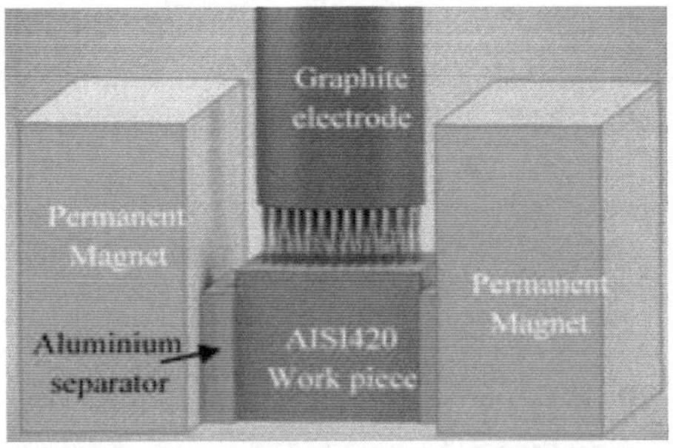

Figure 1. Schematic diagram of MFAEDM.

Table 2. Experiment parameters set up.

Working conditions	Descriptions
Work piece (mm)	AISI 420 Stavax ESR) 30×30×15
Electrode (mm)	Graphite Ø25×40
Dielectric fluid	Kerosene
Depth of cut (mm)	2
Magnetic flux density (T)	0 - 18
Peak current (A)	8 - 24
Pulse duration (μs)	50 - 100

Machining outputs from both conventional EDM and MFAEDM were evaluated and compared by electrode wear rate, EWR (g/min), material removal rate, MRR (g/min) and surface roughness, R_a (µm). Material removal rate (MRR) value for this study was calculated by using a formula shown in equation 1.

$$MRR = \frac{\text{Mass of Material Removed from Part}}{\text{Time of Machining}} \quad (g/min) \qquad (1)$$

Surface roughness measurements are used to assess the surface quality of the machine surface. Once the experiments were completed, sample surface roughness was measured by using surface roughness measuring device (SURFPAK) SJ-301 which is stylus type surface roughness measuring system that has diverse high-precision function feature for ease of operation. Figure 2 shows the visualization equipment of (a) optical microscope, (b) scanning electron microscope (SEM) and (c) 3D profilometry were used to make comparisons of sample surface topography after EDM machining.

Figure 2. (a) Optical microscope, (b) scanning electron microscope and (c) 3D profilometry

3. Results and discussion

During EDM material removal process, a continuous cycle of discharge sparks was created through electrode towards workpiece to erode work material. Theoretically, when EDM is in discharge progression, melting and vaporisation course in the machining area does not only remove the work piece materials but also erodes the electrode materials. Experiment outcomes such as material removal rate (MRR), electrode wear rate (EWR) and surface roughness (SR) were obtained so that responses from conventional EDM and magnetic field-assisted EDM were compared. In this experiment, mass of electrode and work piece material were weighed accordingly for further reflexion. First, MRR between conventional and MFAEDM are compared.

3.1. MRRF

Material removal rate (MRR) figures were gained by calculating sample mass difference before EDM and after EDM machining against machining time. Throughout the investigation, data record shows that machining time reduced when peak current increased. Experiments by some researchers such as Pavan and Sateesh [9], Bains, et al. [12], Qudeiri, et al. [14] and [15] also show the role of peak current in accelerating the machining period while increasing the value of MRR. The data also shows that machining time decreased even lower when MFAEDM was applied. When machining time was thoroughly reduced, MRR increased. MRR for EDM experiments elevated when peak currents were increased from 8 A (0.041 g/min) to 24 A (0.386 g/min) for conventional EDM and 8 A (0.042 g/min) to 24 A (0.454 g/min) for MFAEDM as illustrated in Figure 3. Similar result trend was discovered when the experiments were repeated with higher pulse time value. It is evident in Figure 4 which illustrates MRR comparison between normal and MFAEDM with higher pulse time at 100 μs. Machining improvement can be clearly visualized in both Figure 3 and Figure 4; whereby higher MRR values were recorded for MFAEDM compared to conventional EDM. For example, at peak current of 24 A and 100 µs pulse time MFAEDM recorded MRR 0.551 g/min compared to 0.483 g/min for conventional EDM.

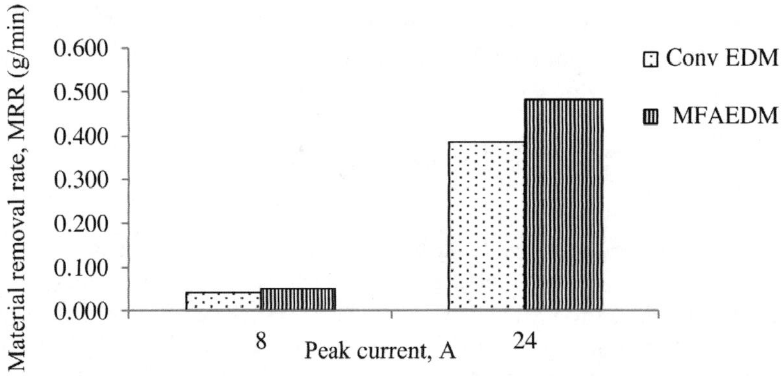

Figure 3. Comparison of MRR between conventional EDM and MFA EDM at 50 µs.

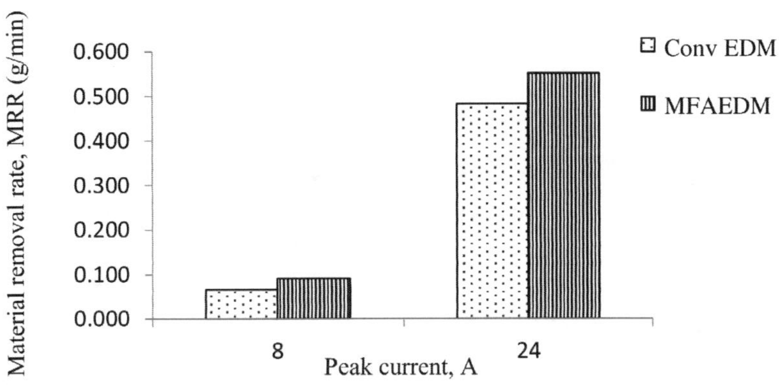

Figure 4. Effect of MRR between conventional EDM and MFA EDM at 100 μs pulse time.

The increment of MRR value produced because higher current led to superior spark and bigger implosion on sample surface [14].The charts also reveal that MRR improved when magnetic field was applied in EDM machining area. Gap condition stability improved when magnets were attached in the EDM machining area by absorbing most debris particles from the spark gap. Hence, it improves sparking cycle which contributes for better EDM machining performance. The resulting magnetic field at the MFAEDM synchronizes any electron and any rotating ion releasing from the core atom to increase the ion concentration thereby increasing the ionization process [3, 12] . The increases in ion density and plasma pressure has improvised the plasma channel to spark down and melt the material more effectively.

3.2. EWR

There is an inter-relationship between the volume of electrode wear and machining time in EDM known as electrode wear rate (EWR). As EDM cycle increases the temperature; the electrode starts to wear [16]. EWR results from EDM process mostly rely on machining parameters such as peak current, pulse time and machining time factors. Besides MRR, EWR is considered as a good tool to measure machining efficiency and to indicate cost-effective process. From the experiments performed, data obtained was calculated to determine EWR of graphite electrode for conventional EDM and MFAEDM comparison.

Figure 5 illustrates the comparison graph of EWR for conventional EDM and MFAEDM at 50 μs. The recorded EWR for graphite electrode was in negative values for both conventional EDM and MFAEDM. The electrode wear rate for conventional EDM was -0.0006 g/min with peak current at I_p, 8 A; whereas EWR value was -0.0058 g/min with peak current at 24 A. Supposedly, high peak current yields high EWR because high electric field strength energy causes greater electrode wear rate [16]. However, it was the opposite with graphite electrode; whereby, as peak current increased EWR value decreased correspondingly. This negative value trend in EWR of graphite has arisen due to a combination of peak current and thermal pyrolysis of carbon layers formation onto graphite electrode surface [17]. Thus the mass of the electrode after machining is higher than before machining causing the result of the EWR calculation to be negative value. Figure 6 shows similar EWR investigation but the experiment was conducted at higher pulse time (100 μs). On the other hand, EWR value was lower for MFAEDM. In MFAEDM, EWR recorded at 24 A (100 μs) was -0.0084 g/min which differs as much as 0.0009 g/min compared to conventional EDM (- 0.0075 g/min).

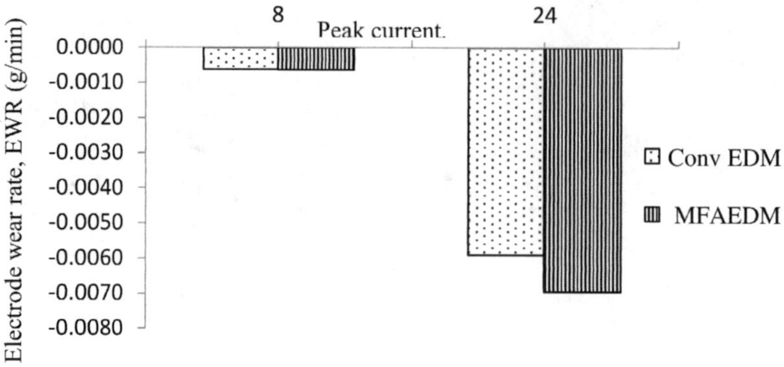

Figure 5. Comparison graph of EWR for conventional EDM and MFAEDM at 50 µs.

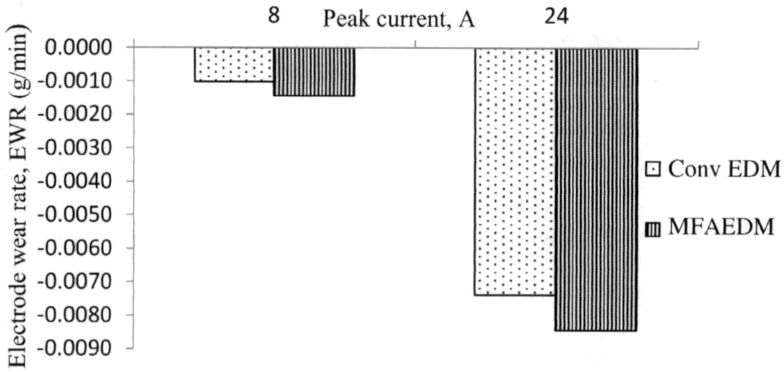

Figure 6. Comparison graph of EWR for conventional EDM and MFAEDM at 100 µs.

3.3. Surface roughness

Surface roughness measurement normally represents surface finish texture state and aesthetic appearance value. EDM machining parameters such as peak current and pulse-on time have prominent effect on sample's surface roughness. Peak current is the parameter that determines spark length and spark explosion magnitudes. Pulse time, which is time exposure (t_{on}) to spark and time for dielectric re-ionization (t_{off}), also corresponds to surface integrity in EDM [5, 18]. Due to the significant characteristic, analysis of sample surface roughness were carried out for conventional EDM and MFA EDM.

Figure 7 and Figure 8 show comparison graphs of surface roughness between conventional EDM and MFAEDM at different peak currents at $50\mu s$ and $100\mu s$, respectively. From both charts, R_a values increased when higher peak current applied. At higher current, spark intensity increased and higher heat energy created at the gap; hence, producing more micro-features on sample surface. This spark state created craters in the sample which in turn determined surface finish. In the case of conventional EDM at 100 µs, surface roughness value (R_a) increased from 6.152 µm to 9.294 µm for peak current of 8 A and 24 A respectively. The lower R_a number represents a small surface roughness. When higher peak current applied, the crater is larger and deeper; as a result, the surface finish is coarse. Meanwhile, if low peak current employed; the craters are small and surface finish is better.

As shown in the both graphs, there is a significant effect of magnetic fields in EDM machining. At the presence of magnetic field, R_a is lower than the value obtained by conventional EDM. At peak current of 8A (50 μs), the difference of R_a value between conventional EDM (5.437 μm) and MFAEDM (4.751 μm) was 12.6%. Followed by peak current of 24A (50 μs), R_a difference value between conventional EDM (7.583 μm) and MFAEDM (5.456 μm) was 28.1%. The sparks produced by the influence of the magnetic field are of better quality due to the higher plasma pressure as well as the squeezing effect of the magnetic field flux so that the discharge crater produced on the surface of the machine is small and shallow. Surface roughness obtained by MFAEDM was lower than conventional EDM which was not only at low current but also at higher current practice. Therefore, modification of the magnetic tool on the EDM gives a smoother surface finish.

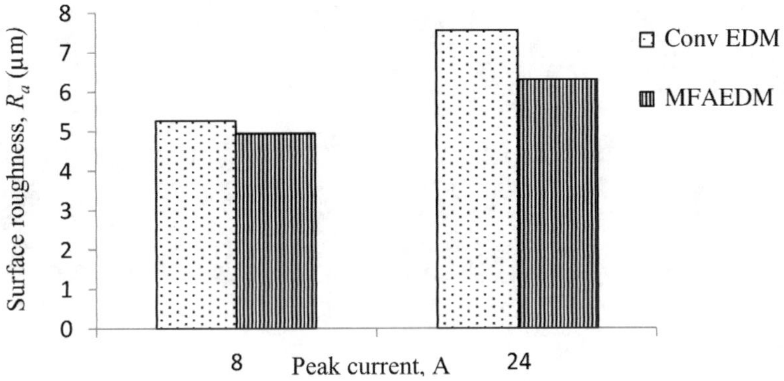

Figure 7. Surface roughness comparison of conventional EDM and MFAEDM at 50 μs.

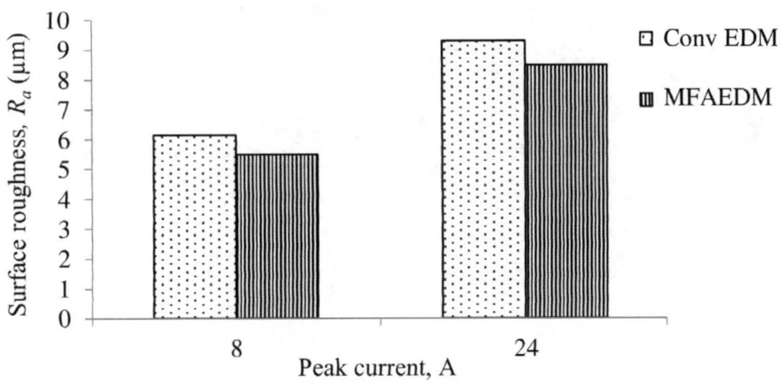

Figure 8. Surface roughness comparison of conventional EDM and MFAEDM at 100 μs.

3.4. Surface topography

Surface topography comparison for conventional EDM and MFAEDM by optical microscope can be observed in Figure 9 where the depth of the spark crater produced by conventional EDM (Figure 9(a)) was deeper than that of MFAEDM whose crater was shallower and wider (Figure 9(b)). While Figure 10 shows the comparison of the both surfaces via SEM images, whereby the bumpier feature and deeper valley attributes can be observed for conventional EDM in Figure 10(a) compared to MFAEDM which has superior surface integrity as witnessed in Figure 10(b). The texture of EDM sample surface is caused by the amount of heat generated by the discharge process which involves melting and vaporisation of sample material followed by rapid cooling [5]. The surfaces were blasted by discharge spark and the crater produced reflects the pattern caused by the spark. Under conventional EDM machining condition, beside the irregular topography the surface shows high amount of tiny loose metal particles which were initially melted due to spark discharges compared to MFAEDM. These small melted metals were unable to be thoroughly washed in dielectric fluid without magnetic field assistance. As a result, metal debris immediately solidified and formed many tiny particles on the machined surface.

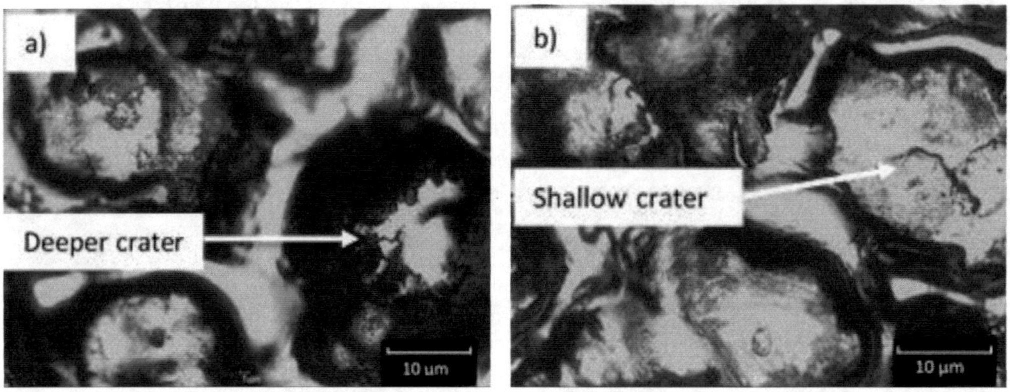

Figure 9. Surface topography of (a) conventional EDM and (b) MFAEDM via optical microscope at 24 A, 100 µs.

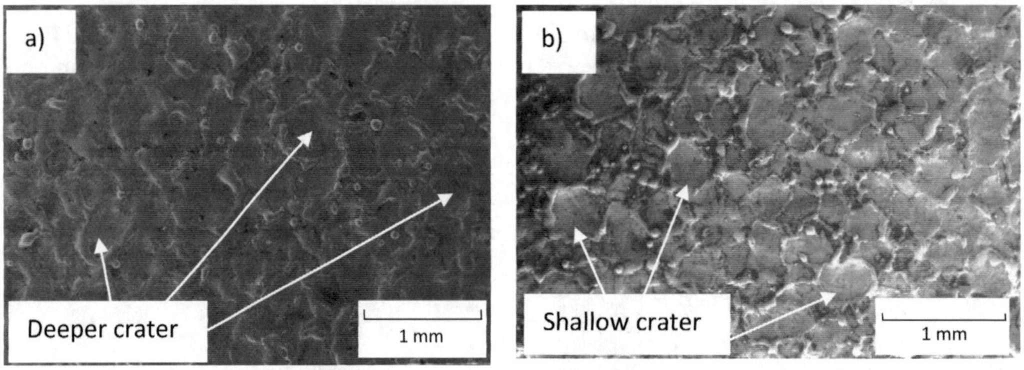

Figure 10. Surface topography of (a) conventional EDM and (b) MFAEDM via SEM: 100× (24 A, 100 µs).

It was evident that MFAEDM had a higher quality surface appearance than conventional EDM as the samples were further examined through 3D profilometry as illustrated in Figure 11. The illustration shows the gap between peak and valley for conventional EDM was 1,111.102 µm which is higher than MFAEDM (838.321 µm). Anyhow these observation of crater surfaces verified that application of MFAEDM has significant improvement on EDM machined surface. It was observed that surface finish quality mostly depended on pulse current and pulse time selection. The combination of discharge spark strength, magnetic field pattern and formation of recast layer influence the surface roughness state [5]. From those observation of microscope, SEM and 3D profilometry evidenced that the crater shape produced by MFAEDM were horizontally smoother because pressure from magnetic fields pattern refined the discharge spark quality by concentrating ionization process in the plasma channel.

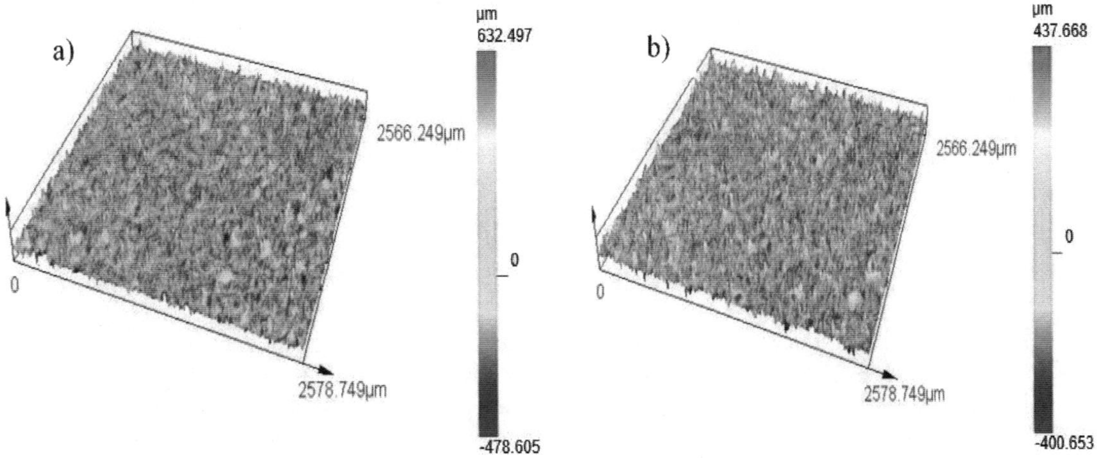

Figure 11. Surface 3D profilometric of (a) conventional EDM and (b) MFAEDM at 5× for 24 A, 100 µs.

4. Conclusion

From this research, MFAEDM clearly provides better MRR and surface quality but in terms of electrode wear rate it is quite difficult to determine which method gives better effect because there is pyrolytic carbon formed on the electrode. This phenomenon occurs when a graphite electrode is used. Surface topography shows that higher peak currents result in poor surface finish as shown by the deeper, wider and irregular shape of the crater. The SEM observations of the micro structure confirm that the craters produced in MFAEDM are less bumpy and shallower than conventional EDM. While 3D profilometry measurement shows that thicker restructuring layer was found in normal EDM surface, while in MFAEDM a thinner restructuring layer is formed. Therefore, the use of MFAEDM is an option for better machining efficiency in terms of MRR, surface roughness as well as EDM machine care because the debris attracted to the magnetic device instead of circulated into the dielectric filter.

Acknowledgments

Authors would like to gratefully acknowledge the financial support from the Ministry of Higher Education Malaysia through FRGS/1/2018/TK03/TATI/03/1. We thank our colleagues from Universiti Malaysia Pahang and University College TATI for providing research equipment and technical expertise support.

References

[1] W. Ming, F. Shen, Z. Zhang, H. Huang, J. Du, and J. Wu, "A comparative investigation on magnetic field–assisted EDM of magnetic and non-magnetic materials," *The International Journal of Advanced Manufacturing Technology,* vol. **109,** pp. 1103-1116, 2020.

[2] A. A. Khan, M. B. Ndaliman, Y. A. Mohammad, M. D. Al-Falahi, H. A. Ibrahim, and M. S. Mustapa, "The Effect of EDM with external magnetic field on surface roughness of stainless steel," 2013.

[3] M. R. Shabgard, A. Gholipoor, and H. Baseri, "A review on recent developments in machining methods based on electrical discharge phenomena," *The International Journal of Advanced Manufacturing Technology,* vol. **87**, pp. 2081-2097, 2016.

[4] A. Gholipoor, M. R. Shabgard, M. Mohammadpourfard, and H. Abbasi, "Comparative study of ultrasonic vibrations assisted EDM and magnetic field assisted EDM processes," *Iranian Journal of Mechanical Engineering Transactions of the ISME,* vol. **21**, pp. 45-64, 2020.

[5] P. Singh Bains, S. S. Sidhu, and H. Payal, "Study of Magnetic Field-Assisted ED Machining of Metal Matrix Composites," *Materials and Manufacturing Processes,* vol. **31**, pp. 1889-1894, 2016.
[6] B. Bojorquez, R. Marloth, and O. Es-Said, "Formation of a crater in the workpiece on an electrical discharge machine," *Engineering Failure Analysis,* vol. **9**, pp. 93-97, 2002.
[7] C. C. Haron, J. Ghani, Y. Burhanuddin, Y. Seong, and C. Swee, "Copper and graphite electrodes performance in electrical-discharge machining of XW42 tool steel," *journal of materials processing technology,* vol. **20(1)**, pp. 570-573, 2008.
[8] Y.-C. Lin, Y.-F. Chen, D.-A. Wang, and H.-S. Lee, "Optimization of machining parameters in magnetic force assisted EDM based on Taguchi method," *Journal of materials processing technology,* vol. **209**, pp. 3374-3383, 2009.
[9] C. Pavan and N. Sateesh, "Taguchi analysis on machinability of Inconel 600 using Copper, Brass, and Copper tungsten electrodes in EDM," *Materials Today: Proceedings,* 2020.
[10] F. Klocke, M. Schwade, A. Klink, and D. Veselovac, "Analysis of material removal rate and electrode wear in sinking EDM roughing strategies using different graphite grades," *Procedia CIRP,* vol**. 6**, pp. 163-167, 2013.
[11] S. P. Arikatla, K. T. Mannan, and A. Krishnaiah, "Influence of Pulse Current and Pulse on Time on Material Removal Rate and Surface Roughness in Electric Discharge Machining of AISI T1 High Speed Steel," 2013.
[12] P. S. Bains, S. S. Sidhu, and H. Payal, "Magnetic field assisted EDM: New horizons for improved surface properties," *Silicon,* vol. **10**, pp. 1275-1282, 2018.
[13] F. Klocke, L. Hensgen, A. Klink, L. Ehle, and A. Schwedt, "Structure and composition of the white layer in the wire-EDM process," *Procedia CIRP,* vol. **42**, pp. 673-678, 2016.
[14] J. E. A. Qudeiri, A. Zaiout, A.-H. I. Mourad, M. H. Abidi, and A. Elkaseer, "Principles and Characteristics of Different EDM Processes in Machining Tool and Die Steels," *Applied Sciences,* vol. **10**, p. 2082, 2020.
[15] F. Alavi and M. P. Jahan, "Optimization of process parameters in micro-EDM of Ti-6Al-4V based on full factorial design," *The International Journal of Advanced Manufacturing Technology,* vol. **92,** pp. 167-187, 2017.
[16] O. B. Abouelatta, "3D surface roughness measurement using a light sectioning vision system," in *Proceedings of the World Congress on Engineering, London, UK*, 2010, pp. 698-703.
[17] S. Abdulkareem, A. A. Khan, and M. Konneh, "Reducing electrode wear ratio using cryogenic cooling during electrical discharge machining," *The International Journal of Advanced Manufacturing Technology,* vol. **45**, pp. 1146-1151, 2009.
[18] A. Gholipoor, H. Baseri, M. Shakeri, and M. Shabgard, "Investigation of the effects of magnetic field on near-dry electrical discharge machining performance," *Proceedings of the Institution of Mechanical Engineers, Part B: Journal of Engineering Manufacture,* vol. **230,** pp. 744-751, 2016.

Performance Evaluation of Titanium Dioxide (TiO2) Nano Cutting Fluids in CNC Turning of Aluminium Alloy (AL319) via Minimum Quantity Lubricant (MQL) Technique

S. Zainal Ariffin[1,] *, A.A.M. Redhwan[1,*,] A. Arifuddin[1,] *, A.R.M. Aminullah[1], M.S. Azim[1], W. Nughoro[1], A.M. Syafiq[1], and A.M. Efendee[1].

[1]Faculty of Engineering Technology, University College TATI (UC TATI), 24000 Kemaman Terengganu, Malaysia. Tel. /Fax +6-09-8601000/+6-09-8635863

*Email: zainal@uctati.edu.my , redhwan323@gmail.com & arifuddin8196@gmail.com

Abstract: Aluminium alloy AL319 is a common alloy that has superior mechanical properties such as hardness and weldability. Nanofluids can be applied to various industrial widely in the aircraft, automotive industry and engineering problems, such as heat exchangers, cooling of electronic equipment, and chemical processes. This research objective focuses on evaluating the performance of Nano cutting fluid mixing titanium dioxide (TiO2) in CNC Turning using material AL319 Aluminium Alloy with MQL technique. The cutting performance is assessed in terms of surface roughness, cutting temperature, and tool wear was investigated. Machining parameters used for spindle speeds of 1000 to 1800 rpm and feed speeds of 0.10 to 0.20 mm / rev are used on CNC turning machines. MQL pressure is constant at 0.5 Mpa, and for parameters of TiO2 Nano liquid with concentrations of three volumes (0.5, 1.0, and 1.5%) was then compared with conventional CNC cutting liquid. Beforehand, TiO2 is diffused in CNC conventional coolant base by using the one-step method. Response surface method (RSM) via Face Centered Design (FCD) was used in designing the experimental use of the variance analysis (ANOVA) to determine which parameters are statistically important. The stability of TiO2 Nano cutting fluid is checked via visual sedimentation. The experiment concluded that the lowest cutting temperature of 28°C and surface roughness of 0.863 µm Ra when TiO2 Nano cutting fluid with 1.5% volume concentration is employed. The TiO2 Nano cutting fluids are recommended in future work to obtain more significant results. The experimental research reveals that the performance of TiO2 Nanofluid in terms of surface roughness, cutting temperature, and tool wear are found to be better compared to dry machining, wet, and MQL machining using conventional cutting fluid. Nanofluids can be considered as the future of heat transfer fluids in various heat transfer applications. They are expected to give better thermal performance than conventional fluids due to the presence of suspended nanoparticles which have high thermal conductivity.

1. Introduction

Turning is a machining process in which the workpiece rotating while the cutting tool moving along the axis. (Abbas et al., 2019) . This process will generate heat and friction around the workpiece. As a solution to this problem, cutting fluid is used to decrease the heat generated. The conventional methods of enhancing the cooling rate have already reached their limits. The use of novel approaches is essential to achieve high-performance cooling and lubrication. There is very little difference in machining performance of the conventional cutting fluid under the MQL technique and under the process of wet machining (Sharma, Tiwari, Dixit, & Singh, 2017). However, the parameter used need to be considered, from the analysis it is noted that the most influential factor for the multi-objective function is feed rate followed by the cutting condition, depth of cut and cutting speed (Viswanathan, Ramesh, & Subburam, 2018). It was proved that the adoption of hybrid lubricating/cooling strategies is a viable alternative to the conventional ones, leading to a drastic reduction of the crater wear as well as good surface integrity given an optimized positioning of the lubricant/coolant nozzles (Sartori, Ghiotti, & Bruschi, 2017).

MQL is minimum quantity lubricant which had been used by many factory and machinist to minimize or reduce the used of coolant. It should be stated that MQL-nanofluid offered promising results through the three studied cases compared to dry and flood approaches. This is mainly attributed to the cooling capabilities of both flood and MQL-nanofluid techniques to reduce the severity of the high-heat generated in the cutting zone. It should be also stated that MQL-Nano-mist does not only offered a promising cooling property but also has a pure tribological effect on the performance of the cutting process(Abbas et al., 2019). Other than that, minimum quantity lubrication is an extensively accepted lubrication system owing to its improved heat removal in the machining areas (Naresh Babu, Anandan, Muthukrishnan, Arivalagar, & Dinesh Babu, 2019). Furthermore, Al2O3–MWCNT–based MQL noticeably reduced the surface roughness and main cutting force when compared with cryogenic CO2–based cooling (Jamil et al., 2019).

Nanofluids provide a potential way to fulfill this requirement. Nanofluid can be defined as a fluid containing nanometer-sized particles. This Nanofluid has high potential as a heat transfer. Because of its potential as heat transfer, it can increase tool life, improving surface roughness, reducing workpiece thermal deformation, improving the surface finish of machining. Titanium dioxide, also known as titanium (IV) oxide or titanium, is the naturally occurring oxide of titanium, chemical formula TiO2. TiO2 prepares as a nanofluid. It will be added to the coolant by using the one-step method. Cutting fluid is a type of coolant and lubricant designed specifically for metalworking processes such as stamping and machining. There are various kinds of cutting fluids, which include oils, oil-water, pastes, gels, aerosols (mists), and air or other gases. They may be made from petroleum distillates, animal fats, plant oils, water, and air, or other raw ingredients. The experimental study clearly reveals that the performance of TiO2 nanofluid in terms of surface roughness, tool wear, cutting force and chip morphology is found to be better compared to dry machining, wet/MQL machining with conventional cutting fluid (Sharma, Tiwari, Singh, & Dixit, 2016). The thermal conductivity, density, specific heat and viscosity of three different nanofluids are compared with each other and found that the addition of nanoparticles into base fluid enhanced its cooling ability without much affecting its viscosity (Sharma, Tiwari, & Dixit, 2016).

This study focuses on evaluates the performance of TiO2 Nano cutting fluid in CNC turning of Aluminum Alloy AL319 via the MQL technique. The cutting performance is surface roughness, cutting temperature, and tool wear was investigated versus feed rate and spindle speed. This study will prove the effectiveness of TiO2 Nanofluid concentration in the turning process through the combination of TiO2 Nano and coolant.

2. Methodology

2.1. Preparation of Nano Cutting Fluid

Titanium dioxide TiO2 Nanofluid dispersed into conventional coolant preparation. The most widely used for preparing Nanofluid is the one-step method. First, titanium dioxide TiO2 used in this method is produced as a wet liquid. Next, it will be measured the volume of TiO2 in 1.50% which is the higher volume concentration Vt% by using a measuring cylinder. Then, measured the volume of distilled water and the volume of coolant will be used according to the calculation. After that, mix distilled water and coolant by using a method of stirring for 30 minutes. Last, mix Nanofluid with that already distilled water and coolant liquid by using a method of stirring within thirty minutes. Lastly, the solution will be placed on a water bath ultra sonicator for 30 minutes to make sure it dissolves completely as shown in Figure 1 below.

Figure 1. Sonication Process

2.2. Experimental Setup

Figure 2 below shows how the test rig was designed and developed. The air pressure was blown to give pressure at the coolant and the coolant will go up along the tube straight to the tool through MQL for lowering the quantity of lubricant. The nanofluid from the MQL nozzle spray mist directly to the workpiece and a cutting tool to reduce the temperature and reduce the tool wear of the tool while improving the surface roughness of the workpiece.

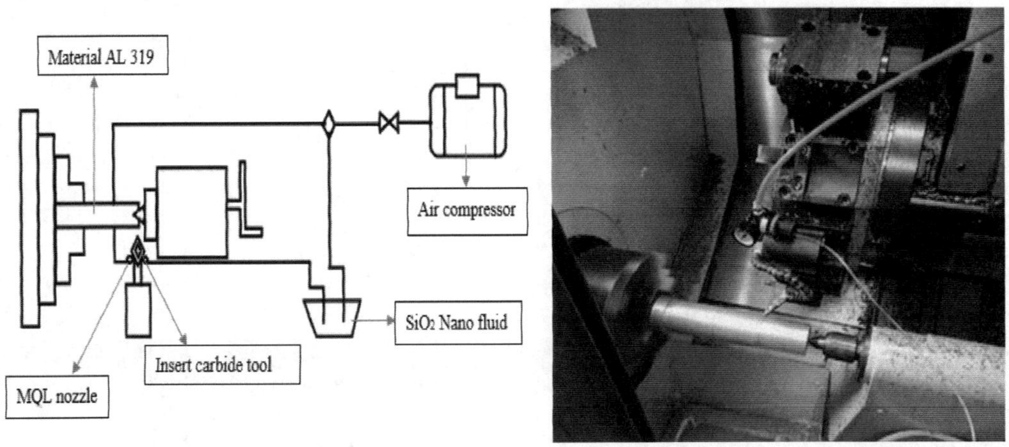

Figure 2. Setup of Test Rig

2.3. Design of Experiment (DOE)

The design of the experiment (DOE) of this project consists of 3 input parameters were measured such as feed rate, depth of cut and cutting speed. From this experiment, 3 responses were analyzed that is temperature, tool wear and roughness surface. The temperature measured using infrared laser thermometer DT 8280. The roughness surfaces the tool that was used to measure are profilometer surface roughness tester SJ-210 and the tool wear were using a machine vision system. RSM method was used to running this experiment. Table 1 shows the design of the experiment by using different levels with different control factors such as spindle speed, feed rate, and concentration of Nanofluid. Spindle speed that uses at 1000, 1400, and 1800 rpm. The concentration of Nanofluid is 0.50, 1.00, and 1.50 Vt%. Feed rate 0.10, 0.15, 0.20 mm/rev.

Table 1. Design of Experiment Using Level

Control Factor	Symbol	Units	Level 1	Level 2	Level 3
Spindle speed	V	RPM	1000	1400	1800
Feed rate	f	mm/rev	0.10	0.15	0.20
Concentration Nanofluid	np	Vt%	0.50	1.00	1.50

2.4. Response Surface Methodology (RSM)

Based on the input parameter and response need to analyze, Table 2 have been designed by using RSM software base on the concentration of Nanofluid, feed rate and spindle speed. The RSM software gives 20 runs of the experiment that needed to be done for this project.

Table 2. Design Layout by RSM

STD	RUN	BLOCK	Volume Concentration (Vt %)	Spindle speed (RPM)	Feed rate (mm/rev)
5	1	Block 1	1.50	1000	0.10
19	2	Block 1	1.00	1400	0.15
11	3	Block 1	1.00	1400	0.10
9	4	Block 1	1.00	1000	0.15
2	5	Block 1	0.50	1800	0.10
14	6	Block 1	1.50	1400	0.15
1	7	Block 1	0.50	1000	0.10
6	8	Block 1	1.50	1800	0.10
20	9	Block 1	1.00	1400	0.15
7	10	Block 1	1.50	1000	0.20
15	11	Block 1	1.00	1400	0.15
4	12	Block 1	0.50	1800	0.20
3	13	Block 1	0.50	1000	0.20
8	14	Block 1	1.50	1800	0.20
18	15	Block 1	1.00	1400	0.15
17	16	Block 1	1.00	1400	0.15
10	17	Block 1	1.00	1800	0.15
12	18	Block 1	1.00	1400	0.20
16	19	Block 1	1.00	1400	0.15

13	20	Block 1	0.50	1400	0.15

3. Results and Discussion

3.1. Stability Tests

Sonication is implemented using an ultra-sonicator instruction to render Nanofluid stable and reduce agglomerate thickness. Each volume concentration Nanofluid was generated in a volume of 100 ml and placed into a sonicator for 30 minutes. Then, all through the measuring cycle, the Nanofluids were very stable. The Nanofluid was prepared with exact measurement showed good stability for several days without any significant sedimentation as shown in Figure 3 and the formula for dilution of nanofluid is shown below.

Figure 3. Stability of Nano Cutting Fluids.

$$\text{Formula } \Delta v = v2 - v1 = v1(\varphi 1/\varphi 2)1 \quad (1)$$

3.2. Machining Results

Based on the detailed result of the experiment that was run 20 times using the RSM method to get the value of the most precision value of surface roughness, cutting temperature, and tool wear. From the result, the lowest value surface roughness (Ra) for 0.50 volume concentration of TiO2 Nanofluid is 1.022 μm when using 1000 rpm and 0.10 feed rate (mm/rev) with the depth of cut 0.50 mm. The highest value surface roughness (Ra) for 0.50 volume concentration of TiO2 Nanofluid is 2.492 μm when using 1000 rpm and 0.20 feed rate (mm/rev) with a depth of cut 0.50 mm. For 1.00 volume concentration of TiO2 Nanofluid the lowest value surface roughness (Ra) is 0.901 μm when using 1400 rpm and 0.10 feed rate (mm/rev) with a depth of cut 0.50 mm. The highest value surface roughness (Ra) for 1.00 volume concentration of TiO2 Nanofluid is 2.268 μm when using 1400 rpm and 0.15 feed rate (mm/rev) with a depth of cut 0.50 mm. For 1.50 volume concentration of TiO2 Nanofluid the lowest value surface roughness (Ra) is 0.863 μm when using 1000 rpm and 0.10 feed rate (mm/rev) with a depth of cut 0.50 mm. The highest value surface roughness (Ra) for 1.50 volume concentration of TiO2 Nanofluid is 3.048 μm when using 1800 rpm and 0.20 feed rate (mm/rev) with a depth of cut 0.50 mm.

3.3. Cutting Temperature

Figure 4 below shown the comparison cutting temperature between three different concentrations 0.50, 1.00, 1.50. The higher value of Nanofluid concentration produces a lower cutting temperature. For the higher contain Nanofluids concentration 1.50 (Vt%), the cutting temperature recorded is 28 °C. while for the 1.00 (Vt%) Nanofluids concentration the cutting temperature was measured is 29 °C. For the 0.50 (Vt%) Nanofluids concentration the cutting temperatures was measured at 30.5 °C and for the pure conventional coolant was give the higher cutting temperature is 32.3 °C. So, it justifies the higher Nanofluids concentration will give the lowest cutting temperature for this experiment. The results also showed the comparison cutting temperature between three feed rate 0.10, 0.15, 0.20. The higher value of feed rate produces a higher cutting temperature. For the higher contain Nanofluids concentration 1.50 (Vt%) which is produced for cutting temperature is 28, 28.5, 29.8 (°C). while for the 1.00 (Vt%) Nanofluids concentration the cutting temperature was measured is 29 °C, 29.5 °C, 30.5 °C. For the 0.50 (Vt%) Nanofluids concentration, the cutting temperature was measured 29.5 °C, 30.2 °C, 30.5 °C, and for true conventional coolant was give the higher cutting temperature is 32.3 °C, 32.8 °C, 33.1 °C. So, it justifies the lowest feed rate will give the lowest cutting temperature for this experiment.

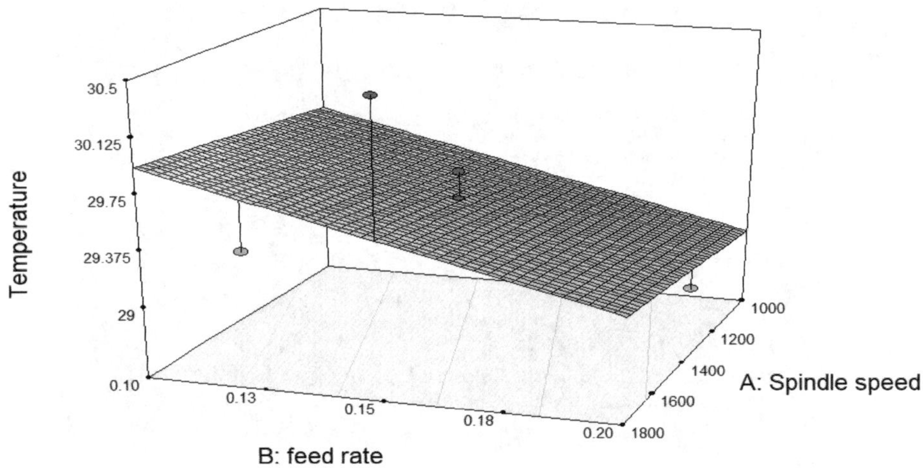

Figure 4. Graphs of Cutting Temperature

3.4. Surface Roughness

Figure 5 below shows the comparison surface roughness between three different concentrations 0.50, 1.00, 1.50 (Vt %). The lower value is the best surface roughness produce by the higher contain nanofluids concentration 1.50 (Vt%) which is produced for Ra is 0.863 μm, while for the 1.00 (Vt%) nanofluids concentration the Ra was measured is 0.901 μm. For the 0.50 (Vt%) Nanofluids concentration the Ra was measured 1.022 μm and for the true conventional coolant was given the higher Ra is 1.886 μm. So, it justifies the higher Nanofluids concentration will give the lowest and the best Ra for this experiment. The experiment then compared to the pure MQL coolant and it proves that a higher concentration of Nanofluid was better the result of surface roughness. The comparative surface roughness between feed rate (mm/rev) is seen from the graph. The lowest feed rate value would yield the best surface roughness which is the lower surface roughness value. The average surface roughness for the lowest feed rate of 0.10, was 0.863 μm. And for the medium feed rate which was an average of 0.15 surface roughness 1.937 μm. Finally, the average surface roughness Ra was 1.833 μm for the highest feed rate of 0.20. This graph shows that feed rate plays important role in measuring surface roughness.

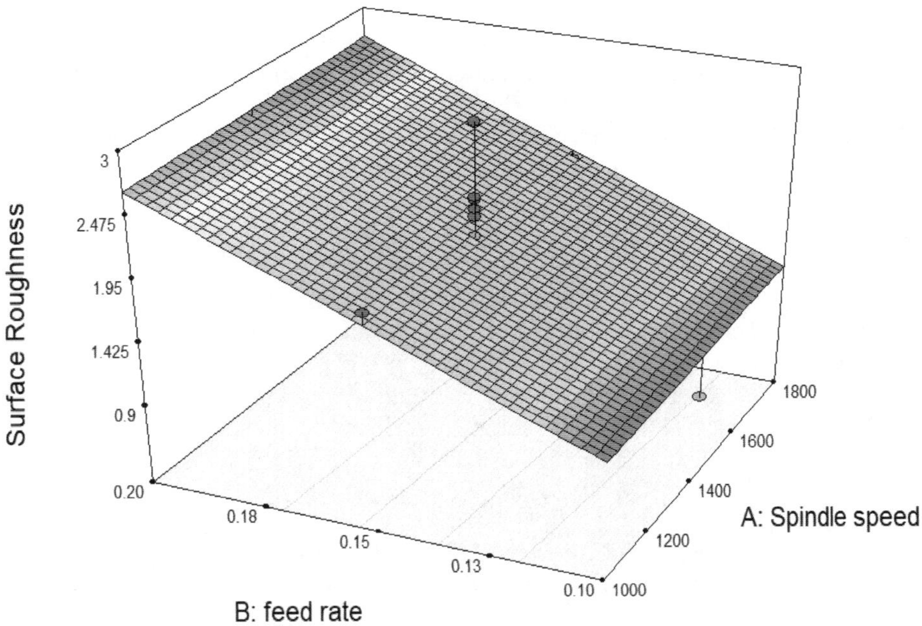

Figures 5. Graphs of Surface Roughness

3.5. Tool Wear

Based on the result of tool wear, in Figure 6 the lowest result was recorded is around 0.0001 grams when using 1.5 volume concentration of Nanofluid and 1000 RPM of spindle speed. It safe to say that, the higher the volume concentration, the better the tool wear and for the spindle speed, the slower the speed the better the tool wear.

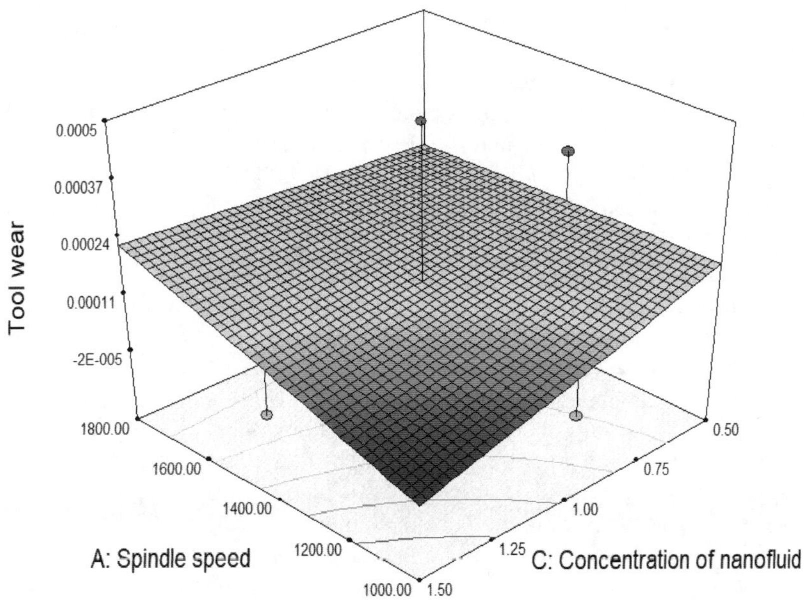

Figure 6. Graphs of Tool Wear

4. Conclusion

In this investigation, the aim was to assess the objective experiment in achieving and the result showed the containing a huge amount of volume concentration TiO2 nanofluid improve the surface quality and cutting temperature from 31°C to 28°C but it cannot justify the tool wear because of the factor that has been told. The following conclusion can be drawn to describe the performance of different concentrations of TiO2 at Al319 in CNC Milling. Recommendations for future experiments should use different techniques and environments, increase the concentration of Nanofluid to get a more accurate result.

i. TiO2 Nanofluids improved surface roughness quality and cutting temperature.
ii. By using a higher volume concentration of TiO2 nanofluid the minimum surface roughness and cutting temperature were obtained.
iii. The best volume concentration of TiO2 nanofluid is 1.50 will result in the best value of surface roughness is 0.863 μm and cutting temperature 28 °C.

Acknowledgments

The authors would like to thank the University College TATI (UC TATI) for sponsoring the work-study under Short Term Grant (STG) with grant STG 9001-1902

References

[1] Abbas, A. T., Benyahia, F., El Rayes, M. M., Pruncu, C., Taha, M. A., & Hegab, H. (2019). Towards Optimization of Machining Performance and Sustainability Aspects when Turning AISI 1045 Steel under Different Cooling and Lubrication Strategies. *Materials (Basel),* ***12*****(18).** doi:10.3390/ma12183023

[2] Jamil, M., Khan, A. M., Hegab, H., Gong, L., Mia, M., Gupta, M. K., & He, N. (2019). Effects of hybrid Al2O3-CNT nanofluids and cryogenic cooling on machining of Ti–6Al–4V. *The International Journal of Advanced Manufacturing Technology,* ***102***(9-12), 3895-3909. doi:10.1007/s00170-019-03485-9

[3] Naresh Babu, M., Anandan, V., Muthukrishnan, N., Arivalagar, A. A., & Dinesh Babu, M. (2019). Evaluation of graphene-based nanofluids with minimum quantity lubrication in turning of AISI D3 steel. *SN Applied Sciences,* ***1*****(10).** doi:10.1007/s42452-019-1182-0

[4] Viswanathan, R., Ramesh, S., & Subburam, V. (2018). Measurement and optimization of performance characteristics in turning of Mg alloy under dry and MQL conditions. *Measurement,* ***120***, 107-113. doi:10.1016/j.measurement.2018.02.018

[5] Sartori, S., Ghiotti, A., & Bruschi, S. (2017). Hybrid lubricating/cooling strategies to reduce the tool wear in finishing turning of difficult-to-cut alloys. *Wear,* 107-114. doi:10.1016/j.wear.2016.12.047

[6] S. Zainal Ariffin, Ahmad Razlan, M.Alias and M Rahman, Experiment of Surface Roughness and Tool Wear on Coolant Condition Technique Using Alloy 319 Used in Automotive Industries. International Journal of Engineering Research & Science (IJOER).ISSN 2395-6992. DOI: 10.25125/engineering-journal-IJOER-OCT-2017-

[7] Sharma, A. K., Tiwari, A. K., Singh, R. K., & Dixit, A. R. (2016). Tribological Investigation of TiO2 Nanoparticle-based Cutting Fluid in Machining under Minimum Quantity Lubrication (MQL). *Materials Today: Proceedings,* ***3*****(6),** 2155-2162. doi:10.1016/j.matpr.2016.04.121

[8] S.Zainal Ariffin, Ahmad Razlan, M. Efende. M.Ali and M.M Rahman, Optimization of Coolant Techique Condition for Machining A319, APCOMS - iMEC 2017 - Journal of Science PP:35

Paper ID 85, The 4th Asia Pacific Conference on Manufacturing System and The 3rd International Conference in Indonesia . 7 & 8 December 2017

[9] Viswanathan, R., Ramesh, S., & Subburam, V. (2018). Measurement and optimization of performance characteristics in turning of Mg alloy under dry and MQL conditions. *Measurement,* ***120***, 107-113. doi:10.1016/j.measurement.2018.02.018

[10] Sharma, A. K., Tiwari, A. K., Dixit, A. R., & Singh, R. K. (2017). The investigation into Performance of SiO 2 Nanoparticle-Based Cutting Fluid in Machining Process. *Materials Today: Proceedings,* ***4*****(2),** 133-141. doi:10.1016/j.matpr.2017.01.006

[11] Chatha, S. S., et al. (2016). "Performance evaluation of aluminum 6063 drillings under the influence of nanofluid minimum quantity lubrication." Journal of Cleaner Production **137**: 537545.

[12] S. Zainal Ariffin, et.al (2016) Effect of Coolant Nozzle Sizes On Turning Aluminum Alloy AL319 Indian Journal of science and technology, Vol **9 (9),** Doi: 10.17485/ijst/2016/v9i9/88712

[13] Gupta, M. K., et al. (2016). "Optimization of machining parameters and cutting fluids during nanofluid based minimum quantity lubrication turning of titanium alloy by using evolutionary techniques." Journal of Cleaner Production **135**: 1276-1288.

[14] Mao, C., et al. (2012). "Investigation of grinding characteristic using nanofluid minimum quantity lubrication." International Journal of Precision Engineering and Manufacturing **13(10):** 17451752.

[15] Sayuti, M., et al. (2014). "Novel uses of SiO 2 nano-lubrication system in hard turning process of hardened steel AISI4140 for less tool wear, surface roughness, and oil consumption." Journal of Cleaner Production **67**: 265-276.

Performance and Emission Measurement of a Single Cylinder Diesel Engine Fueled with Palm Oil Biodiesel Fuel Blends

M Q Rusli[1], M F Muhamad Said[1*], A M Sulaiman[1], M F Roslan[1], I Veza[1], M R Mohd Perang[1], H L N Lau[2], and N S Abd Wafti[2]

[1] Automotive Development Centre, Institute for Vehicle Systems and Engineering, Universiti Teknologi Malaysia, 81310 Johor Bahru, Malaysia
[2] Malaysian Palm Oil Board, Persiaran Institusi, Bandar Baru Bangi, 43000, Kajang, Malaysia
[*] Email: mdfarid@utm.my

Abstract. A high concentration of palm oil biodiesel fuel blend is a strong candidate to substitute conventional diesel fuel (B10) due to its prospect to release fewer emissions. This paper shall present findings when testing palm oil biodiesel fuel blends (B10, B20, and B30) at medium speed (2500 rpm) in a single-cylinder Yanmar L70N diesel engine coupled to an eddy current dynamometer. In this study, a timed burette and an airbox determined the engine's fuel and air consumption, respectively. In contrast, a flue gas analyzer measured the emission released by the fuel blend. The result indicates that as the concentration of palm oil biodiesel increases in the blends (from B10 to B30), there is no significant difference (average less than 5%) between the fuel blends in terms of brake thermal efficiency, brake specific fuel consumption, and air-fuel ratio. On the other hand, the emission results showed a reduction in CO, CO_2, and NO_x when the concentration of palm oil biodiesel increases in the blends. Considering the fuel blend would have no significant difference in the engine performance and its ability to reduce most emission, it is safe to conclude that B30 would be a good alternative for current diesel fuel (B10).

1. Introduction

The automotive industry is popularly accused of why climate change is still an unresolved issue. Many are aware – even preschoolers – that vehicles' emission causes global sufferings such as prolonged drought, sudden flooding, crop failure, and many more. While there have been policies introduced and new technologies deployed to combat it, it is evident that the world needs more, and we are indeed in a race to patch the world before it is too late.

Although many are now concentrating on electric vehicles, EVs, this technology is still a few years away to fully replace vehicles operating by internal combustion engines. EVs still rely on power plants that mainly consume crude oil to power the EVs, and there are still debates on safer ways to dispose of retiring batteries. Contrary, EVs may still damage the environment instead of being its savior. Even when these issues were resolved, massive transports on the sea and even lighter transports on developing countries' roads may still rely on petroleum fuel and the much familiar internal combustion engines.

Since the world may need to cope with internal combustion engines painting the road for a few more years, an immediate solution with smaller changes is necessary to resolve this emission problem. An obvious solution would be to either modify the vehicle or improve what it consumes. The former might be developing a more efficient manufacturing process, improving engine combustion quality, or marketing sustainable vehicles. In contrast, the latter could be the application of biodiesel fuel blends on compressed ignition diesel engines.

Biodiesel fuel blend is a binary mixture of diesel and biodiesel at predetermined concentrations. The application of biodiesel fuel blend for diesel engines is not entirely new that many countries had already begun switching to it. Globally, biodiesel fuel blends exist in varying feedstocks – due to countries' having different resources and economic aspects [1] – and concentration. Furthermore, there are different generations of biodiesels: edible oils, non-edible oils, and waste oils, as the successor solved issues of its predecessor [2]. In recent years, biodiesel fuel blends are no longer binary. There are now tertiary and quarternary blends with either alcohol or vegetable oil mixed with the initial diesel and biodiesel component [3].

From here, there is a need to qualitatively determine and compare the performance of a biodiesel fuel blend with another. The reason is to determine the optimum fuel blend within the broad range of possible feedstocks, components, and concentrations and determine its suitability in an actual engine. Numerous biodiesel fuel blends had blended throughout the years, and researchers rigorously tested them in engines of all shapes and sizes. There were works to compare the performance between biodiesel fuel blends of similar feedstock but with different concentrations, such as performed by Gad et al. [4] and vice versa, as performed by Abed et al. [5]. Another example was comparing different generations of biodiesels' performance as performed by Yesilyurt et al. [3].

Biodiesel fuel blends are named based on the concentration of biodiesel components in the blend. Here, a B30 palm oil biodiesel fuel blend means a blend consisting of 30% palm oil biodiesel and 70% diesel fuel. With the difference in concentration, the fuel blends' properties vary, especially its density, viscosity, calorific value, and cetane number. When quantifying the performance of biodiesel fuel blends, researchers agree that it depends on these four properties. Comparatively, diesel tends to have a lower density, viscosity, and cetane number but has higher calorific value compared to palm oil biodiesel fuel blends.

The Malaysian Palm Oil Board, MPOB, oversees the nation palm oil asset and produces palm oil biodiesel. Along with the Malaysian Government, they are hoping to mandate a higher percentage of palm oil biodiesel fuel blend (B20) and replaces the already commercialized B10. Previously, there had been researching works centering on B20. However, as the year progressed, there had been improvements in the blends' manufacturing process, thus possibly improving its quality and rendering previous results obsolete. Due to this, there is a need to reinvestigate its performance in the engine and emission aspect. Hence, this paper shall present findings in these aspects when testing three blends of palm oil biodiesel fuel blends at different concentrations, namely: B10, B20, and B30 in a single-cylinder Yanmar L70N diesel engine.

2. Methodology

2.1. *Fuel Properties and Handling*

In this paper, three palm oil biodiesel fuel blends were supplied by MPOB with B10 serving as the reference blend. This reference follows B10 as already a commercialized fuel in Malaysia, and the following tests compare the remaining two fuel blends against B10. The fuel blends are binary blends, consisting of only diesel and biodiesel components, and are absent of any other additives. Table 1 lists the properties of the fuel blends, determined using respective ASTM standards. Due to the fuel blends hygroscopic and auto-oxidation nature, the fuel blends were stored in 20 liters oil containers and were under a regulated temperature to avoid fuel degradation. Other than this, operators were made sure to wear latex gloves and thoroughly flush the engine when testing the next fuel blend. This meticulous handling ensured that there is no impurity in the blends, and their quality is maintained.

2.2. *Test Procedures*

The three biodiesel fuel blends were tested on a diesel engine with the specification shown in Table 2. Figure 1 shows the schematic diagram of the test rig. From here, the diesel engine was coupled to a 130-kW eddy current *Magtrol* brake dynamometer to control and maintain the engine speed. The fuel blends were tested at five different loads and at a fixed speed of 2500 rpm, the engine average operating speed. Before the experiment, the atmospheric conditions were first recorded, namely: the atmospheric pressure, temperature, and humidity. Experiments were postponed if the current atmospheric conditions are too different from previous conditions. Data were collected when the engine reached a steady-state condition and is an average of three data under the same operating condition.

The engine was connected to an airbox with a 77-mm diameter orifice and a digital manometer. The water level height in the manometer was used to determine the air consumption rate using Equation 1. Aside from this, a modified burette was installed between the engine and the fuel reservoir. By using the burette and a stopwatch, the fuel consumption was determined. With the air consumption rate and the fuel consumption rate, the Air Fuel Ratio was later determined. Aside from this, the brake specific fuel consumption and the brake thermal efficiency were calculated using equation 2 and equation 3.

Next, to quantify the emission performance, a *Testo* 350 gas analyzer was used to sample the exhaust gas. The gas analyzer was installed with a CO, CO_2, and NO_x emission sensor to measure exhaust gas emissions. The gas analyzer's probe was placed inside the exhaust tailpipe, and an operator began sampling the gas by initializing the analyzer software. This software was installed on a personal computer.

Table 1. Properties of Tested Fuel Blends

Tested Fuel	Density (kg/m^3 40°C) ASTM D7042	Kinematic Viscosity (mm^2/s 40°C) ASTM D7042	Calorific Value (kJ/kg) ASTM D224	Cetane Number ASTM D6890
B10	855.4	4.610	45 634	54.4
B20	855.0	4.524	44 964	57.8
B30	861.2	4.595	44 409	59.0

Table 2. Engine Specification

Displacement	320 Cubic Centimeters
Compression ratio	20:1
No of Cylinder (Bore x Stroke)	Vertical single cylinder (78 mm x 67 mm)
Max Power	4.9 kW/ 3600 rpm
Max Torque	18 Nm / 2400 rpm

$$\dot{m}_a = 64.94 C_d d^2 \sqrt{\frac{h p_a}{T_a}} \quad (1)$$

$$BSFC = \frac{\dot{m}_f}{P_b} \tag{2}$$

$$\eta_c = \frac{P_b}{\dot{m}_f Q_{CV}} \tag{3}$$

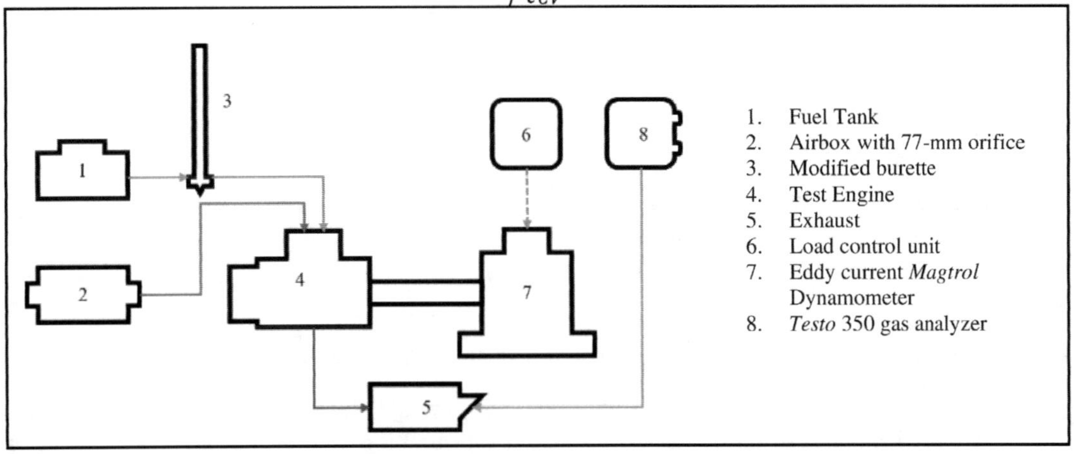

Figure 1. Test Rig Schematic Diagram

3. Results and Discussion

3.1. Engine Performance

The brake thermal efficiency implies the percentage of chemical energy from the fuel blends converted to mechanical energy. Figure 2 shows that as the load increases and is approaching maximum the brake thermal efficiency increases. This is due to the extra heat generated at high load. Figure 2 also shows that as the concentration of palm oil biodiesel increases in the blends, the thermal efficiency decreases as agreed by Singh et al. [6]. They argued that the decreasing efficiency is due to the fuel blends decreasing calorific value (heat produced in complete combustion) as the biodiesel concentration in the blends increases. The result agrees with this statement since the calorific value of tested fuel blends does decrease with increasing palm oil biodiesel in the fuel blends. B10 having the highest calorific value generated the highest thermal efficiency whereas B30 having the lowest calorific value generated the lowest thermal efficiency. On the other hand, Teoh et al. [7] suggested that rich oxygen molecule promotes complete combustion, thus improving the brake thermal efficiency. This reason could be why B30 – which contains richer oxygen molecule than B20 – overtook B20 at mid load operation.

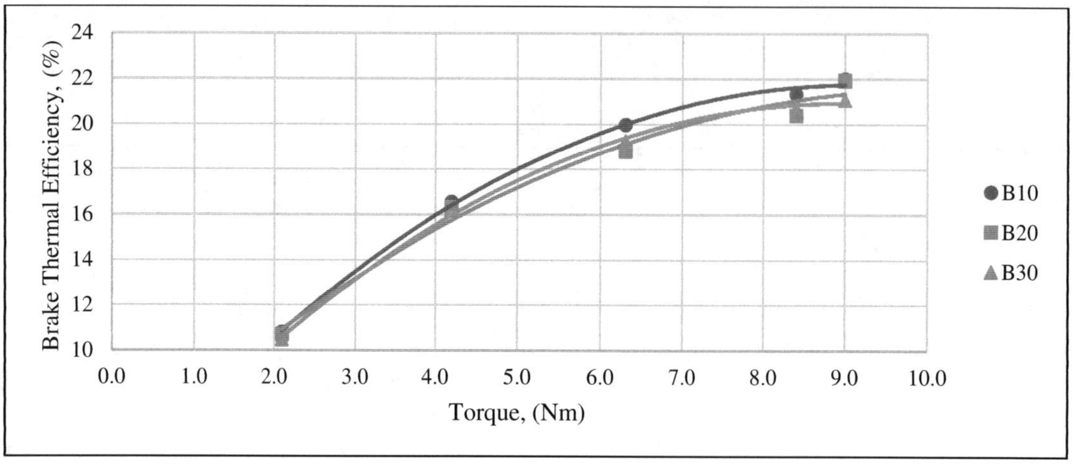

Figure 2: Brake Thermal Efficiency of Palm Oil Biodiesel Fuel Blends

Figure 3 shows the brake specific fuel consumption, BSFC of palm oil biodiesel fuel blends. The figure shows that as the load increases, the brake specific fuel consumption decreases. Figure 3 also shows that the brake specific fuel consumption increases as the biodiesel concentration increases in the blend. A higher concentration of biodiesel fuel blends has a lower caloric value, which results in an increase in BSFC. Engines using these fuel blends would therefore consume more fuel to maintain the load. This generally means for a fixed volume of the biodiesel fuel blends, engines fueled with B10 would travel further than engines fueled with B30. Referring to equation 2, BSFC is directly proportional to the fuel consumption rate.

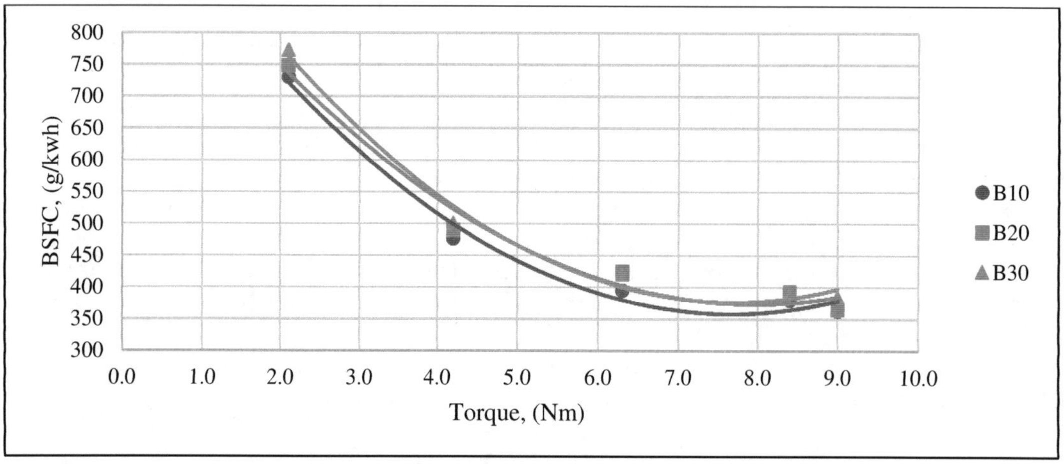

Figure 3: Brake Specific Fuel Consumption of Palm Oil Biodiesel Fuel Blends

The Air Fuel Ratio is the ratio of air and fuel consumption. A diesel engine is commonly operated under a lean condition where more air is consumed than fuel. Figure 4 shows that the engine did operate in a lean condition, and the AFR decreases with increasing engine load. The figure also shows that B10 has the highest AFR, while B30 is the lowest. B30 has high oxygen content, which further reduces the AFR since the engine requires fewer external air for the combustion process. This shows that a higher concentration of biodiesel fuel blends helps to promote better combustion as compared to neat biodiesel fuel blends.

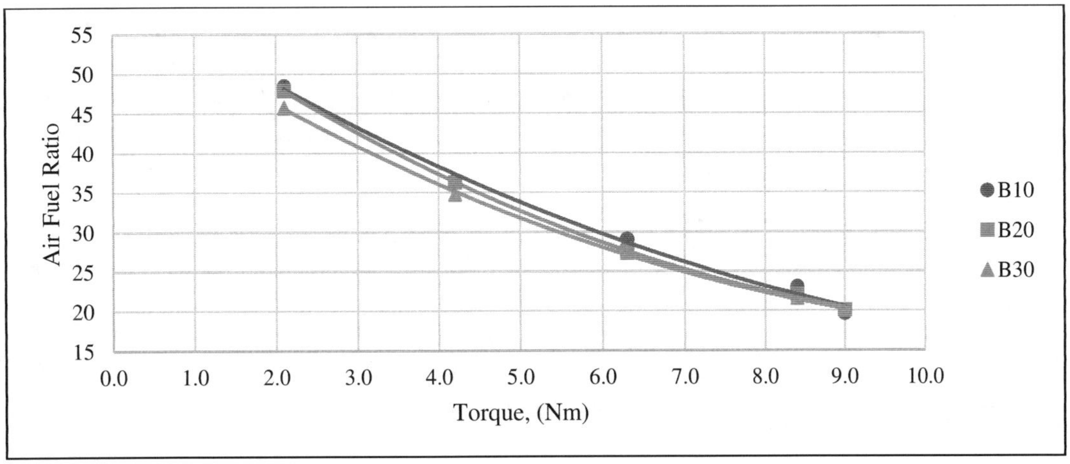

Figure 4: Air Fuel Ratio of Palm Oil Biodiesel Fuel Blends

From these figures, it is evident that the blends' performance differences are not significant with each other. Table 3 shows the percentage difference of the different fuel blends with B10 as the reference fuel. The percentage difference is at most 7.67%. This small percentage difference denotes that if an engine switches fuel blends, the performance difference would not be too noticeable. The results obtained could apply to multi-cylinder engines under a similar test condition.

Table 3: Engine Performance Percentage Difference Against B10

	B20			B30		
Torque (Nm)	**Air Fuel Ratio (%)**	**Brake Specific Fuel Consumption (%)**	**Thermal Efficiency (%)**	**Air-Fuel Ratio (%)**	**Brake Specific Fuel Consumption (%)**	**Thermal Efficiency (%)**
2.1	-1.18	2.48	-0.97	-5.59	5.85	-2.92
4.2	-0.64	2.98	-1.45	-4.61	5.00	-2.14
6.3	-5.33	7.67	-5.74	-6.05	6.67	-3.67
8.4	-4.32	3.29	-4.33	-6.49	3.01	-2.87
9.0	2.34	0.63	-0.24	2.19	6.06	-4.17

3.2. Emission Performance

Figure 5 shows an increase in the percentage of carbon monoxide, CO released with the increased engine load. However, with the increasing concentration of palm oil biodiesel, fewer CO were released, similar to Saravanan et al. [8]. The formation of CO is due to incomplete combustion [7]. The results showed that a higher blend of biodiesel helps to reduce CO. In times, Saravanan et al. [8] suggested that extra oxygen molecules might react and generate carbon dioxide, CO_2. Like Figure 5, the percentage of CO_2 in Figure 6 increases with an increasing load but decreases with the increase in palm oil biodiesel concentration. Anwar et al. [9] implied that the increase in CO_2 emission is due to the fuel blends' high oxygen content and its high cetane number. This result agrees with Anwar et al. [9] since B30 has the highest cetane number of 59.0 and generates the fewest CO_2 at all loads. The CO and CO_2 content agree with the oxygen, O_2 content released by the palm oil biodiesel fuel blends. From Figure 7, at high load, the O_2 content is low. Referring to Figure 5 and Figure 6, it is observed that the CO and CO2 content is high at high load. Since the total number of oxygen molecules is fixed, fewer O_2 molecules were generated since more oxygen molecules generated CO and CO_2.

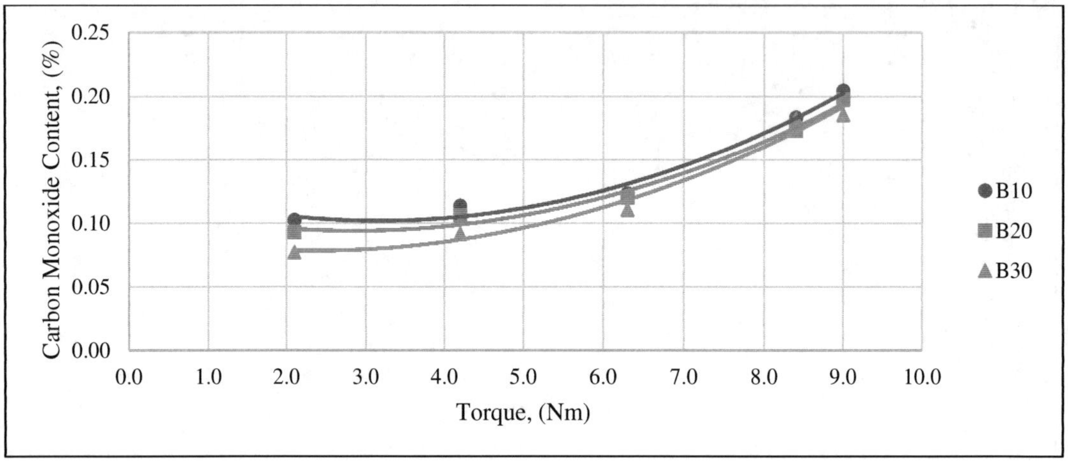

Figure 5: Carbon Monoxide, CO Content of Palm Oil Biodiesel

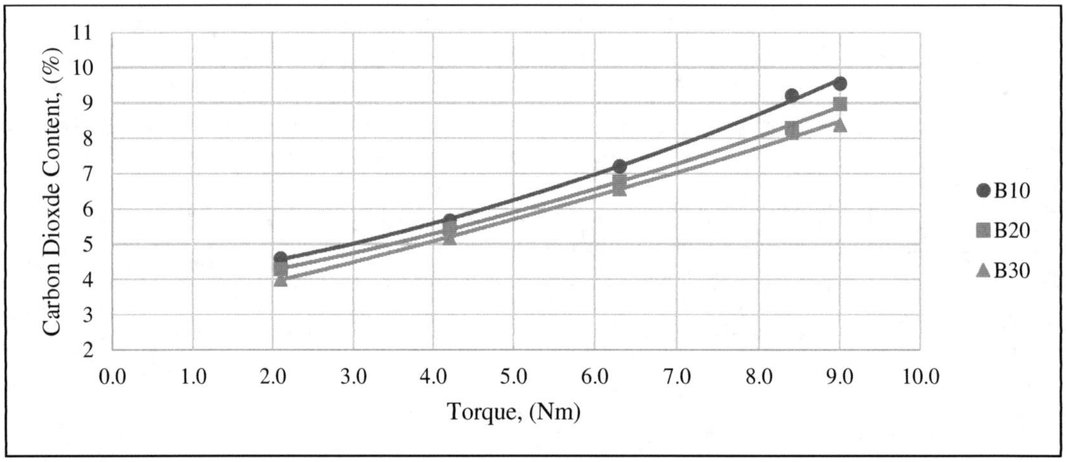

Figure 6: Carbon Dioxide, CO_2 Content of Palm Oil Biodiesel

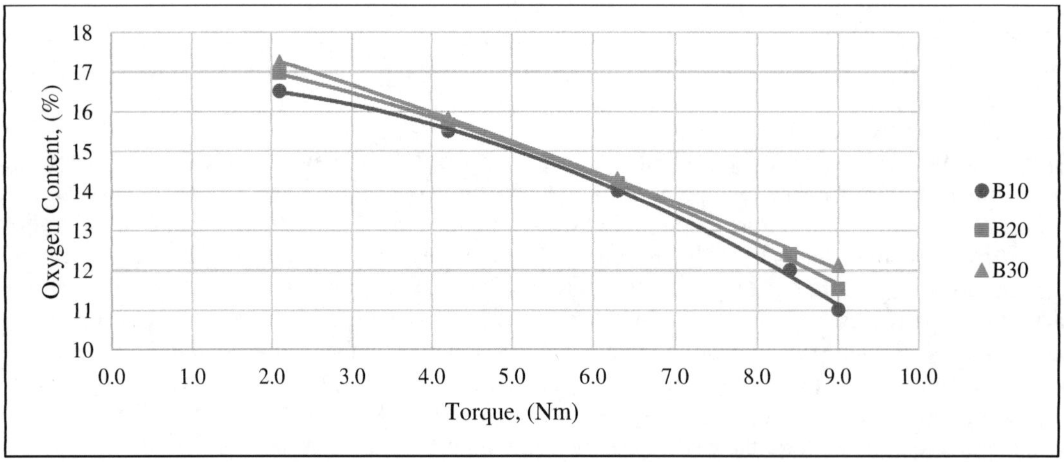

Figure 7: Oxygen, O_2 Content of Palm Oil Biodiesel

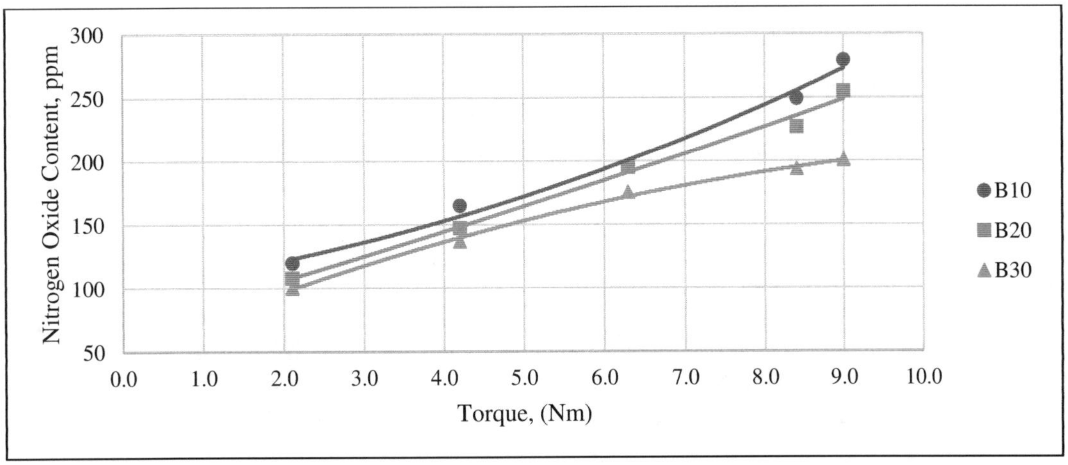

Figure 8: Nitrogen Oxides (NO_x) Content of Palm Oil Biodiesel

For nitrogen oxides in Figure 8, the emission released increases with the increased engine load but decreases with the increasing concentration of palm oil biodiesel. Contrary to common findings, the result showed that a higher concentration of biodiesel fuel blend produces fewer NO_x compared to a lower concentration of biodiesel fuel blends. The formation of NO_x is directly dependent upon the inside temperature of the combustion chamber [10]. Shrivastava et al. [10] implied that a higher concentration of biodiesel fuel blend might have a lower combustion temperature, which is ascertained by the lower exhaust gas temperature. Since the formation of NO_x depends on the temperature phenomenon, the lower combustion temperature causes fewer generation of NO_x.

From these figures, it is evident that B30 generated the fewest emission. From Table 4, B30 at the highest load produces 9.11% fewer carbon monoxide, 12.25% fewer carbon dioxide, and 28.10% fewer oxides of nitrogen compared to the baseline fuel B10. At low load, a similar reduction trend is also observed. At mid load operation, the percentage difference is not uniform due to the inconsistent heat generated in the engine which affects the generation of emission.

Table 4: Emission Performance Percentage Difference Against B10

Torque (Nm)	B20			B30		
	CO (%)	CO_2 (%)	NO_x (%)	CO (%)	CO_2 (%)	NO_x (%)
2.1	-9.19	-6.54	-9.92	-24.34	-12.85	-16.33
4.2	-6.52	-3.89	-10.79	-19.12	-8.83	-17.09
6.3	-2.84	-6.10	-1.54	-10.14	-9.15	-10.21
8.4	-5.57	-9.99	-9.04	-0.49	-11.51	-22.37
9.0	-3.28	-6.07	-8.75	-9.11	-12.25	-28.10

4. Conclusion

This paper determined the performance and measured the emission released by three palm oil biodiesel fuel blends from MPOB. Malaysia had mandated the usage of B10 for diesel engines and is currently mandating B20 to replace B10. Based on the results, the following conclusions were made.

- B30 is an acceptable successor to replace B20 in the future.

- The performance difference of B30 is not significant against B10 and B20. The percentage differences for air-fuel ratio, brake specific fuel consumption and thermal efficiency is an average less than 5%.
- The brake specific fuel consumption of B30 is slightly higher than B20 and B10.
- However, B30 compensates with the significant reduction of CO, CO_2, and NO_x compared to B20 and B30.

Indeed, mandating B30 for diesel engines would effectively reduce the national greenhouse gas released. When comparing the tested palm oil biodiesel fuel blends and its previous version, the tested fuel blends' quality had improved. This paper recommends further identifying and investigating the decisive factors that improved the quality of the fuel blends. This paper also recommends testing the fuel blends at different conditions and proving the fuel blend's lower combustion temperature due to the decreasing trend of nitrogen oxide emitted as the concentration of palm oil biodiesel increases in the fuel blend.

Acknowledgment

This study was supported by the research grant number R.J130000.7309.4B405 - Impact of Biodiesel Fuels (B30) on Selected Diesel Engine and Injector Spray Quality. This study was collaborated and sponsored by the Malaysian Palm Oil Board, MPOB.

References

[1] Sakthivel R, Ramesh K, Purnachandran R and Mohamed Shameer P 2018 A review on the properties, performance and emission aspects of the third generation biodiesels *Renewable and Sustainable Energy Reviews* **82** 2970-92

[2] Singh D, Sharma D, Soni SL, Sharma S, Kumar Sharma P and Jhalani A 2020 A review on feedstocks, production processes, and yield for different generations of biodiesel *Fuel* **262**

[3] Yesilyurt MK, Aydin M, Yilbasi Z and Arslan M 2020 Investigation on the structural effects of the addition of alcohols having various chain lengths into the vegetable oil-biodiesel-diesel fuel blends: An attempt for improving the performance, combustion, and exhaust emission characteristics of a compression ignition engine *Fuel* **269**

[4] Gad MS, El-Araby R, Abed KA, El-Ibiari NN, El Morsi AK and El-Diwani GI 2018 Performance and emissions characteristics of C.I. engine fueled with palm oil/palm oil methyl ester blended with diesel fuel *Egyptian Journal of Petroleum* **27(2)** 215-9

[5] Abed KA, Gad MS, El Morsi AK, Sayed MM and Elyazeed SA 2019 Effect of biodiesel fuels on diesel engine emissions *Egyptian Journal of Petroleum* **28(2)** 183-8

[6] Singh M and Sandhu SS 2020 Performance, emission and combustion characteristics of multi-cylinder CRDI engine fueled with argemone biodiesel/diesel blends *Fuel* **265**

[7] Teoh YH, Yu KH, How HG and Nguyen HT 2019 Experimental Investigation of Performance, Emission and Combustion Characteristics of a Common-Rail Diesel Engine Fuelled with Bioethanol as a Fuel Additive in Coconut Oil Biodiesel Blends *Energies* **12(10)**

[8] Saravanan A, Murugan M, Sreenivasa Reddy M and Parida S 2020 Performance and emission characteristics of variable compression ratio CI engine fueled with dual biodiesel blends of Rapeseed and Mahua *Fuel* **263**

[9] Anwar M, Rasul MG and Ashwath N 2020 A pragmatic and critical analysis of engine emissions for biodiesel blended fuels *Fuel* **270**

[10] Shrivastava P, Verma TN and Pugazhendhi A 2019 An experimental evaluation of engine performance and emisssion characteristics of CI engine operated with Roselle and Karanja biodiesel *Fuel* **254**

Development of Android Based Application for Monitoring Engine Lubricant Condition in Four Wheeled Vehicle and Driving Behaviour Effects on Engine Lubricant Condition

H S Nugroho[1,*], M A Dewantoro[1,*] and D E Atmaja[1,*]

[1] Mechanical Department, Faculty of Engineering, Universitas Indonesia, Depok, Jawa Barat 16424, Indonesia

*Email: henky.suskito@eng.ui.ac.id , m.akhadewantoro@gmail.com, david.edika@ui.ac.id

Abstract. Engine lubricant is a liquid chemical that is given between two moving objects to reduce the frictional force that occurs. Engine lubricant needs to be replaced periodically to maintain the durability of four-wheeled vehicles. Manufacturers of four-wheeled vehicle manufacturers have given the lubricant replacement period to consumers, but this time period is only a reference. The purpose of this research is to study the calculation method to find the right time to replace the lubricant along with its parameters, study the relationship and characteristics between engine lubricant temperature with coolant temperature, and study the relationship of driving behaviour to decrease engine lubricant conditions. This research was carried out by taking data on engine lubricant temperature and coolant using ANCEL and then analysed it to get the formula for determining engine oil temperature and found the mean absolute error of 0 to 3.60. This research is done by testing the driving behaviour of eco, normal, and sport as far as 300 km for each driving behaviour. In this study, engine speed and coolant temperature data are taken through OBD II and then processed using Raspberry Pi into RPS and engine lubricant temperature then further processed by the backend then the data is sent to Android. On the Android application, the output of the data processing results is displayed as a percentage of engine lubricant conditions, engine lubricant remaining distance, and engine lubricant remaining time. The test results show that in sports driving behaviour, the condition of engine lubricants decreased the most at 3.9% followed by normal at 3.18% and the lowest decreased was eco with 2.39%.

1. Introduction

Indonesia experienced the number increase of four-wheeled vehicle users in 2018 that reached 16.44 million drivers, 2.2 times compared to 2008 [1]. Along with these increases, the condition of four-wheeled vehicles in Indonesia developed in terms of production and user adjustments which also occurred all over the world such as three regions, Europe, America, and China and tended to pay more attention to drivers, especially the 'modern' generation with an interest in digital technology [2].

Internet of Things as a term that supports the development of world technology provides opportunities for the automotive sector [3] which results in one in nine survivors of road accidents in developing countries due to the informative ability of four-wheeled vehicles [4]. Around the world, the benefits provided by the implementation of IoT in the automotive sector had an impact on reducing maintenance costs by around 10% - 40% and increasing vehicle life cycles by around 3% - 5% [5]. The change in driver characteristics to 'modern' is in line with the market survey themed "Global Executive

IOP Conf. Series: Materials Science and Engineering 1068 (2021) 012021 doi:10.1088/1757-899X/1068/1/012021

Survey" which said that 58% of four-wheeled vehicle consumers had the appeal of digital devices in accessing vehicle data remotely, checking and diagnosing, and scheduling maintenance activities [6] so that the use of IoT in maintenance activities can provide many benefits.

Vehicle maintenance activities were important to operate in optimal conditions [7] in terms of safety, performance, and comfort. Component that had been identified and considered important to be the subject of this study was engine lubrication systems [8] which needs to be replaced periodically according to vehicle manufacturers recommendation to maintain the durability of engines. The period for changing the lubricant could be influenced by many things, such as engine speed and engine lubricant temperature [9]. The problem with this theory is that only certain vehicles have access to the engine lubricant temperature reading, so it is necessary to find another approach to obtain the engine lubricant temperature value. Therefore, IoT application was designed with Raspberry Pi 3 and OBD II [10] as data collectors, cloud as backend, and Android as data visualization so that users could determine the condition of engine lubricants in real time and calculate the correct timing for replacing engine lubricant. The research also conducted tests to determine the effect of driving behaviour on the condition of engine lubricant so that four-wheeled vehicle drivers can find out how to drive properly so that the quality of the engine lubricant condition can be maintained.

2. Methodology

2.1. Model Calculation Design

Based on test data performed using a dynamometer, it was found that engine lubricant temperature and engine rotation speed (RPM) can affect engine durability and/or engine lubricant degradation (Figure 1), In addition, based on research conducted on the relationship between lubricant temperature and lubricant life, a parabolic graph was obtained [9].

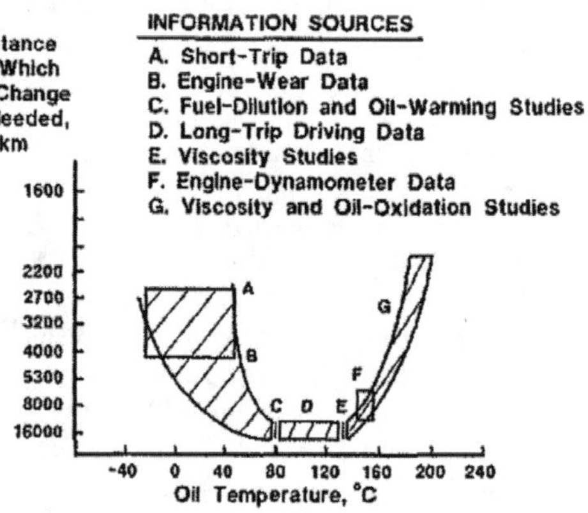

Figure 1: Parabolic relationship between lubricant temperature and lubricant life (Schwartz, 1987)

To increase safety in calculating the lubricant change interval, a constant is made based on the graph. These constants are divided into three parts, namely the constant when the lubricant temperature is below the ideal temperature (k1), at the ideal temperature (k2), and above the ideal temperature (k3).

$$\frac{1}{(distance\ to\ oil\ change)} = k_1(T_{oil} - T_C)^2 \quad (1)$$

$$\frac{1}{(distance\ to\ oil\ change)} = k_2 \quad (2)$$

$$\frac{1}{(distance\ to\ oil\ change)} = k_3(T_{oil} - T_E)^2 \tag{3}$$

Thus, from the distance, oil-aging rate can be calculated with penalty factor, effective revolution, and revolution left.

$$Penalty\ Factor = \frac{distance\ to\ oil\ change\ at\ ideal\ temperature}{distance\ to\ oil\ change\ at\ given\ temperature} \tag{4}$$

$$Effective\ Revolution = Engine\ Rev. \times Penalty\ Factor \tag{5}$$

$$Revolution\ Left = Engine\ Rev.Max - Effective\ Revolution \tag{6}$$

The maximum revolution value was revolutions that can occur throughout the life of the engine lubricant. Nissan Juke engine lubricant must be replaced every 10,000 km [11]. Therefore, the maximum revolution value in this study is the number of engine revolutions that occur to reach 10,000 km with a mixture of various driving behaviours.

In this research, because the programming carried out in this study will process data per second, the value of the engine speed taken from the vehicle (RPM) is converted into the number of engine revolutions per second (RPS) then multiplied by the penalty factor so that it becomes an Effective Revolution. To get the percentage of the latest engine lubricant condition, a comparison of the revolution left value with the maximum revolution value is carried out.

$$\%\ Lubricant\ Condition = \frac{Revolution\ Left}{Engine\ Rev.Max} \times 100\% \tag{7}$$

Not only in the from of percentage, the result could be shown in days (full driving) and kilometers.

$$Days\ Left = \frac{Revolution\ Left}{(Rev.Left\ per\ Second_{t-1} - Rev.Left\ per\ Second_t) \times 86400} \tag{8}$$

$$KM\ Left = \%\ Lubrication\ Condition \times 10000 \tag{9}$$

In order to obtain lubricant temperature data, it is necessary to convert obtainable data from OBD II such as coolant temperature. It had correlation to lubricant temperature as it followed the coolant temperature value with a bit elasticity and damping [12]. In this research, these correlations were obtained by calculating conversion equation using polynomial with the least MAE. To convert the coolant temperature data, team used ANCEL (non-real-time condition) as lubricant temperature standard and infrared thermometer (BENETECH GM320) to verified real lubricant temperature value.

After conducting a literature study and determining the formula to be used, the modeling was made as a pre-liminary test before testing and taking real data. Modeling is made using the Simulink application (Figure 2). This modeling is done by entering the predetermined formulas along with the required data.

IOP Conf. Series: Materials Science and Engineering 1068 (2021) 012021 doi:10.1088/1757-899X/1068/1/012021

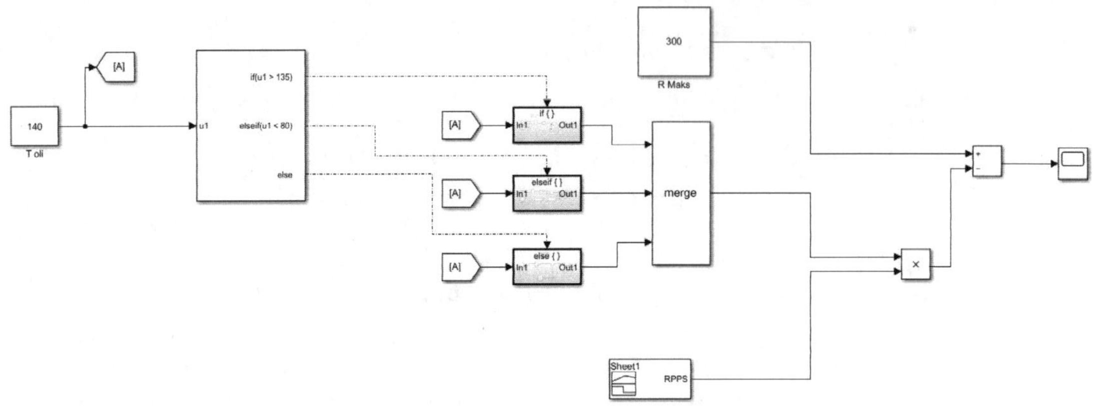

Figure 2 . Calculation Modeling Using Simulink

2.2. System and Application Design

Vehicle monitoring applications were built by designing both the hardware and the software. This system contained three main subsystems, the data collecting device, the data processing algorithm, and the data visualization user interface (Figure 3).

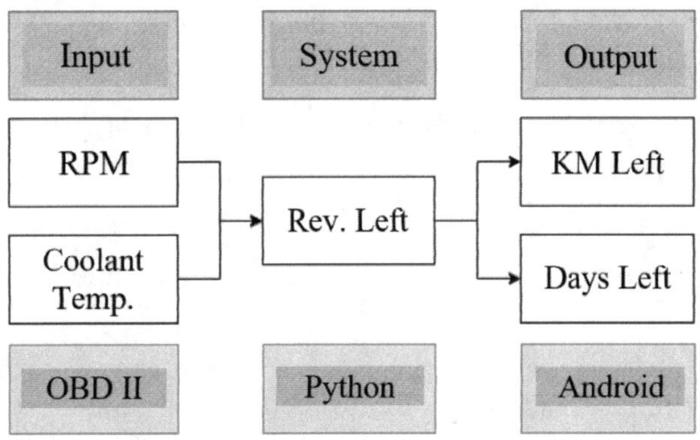

Figure 3. Application Block Diagram

The first two subsystems were built using Python language. The data collecting device was designed using OBD II and Raspberry Pi 3 that connected directly from the vehicle with an embedded program to collect parameter values, empirically calculate the component condition and send it to the cloud [13]. The data processing algorithm was embedded in the cloud using the Django framework and PostgreSQL as the database storage (Figure 4). In this framework, the data were being visualized using Android applications that already common in people nowadays.

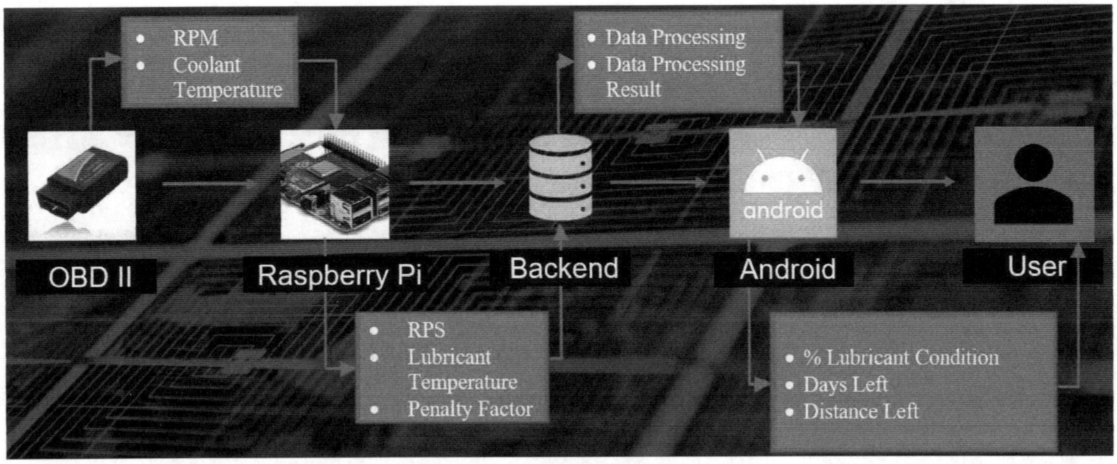

Figure 4. Program Flow

2.3. *Experiment Design*

In this research experiment, the test drive was been done using three driving behaviour to look for difference of lubricant condition for each behaviour. The driving behaviours were divided into three stage that determined by accelerator position and coasting time within 1 minute which being calculated from MSA calculation. The results were indices representative driver could maintain the behaviour separately with 14.23% R&R percentage and the behaviour can be differentiated into three behaviour, Eco, Normal, and Sport perfectly (Table 1).

Table 1. Driving Behaviour Parameters

Driving Behaviour	Accelerator Position (%)	Coasting Time (min/km)
Eco	<12 (16)	>1.8
Normal	20 (16) – 25 (27)	1.3 (1) – 1.6 (1.8)
Sport	>30 (27)	<1

From above parameters, the test was designed within 300 km distances for each behaviour with Nissan Juke 2015 as observed vehicle. After each behaviour, the components were replaced to keep the lubricant's condition of every driving behaviour.

3. Result and Discussion

3.1. *Lubricant and Coolant Temperature Conversion*

Three times data collection of coolant and engine lubricant temperature was carried out to determine the differences and characteristics of the temperature of the coolant and engine lubricant (Figure 5).

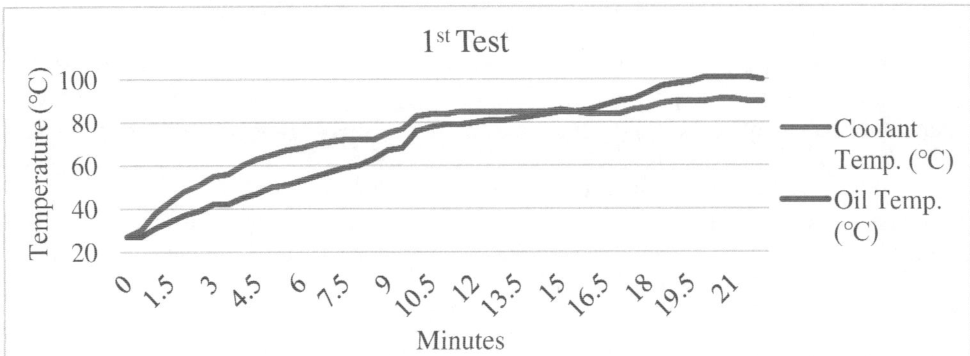

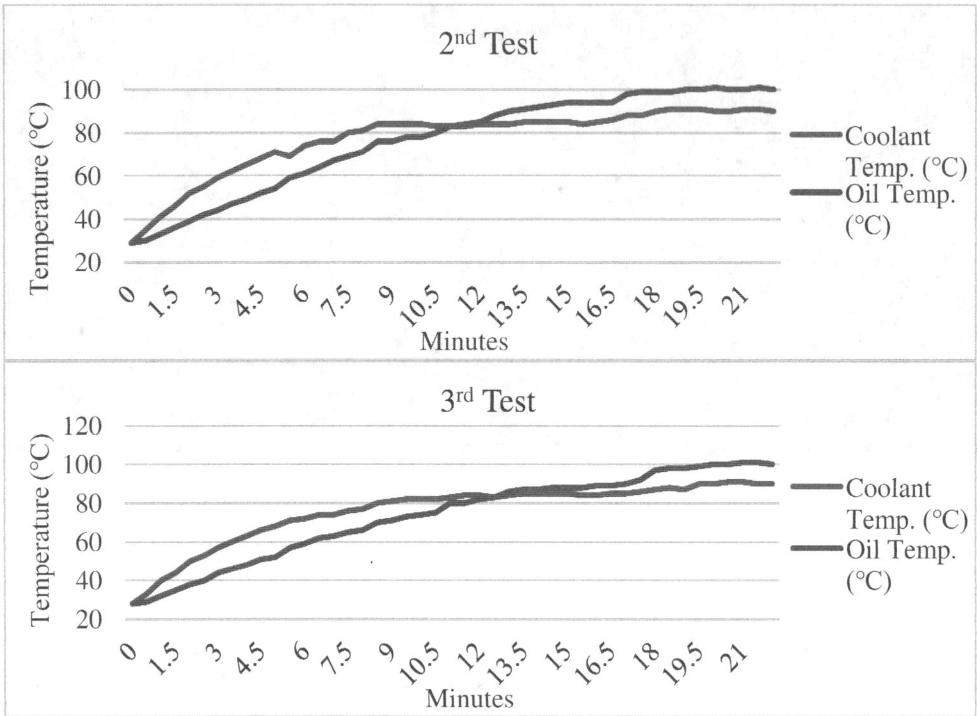

Figure 5. Three Correlation Result of Lubricant and Coolant Temperature

The verification of the results was being done during off-condition engine and driving. The average difference of the temperature was 0.4ºC or 1.3% error during off-condition engine and 1.4ºC or 1.4% error during driving. The conversion equation gave nearly predicted lubricant temperature from coolant temperature with MAE around 0ºC – 3.6ºC.

From the three graphs, in the first experiment, the temperature of the engine lubricant took longer to overtake the coolant temperature, which was 15.5 minutes. Meanwhile, in the second and third experiments, the meeting point of the two temperatures was relatively faster, namely the 11th minute for the second experiment and the 12.5th minute for the third experiment. This happens because in the first experiment, the test was carried out in the morning when the ambient temperature was lower, while the second and third experiments were carried out at noon when the ambient temperature was higher. Another factor that influences this is that in the first experiment, the test was carried out with the condition of the vehicle at rest in place, while in the second experiment, the test was carried out with the condition of the vehicle being driven immediately as soon as the engine was started, and in the third experiment, the test was carried out with a half test condition carried out when the vehicle does not move then half again while the vehicle is being driven. The results of this test support the theory put forward by Phillips and Termo Industries which states that the vehicle engine will heat up faster (optimum working temperature) when driven slowly after starting. However, the three graphs have something in common, namely the meeting point of coolant and engine lubricant temperature occurs in the temperature range 83 °C to 85 °C.

3.2. Constant and Revolution Maximum Calculation

Determination of constant values is done by performing calculations using various data in predefined formulas (Table 2).

Table 2. Lubricant Constant Value

Constant	Value
K_1	1.81536E-07
K_2	2.22439E-07
K_3	2.66662E-08

The vehicle (Nissan Juke 2015) requires a replacement of engine lubricant every 10,000 km [11]. In this study, the maximum revolution value is obtained from data collection of eco, normal, and sport driving behaviour which is carried out as much as 300 km per one driving behaviour. The three data are combined and converted into 10,000 km (Table 3).

Table 3. Engine Revolution Total Each Behaviour

Driving Behaviour	Engine Revolution Total
Eco (300 km)	556761.4
Normal (300 km)	704305.6
Sport (300 km)	962652.7
Mean	741239.9

Thus, the maximum revolution obtained was 24707996.67 revolution.

3.3. *Application Program*

The built system that consisted of three main systems were each simulated to do exchanged data information through either internet or USB in laptop. The simulation was resulting in good real-time condition with each second accurate data (Figure 6).

Figure 6: Android User Interface with Output Data Information

3.4. *Engine Lubricant Condition Experiment*

In this subsection, the result condition of engine lubricant using empirical calculation and shown result on application were described for each driving behaviour in the end of session for 300 km. For Eco behaviour, the result values were shown on Table 4.

Table 4. Eco Behaviour Experiment Result

Information	Empirical Calculation	Application Shown Result
Revolution Reduction	590832.44	590832.44
% Lubricant Condition	97.61%	98%
Distance Left (km)	9760.87	9761
Time Left (days)	13.11	13

For Normal behaviour, the result values were shown on Table 5.

Table 5. Normal Behaviour Experiment Result

Information	Empirical Calculation	Application Shown Result
Revolution Reduction	786729.79	786729.79
% Lubricant Condition	96.82%	97%
Distance Left (km)	9681.59	9682
Time Left (days)	10.03	10

For Normal behaviour, the result values were shown on Table 6.

Table 6. Normal Behaviour Experiment Result

Information	Empirical Calculation	Application Shown Result
Revolution Reduction	962652.70	962652.70
% Lubricant Condition	96.10%	96%
Distance Left (km)	9610.39	9610
Time Left (days)	6.37	6

From the results of the verification of the results of the application, it was found that the results of data processing and the results of data processing through the application showed the same results, only on the Android interface, the results shown did not use decimals so the results were rounded to the nearest unit. This rounding causes a maximum error of 0.5% which is caused by rounding on the percentage of engine lubricant condition.

Of the three driving behaviour test results, testing with eco-driving behaviour recorded the smallest decrease in engine lubricant condition compared to normal and sport, which was 2.39%. This is because, during testing of eco driving behaviour, the number of engine revolutions required to drive 300 km is not too large. From the test results of eco-driving behaviour, the average engine rotation speed of the vehicle is only 1,203 revolutions per minute (rpm).

In contrast, of the three driving behaviours, tests with sport driving behaviour recorded the largest decrease in engine lubricant condition. From the tests carried out as far as 300 km, there was a decline in the quality of the engine lubricant due to sports driving behaviour by 3.9%. The average engine rotation speed during the sport driving behaviour test is 2589 rpm.

The normal driving behaviour records a decrease in the quality of the engine lubricant in the middle between eco and sport driving behaviour. From the test results of normal driving behaviour, the decline in the quality of engine lubricants was recorded at 3.18%. The average engine rotation speed in the normal driving behaviour test is also between eco and sport, which is 1483 rpm.

From the three test results, it can be stated that a person's driving behaviour while driving a vehicle can affect the condition of the engine lubricant. The deeper the accelerator lever is pressed, the higher the engine rotation speed and the faster the engine lubricant condition decreases. In addition, one of the parameters in driving behaviour is coasting driving. The more coasting frequency during driving, the more efficient the consumption of engine lubricant will be. This can be proven in the results of this test where testing for eco-driving behaviour has decreased the least amount of lubricant consumption.

The results of this test also prove the correctness of the factors that influence the condition of engine lubricants proposed by Nissan. From the factors put forward by Nissan, there are two factors that can be proven in this test, namely driving briefly with cold temperatures and using vehicles with continuous idling. This is because the engine lubricant operates outside the optimum temperature and the short driving time makes the engine lubricant temperature does not reach the optimum temperature. In addition, continuous idling causes the condition of the engine lubricant to be continuously consumed without the movement of the vehicle.

4. Conclusion

The conclusion of the study was as follows:

- The engine lubricant temperature always follows the coolant temperature without a definite difference and the time it takes for the engine lubrication and coolant temperatures to meet (temperature meeting point) was influenced by the outside temperature of the vehicle and the test method;
- The meeting point for coolant and engine lubricant temperature was always in the range 83 °C - 85 °C;
- Parameters to determine the condition of the engine lubricant were engine lubricant temperature and engine rotation speed;
- Accelerator position and coasting frequency, which were driving behaviour parameters, affected engine lubrication deterioration;
- The designed application could calculate engine lubricant condition in percentage, days, and distance output.

Acknowledgement

The authors would like to recognize the research team for activity support, and Mechanical Engineering Department, Universitas Indonesia for technical and administration support.

References

[1] BPS. (2018). Perkembangan Jumlah Kendaraan Bermotor Menurut Jenis. Retrieved from https://www.bps.go.id/linkTableDinamis/view/id/1133
[2] Felix Kuhnert, C. S. a. A. K. (2018). *Five trends transforming the Automotive Industry* PricewaterhouseCoopers GmbH Wirtschaftsprüfungsgesellschaft.
[3] Tenghong, L., et al. (2012). Research on the Internet of Things in the Automotive Industry.
[4] Association, G. (2014). Understanding the Internet of Things (IoT)
[5] Rahul Rishi, R. S. (2019). *Future of IoT*. India: Ernst & Young Associates LLP.
[6] KPMG. (2017). Global Automotive Executive Survey 2017.
[7] SELIN DÖNMEZ, M. Z. (2016). *Analyzing the Value of Vehicle Maintenance Activities.* (Master), Chalmers University of Technology, SE-412 96 Göteborg, Sweden.
[8] Lathan, C. (1998). *Construction Mechanic Basic* (Vol. 1). United States: NAVEDTRA 14264.
[9] Schwartz, S. E., & Smolenski, D. J. (1987). Development of an Automatic Engine Oil-Change Indicator System.
[10] Manalu Sonya, R., Moniaga, J., Hadipurnawan Dionisius, A., & Sahidi, F. (2017). OBD-II and raspberry Pi technology to diagnose car's machine current condition: study literature. *Library Hi Tech News, 34***(10),** 15-21. doi:10.1108/LHTN-06-2017-0041
[11] Nissan Juke Owner's Manual (2015)
[12] Eriksson, M. (2007). *The Oil Temperature's Effects on Engine Oil Degradation in Trucks.* KTH Industrial Engineering and Management.
[13] Baek, S.-h., & Jang, J.-W. (2015). Implementation of integrated OBD-II connector with external network. *Information Systems, 50*, 69-75. doi:10.1016/j.is.2014.06.011

IOP Conf. Series: Materials Science and Engineering 1068 (2021) 012022 doi:10.1088/1757-899X/1068/1/012022

Smart driving assistance system using Raspberry Pi and actuator networks

Saiful Anwar Che Ghani[1*], Maurice Kettner[2], Ferhat Aslan[2], Fatimah Dzharuddin[3], Norazlianie Sazali[3]

[1]Automotive Engineering Centre, Universiti Malaysia Pahang, 26600 Pekan, Pahang, Malaysia
[2]Institut für Kälte-, Klima- und Umwelttechnik, Karlsruhe University of Applied Sciences, Germany
[3]Department of Mechanical, College of Engineering, Universiti Malaysia Pahang, 26600 Pekan, Pahang, Malaysia

*Email: anwarcg@ump.edu.my

Abstract. Testing and certification of vehicles in succession to their development are usually achieved on chassis dynamometer testbeds. This paper proposes the implementation of a smart real-time driving assistance system using cloud-computing, embedded tools and motorised actuator for vehicles with manual transmission powertrain under testing. For clutch actuation of the vehicle, a robotic leg actuator connected to a high torque stepper motor is controlled by Python code programmed on Raspberry Pi 4, linked through a smart network for remote controlling of the system via a user interface dashboard and digitalized working system so as to employ data connectivity through cloud-computing. By applying this network structure, the implementation functionality of smart driving assistance system using free cloud infrastructure is investigated, when the hardware is setup in Malaysia, whilst the control is from Germany. The framework that governs this prototype involves different domains of engineering such as data processing, manufacturing, mechanics, networking, embedded systems and actuators is able to be utilised in smart driving assistance system for car tested on a chassis dynamometer with latency of less than 1 s.

1. Introduction

Normally, chassis dynamometers are employed to test and certify of newly developed vehicles when it's completely assembled, usually for assessing the different aspects of the engine and powertrain performance [1], [2]. Noise emission, exhaust emission, engine performance, fuel consumption and electromagnetic compatibility are among the common parameters investigated in the test trials [3]. Recently, chassis dynamometer testbeds are more preferred than the test runs on the road due to economical factor, ease of operation, reliability, repeatability and the most tempting to the research and industry with the advancement of technology is the possibility for fully automation [4]. Furthermore, chassis dynamometers also allow for transient trials under varied and pre-set environmental conditions when the tests are conducted inside a climate or high-altitude chamber [5]. In addition, the integration of multiples devices such as emission measurement devices, load control and data acquisition is much easier with chassis dynamometer as compared to road tests [6].

Test trials are normally automated by employing a robotic device, which replaces the functions of a driver [4], [7]. The tasks of the robotic device are commonly for operating the brake, controlling the force given to the accelerator and the engagement of the clutch pedal, and to the extent of manipulating the gear lever to switch gears. Normally when in operation, a pre-defined velocity profile and an associated gear shifting strategy must be pursued for automated test trials on the chassis dynamometer testbed, in which the challenges for control engineering lie in the consideration of actuators as manipulated variables in a closed loop of the vehicle and a driver controller. The driver controller has to be developed to handle different tasks that constitute of several overlying controllers dealing with routines such as drive away, acceleration, gear shifts, deceleration and stopping. The control of the entire remote routine can be addressed by implementing the Internet-of-Things (IoT) approach. This may be beneficial for the development team to access the test facility without the need to be physically present at the testbed. This is important to allow for cross-country collaboration which is crucial to technological innovation which is the main drivers for change in this sector.

This study proposes a conceptual framework to implement a smart real-time driving assistance system using cloud-computing, transport mechanism, embedded tool and motorised actuator for vehicles with manual transmission powertrain under test. The embedded tool is used as client to acquire signals from cloud broker and manipulate the signals to control the movement and speed of the robotic leg to engage and disengage clutch pedal. The test trials on hardware setup in Asia are conducted by exchanging data with data IoT agent in Europe.

2. Methodology

2.1. Message Queuing Telemetry Transport (MQTT) Messaging Protocol

The transport mechanism that is implemented in this work uses the Message Queuing Telemetry Transport (MQTT) messaging protocol which allows the system to reduce the total amount of data sent using a well-known paradigm called publish-subscribe. For a system with restricted equipment, low bandwidth, high latency and reliable networks such as chassis dynamometer in an automotive lab, MQTT protocol is suitable since it is a very simple, lightweight information transmission mechanism [8]. Since MQTT is designed to minimize network bandwidth and equipment resource requirements, while ensuring reliability and a certain degree of delivery guarantee at the same time, the use of MQTT with the driving assistance system that is delay-sensitive and uses little battery power is considered ideal.

The devices that are used to send data can publish them on a specific topic whereas devices or services that are mutually interested in the information can subscribe to the topics of interest. In between these two entities there is another entity called the broker which is responsible for delivering message updates to the subscriber. This framework as depicted in Figure 1, constitutes of components, the GUI Browser Cayenne, as the publisher, local supervisor and Raspberry Pi as the subscribers and the HiveMQ in the cloud as the broker. The communication protocol for the Raspberry Pi that acts as the MQTT-client is written in Python programming language. The Python MQTT client class was implemented which provides the required functions to publish messages and subscribe to topics. Methods from the paho mqtt client class such as connect, disconnect, subscribe, unsubscribe and publish were used within the code.

The Raspberry Pi 4 used in this project needs to subscribe to the topics that are created in the communication protocol to be able to send and receive data, namely target speed, clutch position and send back the On-board Diagnostics (OBD) data such as the actual speed and Revolutions per Minute (RPM). The MQTT signals manipulated by Raspberry Pi will be used to control a stepper motor, which is connected to the clutch pedal, using a circuit shown in Figure 2.

2.2. Robotic Leg

Robotic leg mechanism is driven by a high torque stepper motor. Stepper motor model Nema 34 is used in this project to overcome the torque, τ required in pushing the clutch pedal to the maximum point. The measurement and calculation of the torques when the clutch pedal is pushed at different angles are listed in Table 1. F_m is the measured forces perpendicular to the clutch pedal acquired when the pedal clutch is pushed to the maximum, whereas F_x and F_y are two component vectors of forces resolved into two components. Lever angle and pushrod length are the geometrical parameters of the linkages for the kinematics to transfer the force from stepper motor to the clutch pedal imitating the gait of a human driver.

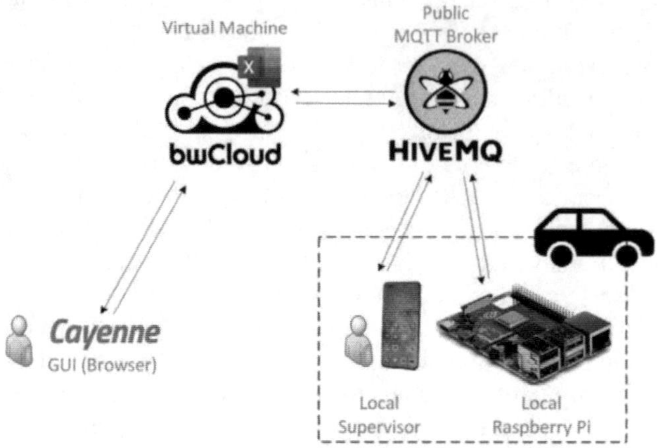

Figure 1. Communication protocol framework.

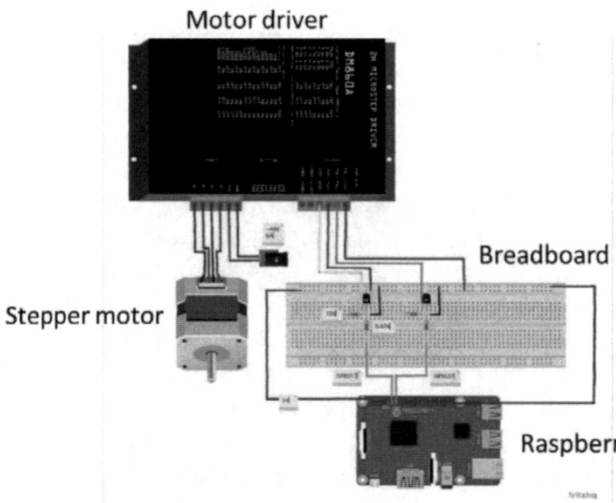

Figure 2. Schematic diagram of stepper motor circuit controlled by Raspberry Pi 4.

Table 1. Measurement and calculation of the torque, τ when the clutch pedal is pushed at different angles.

Pedal angle (°)	F_x (N)	F_y (N)	Lever angle (°)	Pushrod length (mm)	F_m (N)	τ (Nm)
30	129.90	75.00	60	175	173	30.28
40	114.90	96.42	50	136	175	26.52
45	106.06	106.06	45	124	212	26.29
60	75.00	129.90	30	101	300	30.30

2.3. *Real-time monitoring*

Real monitoring of the developed driving assistance system is realised with the development of web-based graphical user interface (GUI) Cayenne. The dashboard that is produced by drag-and-drop function contains slider and meter for clutch position control and vehicle speed, respectively as shown in Figure 3.

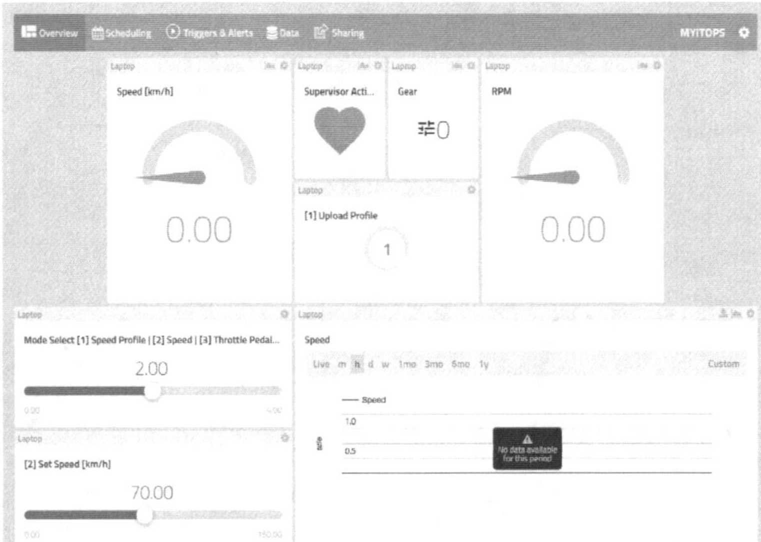

Figure 3. Cayenne Dashboard.

In order to equip the chassis dynamometer with the driving assistance system, a platform as shown in Figure 4 should be fabricated to accommodate all sub-systems, namely electronic enclosure containing Raspberry Pi 4 and electronic components, motor mechanism (stepper motor powered by 110 VDC with gearbox of 10:1 ratio), and linkages (push rod and pedal clamp). The hardware platform is assembled to provide retrofit capabilities of the system in the manual transmission powertrain vehicle without requirement for major modification that can damage the physical of the car.

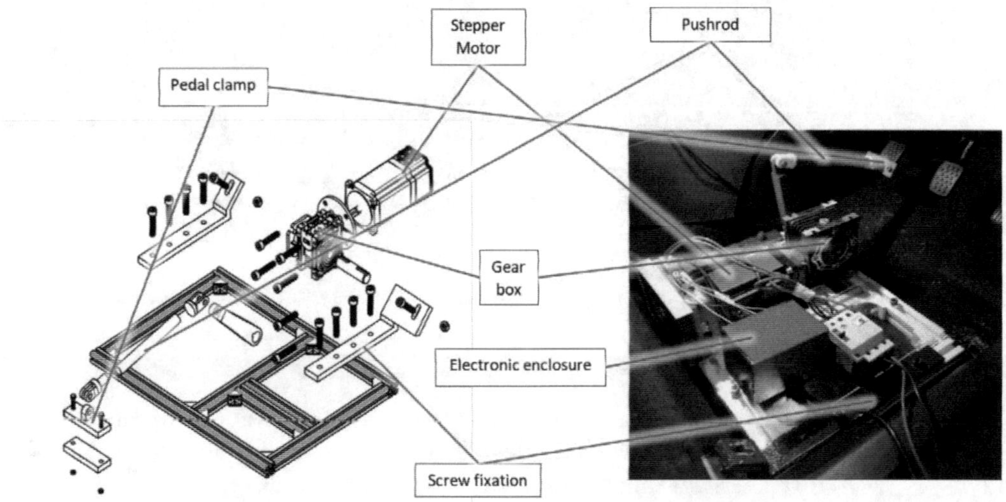

(a) (b)

Figure 4. The smart driving assistance system in (a) CAD exploded view and (b) actual device.

Table 2 provides the specifications of the test vehicle which is a Perodua Myvi 1.3 L manual transmission car. The car has a reference mass of 955 kg and a maximum mass of 1,015 kg. With total displacement of 1298 cc the rated maximum power and torque outputs were 69 kW @ 6000 rpm and 121 Nm @ 3200 rpm, respectively.

Table 2. Formatting sections, subsections and subsubsections.

Vehicle model	Perodua Myvi, 5-speed manual gearbox, front wheel drive
Manufacturer year	2017 (standard gasoline 95 RON)
Reference/ maximum mass	955/ 1,015 kg
Engine type	1.3-liter 1NR-VE
Rated maximum power	69 kW (93 hp)
Rated maximum torque	121 N·m
Number of cylinders	In-line 4 cylinders

2.4. Case study

The experiments were conducted in the UMP Automotive Excellence Centre (AEC) in Pekan, Pahang, Malaysia where the chassis dynamometer was certified to European standards. The test vehicle was tied on a 17.8-inch-roller Mustang Dynamometer. In this conceptual study, the vehicle was driven under no driving cycle conditions, since the aim of the study is merely to investigate the functionality of driving assistance system fitted in the vehicle under test on chassis dynamometer. Figure 5 illustrates experimental setup on Mustang chassis dynamometer testbed. Licensed driver is hired to perform switching and acceleration in the experimental trials. Investigation of latency in communication protocol between MQTT clients and MQTT-broker during the experiment is recorded.

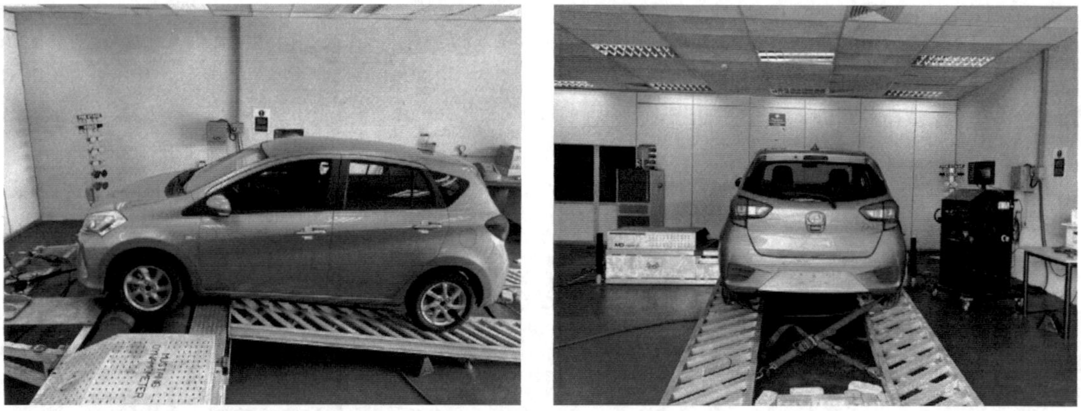

(a) (b)

Figure 5. Experimental setup on Mustang chassis dynamometer testbed.

3. Results and discussion

The experimental results are presented in two subsections, namely (1) the functionality of communication protocol framework and (2) the reliability of the driving assistance system in conducting the driving cycle test on chassis dynamometer testbed.

The performance of the MQTT has been analysed through the latency performance of the subscriber node in broker end for MQTT to highlight efficiency of the communication protocol proposed. The result is shown in Table 3.

Table 3. Experimental trials for the entities in cloud-based communication protocol.

Entity	Device	Location
Publisher	Cayenne	Karlsruhe, Germany
Subscriber	Raspberry Pi 4	Pekan, Malaysia
Broker	HiveMQ	BW Cloud
Latency	765 ms	

The experimental result shows that MQTT communication protocols written in Python language can perform the gear pedal pushing routine with latency of less than 1 s, which is similar to published results by Mukherjee, Dey and De [9] that obtained 30% improvement of average message delivery performance in case of enhanced MQTT-SN (sensor network) and 17% of improvement for enhanced MQTT with latency of 15.5 ms. The data displayed on Cayenne web-server dashboard also shows that reliable information pertaining the speed profile, speed, throttle pedal, clutch position can be obtained accurately in real time.

Test trials to retrofit the driving assistance system inside the tested car resulted the position and reached distance of the push rod as shown in Figure 6. The fixation of the hardware platform is designed based on finite element analysis to study the deformation of the platform when a force is applied axially at the end of the push rod (which is attached to the foot pedal) and the hardware is fixed at different locations along the hardware frame that are aligned with the existing fixations of driver seat. Experiment shows that if at standby condition the position of push rod is 0°, so at midway is 6° and at the end point is 52°. Whereas the reach distance for the push rod in the pushing process is 0 mm, 105.2 mm and 210.4 mm at the standby, mid-point and end-point, respectively.

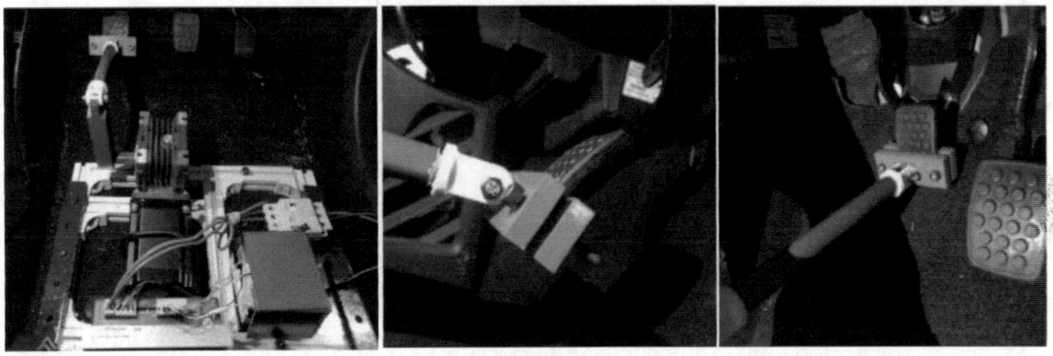

(a) (b) (c)

Figure 6. Retrofitting of driving assistance system (a) for clutch control with (b) initial position and (c) end position when the clutch is fully engaged.

4. Conclusion

In this paper a conceptual framework to implement a smart real-time driving assistance system for clutch pedal control using MQTT communication protocol, python, Raspberry Pi 4 and push rod linked to geared stepper motor for vehicles with manual transmission powertrain under chassis dynamometer test. The results show that MQTT communication protocol for MQTT-clients in different continents communication, namely Asia and Europe can be realised with high latency (< 1 s). The retrofitting of the system for use in a manual transmission power train vehicle for emission study or engine performance assessment is effortlessly being done using the existing fixation for the driver seat in the vehicle. In the future, motorised system for accelerator can be utilized to achieve fully IoT driving assistance system for vehicle under test on chassis dynamometer. In addition, real-time implementation of MQTT for hardware, software and On-board diagnostic (OBD) can be done to justify the system's efficiency.

Acknowledgements

The authors are grateful to the Baden Württemberg Stipendium Plus and Universiti Malaysia Pahang, Malaysia (www.ump.edu.my) for financial supports given under RDU192702, UIC191513, PGRS190305 and iTOPS grant (https://www.bw-stipendium.de/en/the-program/news/current-article/german-malaysian-teamwork/). The authors are indebted to Danial Haris Limi Hawari, Abdul Fattah Mohamad, Nur Hanna Khairul Anuar, Arzielah Ashiqin Alwi, Johannes Neckel, Jessica Schmidt, Carsten Pokluda and Jannik Lassen for the system fabrication and constructive discussion. Furthermore, the authors would like to express the gratitude to the experts in Faculty of Manufacturing and Mechatronics Engineering Technology (FTKPM), Faculty of Mechanical and Automotive Engineering Technology (FTKMA) and Automotive Engineering Centre (AEC) for the technical supports in this project.

References

[1] L. Chen, Z. Wang, S. Liu, and L. Qu, "Using a chassis dynamometer to determine the influencing factors for the emissions of Euro VI vehicles," *Transp. Res. Part D Transp. Environ.*, vol. **65**, pp. 564–573, 2018, doi: https://doi.org/10.1016/j.trd.2018.09.022.

[2] C. Huang and Zhou L, "Evaluation of emission factors for light-duty gasoline vehicles based on chassis dynamometer and tunnel studies in Shanghai, China," *Atmos. Environ.*, vol. **169**, pp. 193–203, 2017, doi: https://doi.org/10.1016/j.atmosenv.2017.09.020.

[3] G. Triantafyllopoulos, A. Dimaratos, L. Ntziachristos, Y. Bernard, J. Dornoff, and Z. Samaras, "A study on the CO2 and NOx emissions performance of Euro 6 diesel vehicles under various chassis dynamometer and on-road conditions including latest regulatory provisions," *Sci. Total Environ.*, vol. **666**, pp. 337–346, 2019, doi: https://doi.org/10.1016/j.scitotenv.2019.02.144.

[4] N. Euler-Rolle, C. H. Mayr, I. Škrjanc, S. Jakubek, and G. Karer, "Automated vehicle driveaway with a manual dry clutch on chassis dynamometers: Efficient identification and decoupling control," *ISA Trans.*, vol. **98**, pp. 237–250, 2020, doi: https://doi.org/10.1016/j.isatra.2019.08.021.

[5] K. Shen, long Chang, H. Chen, Z. Zhang, B. Wang, and Y. Wang, "Experimental study on the effects of exhaust heat recovery system (EHRS) on vehicle fuel economy and emissions under cold start new European driving cycle (NEDC)," *Energy Convers. Manag.*, vol. **197**, p. 111893, 2019, doi: https://doi.org/10.1016/j.enconman.2019.111893.

[6] E. S. Mohamed, "Development and performance analysis of a TEG system using exhaust recovery for a light diesel vehicle with assessment of fuel economy and emissions," *Appl. Therm. Eng.*, vol. **147**, pp. 661–674, 2019, doi: https://doi.org/10.1016/j.applthermaleng.2018.10.100.

[7] C.-W. Hong and T.-W. Shio, "Fuzzy control strategy design for an autopilot on automobile chassis dynamometer test stands," *Mechatronics*, vol. **6**, no. 5, pp. 537–555, 1996, doi: https://doi.org/10.1016/0957-4158(96)00010-4.

[8] X. Liu, T. Zhang, N. Hu, P. Zhang, and Y. Zhang, "The method of Internet of Things access and network communication based on MQTT," *Comput. Commun.*, vol. **153**, pp. 169–176, 2020, doi: https://doi.org/10.1016/j.comcom.2020.01.044.

[9] A. Mukherjee, N. Dey, and D. De, "EdgeDrone: QoS aware MQTT middleware for mobile edge computing in opportunistic Internet of Drone Things," *Comput. Commun.*, vol. **152**, pp. 93–108, 2020, doi: https://doi.org/10.1016/j.comcom.2020.01.039.

Design optimization of passenger SUV's crash box and bumper beam by using finite element method

Ahmad Yunus Nasution[1,2]* , Mohd Ruzaimi Mat Rejab[2] , Quanjin Ma[2] , Mohamad Ardy Firmansyah[1]

[1] Mechanical Engineering Department, Engineering Faculty, Universitas Muhammadiyah Jakarta, Cempaka Putih, 10510, Indonesia
[2] Faculty of Mechanical and Automotive Engineering Technology, University Malaysia Pahang, 26600 Pekan, Pahang, Malaysia

*Email: ahmad.yunus@ftumj.ac.id

Abstract. The accident cases with front crash type occupy the largest data statistics with 7,372 cases. Theoretically, the accident cases which involve passenger cars, kinetic energy is absorbed by the complex system. Some components which are included in the system are crash box and bumper beam. The main purpose of this research is to obtain the absorption of kinetic energy when the accident happened, types of deformation, and optimization in the existing designs of crash box and bumper beam from vehicle unit. Finite element method combines with analytical value are used in the simulation. Whereas, the software used in solid modelling is SolidWork and the numerical analysis used in this research is Abaqus / Explicit. The average reaction force through simulation is obtained by averaging the results of curve plotting, while the average reaction force is obtained through formula analysis by taking material property and dimension data and then inputting it in the calculation. From the simulation, energy absorbed is 9,912 Joule from the whole original structure. The energy absorbed is less than the crash box work which is 14,066 Joule within an error value of 22 %. This is caused by the bending moment which is emerged by bumper beam. Then, optimization is done by increasing lateral lengths of bumper beam with 20 mm, 15 mm and 10 mm, therefore energy absorption increased with 20,362 Joule, 31,886 Joule and 16,348 Joule, respectively.

1.Introduction

Motor vehicle is a means of road transportation which is most popularly used by Indonesian people. In its operation, passenger vehicle also has risk of accident, several cases of accidents such as head-on collision, side collision, single-vehicle accident and so on. Head-on collision cases are on the highest position with 7,372 cases from the total of 27,265 cases [1]. When head-on collision happens, there is energy absorption management which involves a component called crash box and bumper beam whose function is to absorb the energy from the impact when a head collision happens. In the Asean NCAP 2017 [2], vehicle got the predicate of a 5-star car. With the result of good and adequate for the driver and good for the passenger next to driver [3]. In this research, the author will try to present crashworthiness analysis specifically for crash box and bumper beam part with finite element method also the possibility for design optimization of those components.

When there is a front crash there is energy absorption management, one of which involves components called crash boxes and bumper beams. The installation of crash boxes and bumper beams on vehicles is part of passive safety. Crash box and bumper beam is a passenger car part located behind the plastic bumper and in front of the front rail/chassis. Tasked with receiving impact energy in the event of a front crash. Crash boxes and bumper beams are not alone in absorbing hit energy, but assisted by other components such as chassis, front body and so on [4].

Characteristics of energy absorption hit on the crash box can be said to be unique, because it absorbs energy by buckling or bending lengthwise in the direction of the elongated axis of the vehicle. Research on Crashworthiness has grown rapidly in recent years in Indonesia, most conducted by educational institutions such as high-profile institutions [5]. Research on Crashworthiness requires complex facilities and infrastructure and requires a large cost investment, can be done by physical testing by reconciling actual accident cases and can also be done by numerical approach method. Crashworthiness is intended to ensure passenger safety in the event of a collision can also be intended for vehicle certification activities. In this study, the authors present a special crashworthiness analysis on the crash box and bumper beam with element method up to or finite element method and possible design optimization on the component.

Energy Absorption Theory is energy given at the time of an accident where type of mechanical energy is more precisely kinetic energy, and kinetic energy is energy stored in an object with the mass of M at the speed of v. When two structures collide, the stored energy must be transformed into another form. Another form in question is a change in the shape or deformation of the structure of the vehicle. Deformation can be in the form of long inactivity or shortening of a structure. The increase in length when the impact load attracts a structure and shortens when the load is the impact suppresses the structure. The equation below is a form of energy immortality, in which kinetic energy is transformed into structure shortening (δ^2_{max}) [6]. Where the energy or work is defined as in the formula below:

$$\frac{Mv^2}{2} = \frac{EA\delta^2_{max}}{2L} \quad (1)$$

To get δ^2_{max} the formula is simplified to:

$$\delta_{max} = \sqrt{\frac{Mv^2L}{EA}} \quad (2)$$

The formula is applied when the object or test specimen is in the form of a rod or does not have a hollow. While in Structural Impact, in the structure of crash boxes square type/square has a bending value of plastic moments per unit length of

$$M_0 = \frac{\sigma_0 H^2}{4} \quad (3)$$

Where :
M_0 = Bending moment of plastic per unit length
H = Crash box thickness (mm)
σ_0 = Yield stress (MPa)

$$\frac{P_m}{M_0} = 52.22\left(\frac{C}{H}\right)^{1/3}\left\{1 + \left(\frac{0{,}33\,V_0}{CD}\right)^{1/q}\right\} \quad (4)$$

Pm = The average of force used to deform the crash box on a certain length
V_0 = The average length of crash box's sides (the median of crash box's length and breadth)
C = The speed when the collision happen
D and q : Predetermined constant of equation.

2. Materials and Research Method

The materials used in this study were crash boxes and bumper beam RE1 (4x2) M/T, with the name in the products catalogue is Beam Comp Front Bumper, part number 71130-SWA-000ZZ [7]. Crash box and bumper beam is a spare part assembly or assembled spare part. If it is decomposed, there are a total of 11 parts, 10 parts are symmetrical right and left parts and 1 connecting part is a bumper beam. The assembly of the 11 parts is using welding. Part assembled crash boxes and bumper beams are seen in the Figure 1. In the figure, there are two front and rear bumper arrangements, includes mountings, plastic body, number plate mounting and radiator mounting.

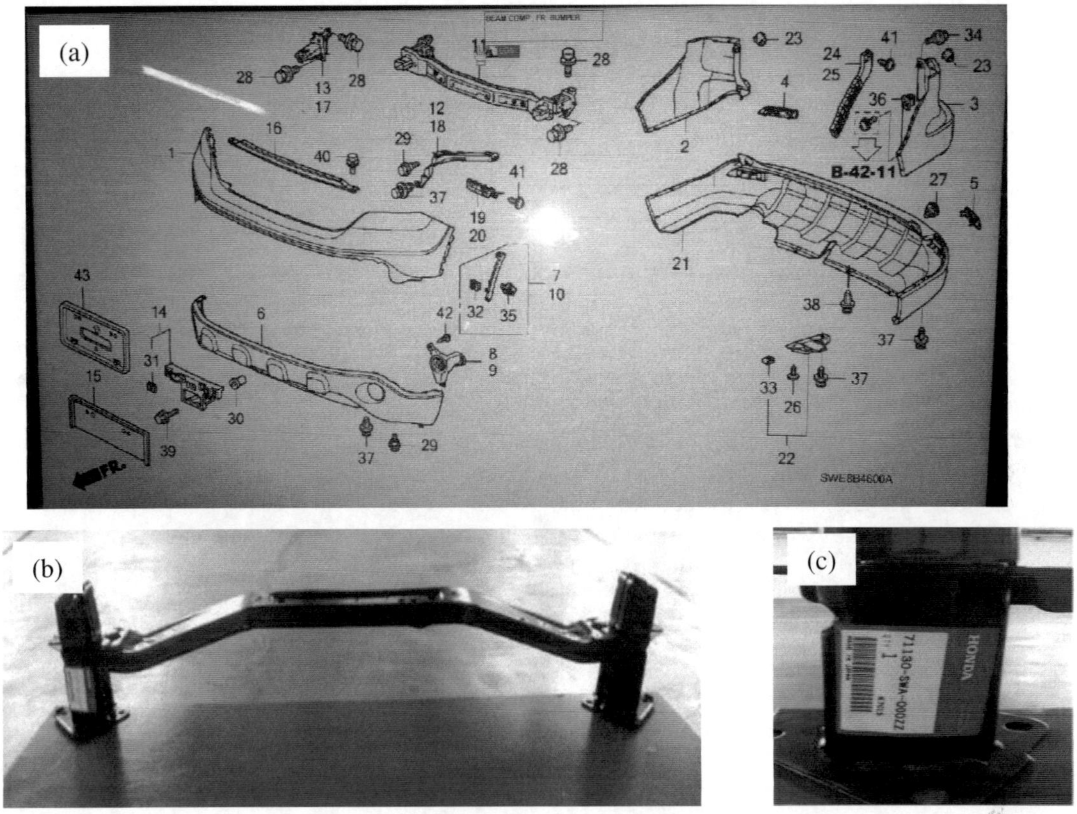

Figure 1. Crash box and bumper beam assembly : (a) assembly diagram; (b) assembly spare part; (c) spare part number.

The material data used for the structure of crash boxes and bumper beams are tabulated in the Table 2:

Table 1. Materials and thicknesses used in crash box and bumper beam [8]

No	Part Name	Materials	Thickness (mm)
1.	Bumper Beam	JSC 440W	1.4
2.	Front Crash Box Rear Crash Box	JSC 780Y	1.8
3.	Right Support Left Support	JSC 780Y	2.2

The data in Table 2 are property materials in materials used for bumper beam structure and crash box structure of JSC 440 W and JSC 780 Y.

Table 2. Mechanical properties JSC 440W and JSC 780Y

IOP Conf. Series: Materials Science and Engineering 1068 (2021) 012023 doi:10.1088/1757-899X/1068/1/012023

No	Material Properties JSC440W		Value	Unit
1.	Density		7800 7.8×10^{-9}	Kg/m^3 $Tonne/mm^3$
2.	Elasticity	Young's Modulus	270 2.7×10^5	GPa MPa
		Poisson's Ratio	0.3	-

No	Material Properties JSC 780Y		Value	Unit
1.	Density		7800 7.8×10^{-9}	Kg/m^3 $Tonne/mm^3$
2.	Elasticity	Young's Modulus	400 4×10^5	GPa MPa
		Poisson's Ratio	0.3	-

In this research the authors used several data collection techniques as mentioned in the points below. Data collection techniques used by the authors in the study are (see Figure 3):

1. Direct Observation Technique, this is done directly by the author when measuring dimensions and geometry on the research object of this final task, namely crash box and bumper beam.
2. Document Engineering is performed when the author searches for property material references.
3. Interview techniques will be conducted to the holder of vehicle brands that crash boxes and bumper beam beams will be analyzed in this research.

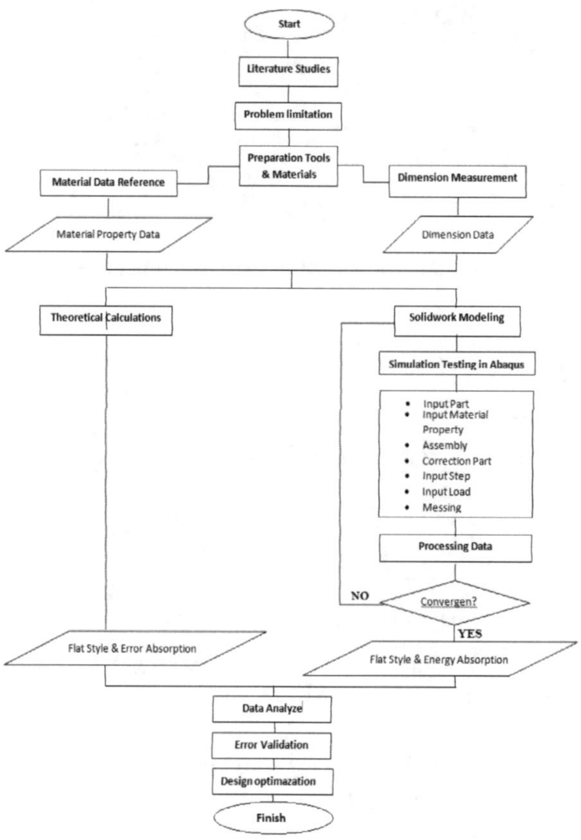

Figure 3. Research flowchart

While the data management techniques used by the authors in this research are by the use of number management software or commonly called Microsoft Excel and other programs needed in data processing. Technical implementation of data collection in this research is as follows:

1. Measurement of dimensions of crash boxes and bumper beams using dimension measuring instruments such as what has been described in the tool and material parts.
2. The shape of the crash box is getting smaller towards the front of the vehicle, causing the area of the cross-section is not the same, will be taken cross-sectional area and determined the area of the cross-section average, namely the middle value of both ends of the profile crash box.
3. Sampling displacement, initial speed, initial acceleration, reaction force and energy absorption are done by setting the research time step and iteration time and will be done automatically by CAE software. For example, we will examine a collision for 6 milliseconds, while the iteration we input is 50. Then sampling data will appear at 0 s, 1ms, 2ms, 3ms and so on up to a full time of 6ms.

3. Results and Discussion

The modelling conducted on Solidworks software obtained four components which were assembled in an assembly [9]. The assembly is as shown in the Figure 4 below:

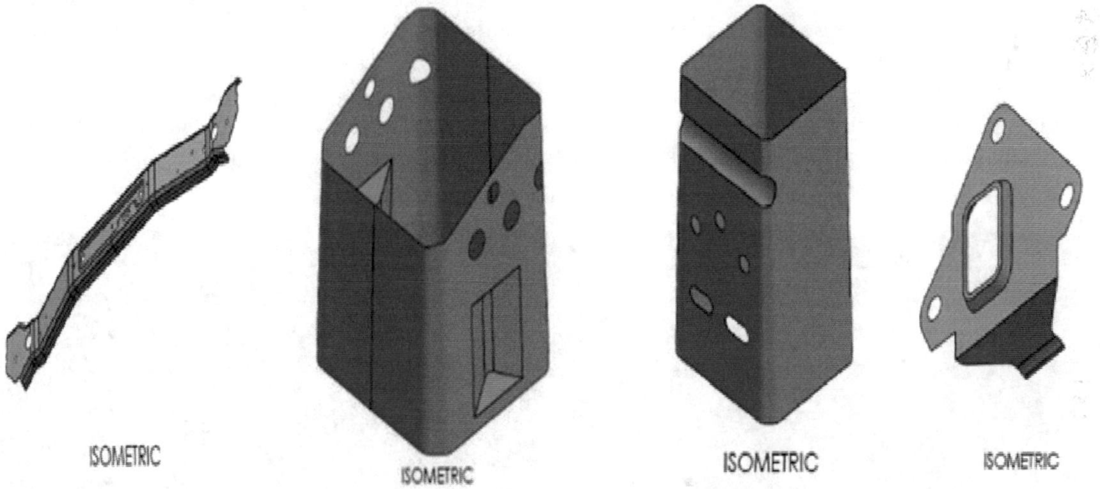

Figure 4. The result of Crash Box and Bumper Beam modelling by using Solidworks software.

The material used for Bumper Beam is JSC 440W with 1.4 mm thickness and for Crash Box is JSC 780Y with 1.8 mm thickness [10]. Results from the tensile test are shown in Figure 5.

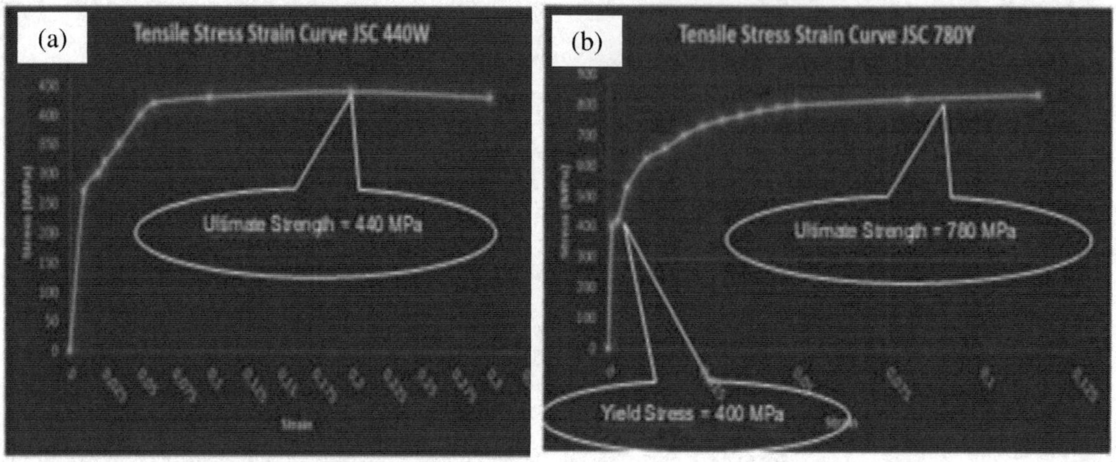

Figure 5. Tensile stress-strain curve: (a) material JSC 440W; (b) material JSC 780Y.

After the simulation was carried out on Abaqus/Explicit then it was found out that the total structural deformation was as far as 95.33 mm and average reactional force as many as 103,765 Newton. The visualization is as shown in the Figure 6.

Thus it is expected that the crushing energy absorbed by the whole structure of crash box and bumper beam as many as 9,912 Joule. It can be seen from the result of the simulation that the type of deformation that happened to the crash box was buckling type and the deformation that happen to the bumper beam was bending [11-15].

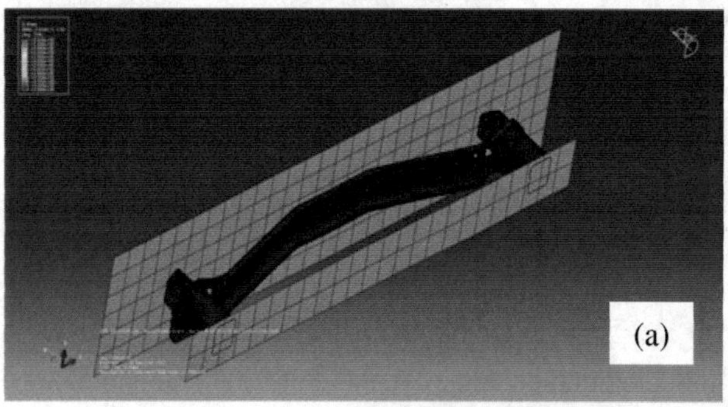

Initial Conditions Before Optimization Structure Deformation

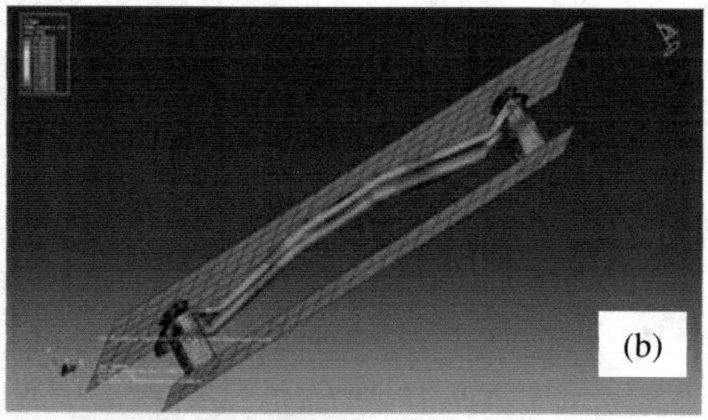

Current Condition of Deformation of Optimization Structure

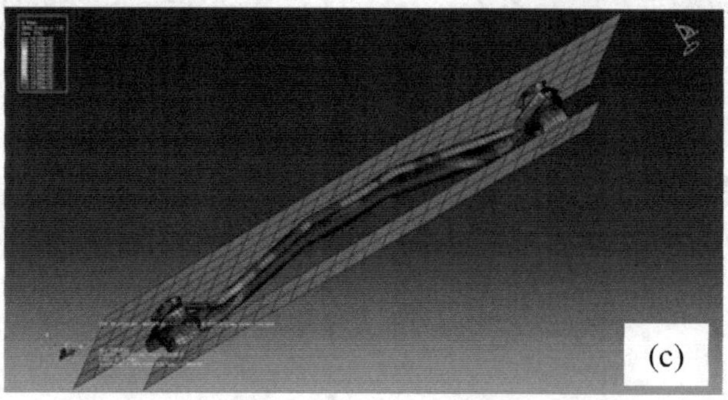

Final Condition After Optimization Structure Deformation

Figure 6. The visualization Crash Box and Bumper Beam by using Abaqus/Explicit : (a) condition before optimization; (b) current condition optimization; (c) after optimization

While the result of manual calculation from the work of the crash box only generated 14,066 Joule of absorbed crushing energy. This is considered as a handicap when the crash box is working on its own the performance of energy absorption is reduced by the bumper beam [16]. To find out the error value that happens, the author conducted a simulation test partially on the crash box. The result of the simulation indicated the amount of energy absorption of 18,260 Joule, with the error value of 22%.

The error value happens because the author omitted the crash initiator and stiffness which can be found in the crash box. Therefore in the analytic calculation by using the formula, crash box is considered as thin-walled square tube. The analysis can be seen briefly in Table 3.

Table 3. The result of bumper beam optimization variation

Analysis Type	Object of analysis	EA Performance (J)	Error Value
Simulation	Crash Box and Bumper Beam	9,912	Not Compared
Simulation	Crash Box	18,260	22 %
Calculation	Crash Box	14,066	

Next, the author conducted design optimization on the bumper beam. This optimization was based on the front-most point of the bumper beam in the original structure is a bit more backward compared to the structure crash box and bumper beam of other vehicles.

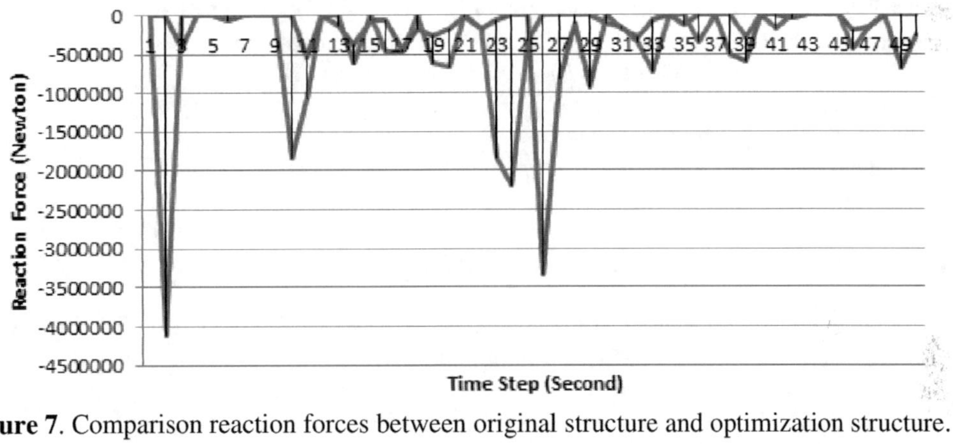

Figure 7. Comparison reaction forces between original structure and optimization structure.

The optimization was conducted with 3 extension variations which are 10 mm, 15 mm and 20 mm, with displacement variation and Mean Crushing Force variation. The optimization showed increasing in the value of crushing energy absorbed by the structure. The result is as shown in the Table 4 below.

Table 4. Simulation result on design optimization variation.

No	Optimization Variation (mm)	Mean Crushing Force (N)	Displacement (m)	Absorbed Kinetic Energy (J)
1	10	179,660	0.09099	16,348
2	15	350,636	0.09094	31,886
3	20	223,100	0.91270	20,362

The result of the above optimization indicated the increase in kinetic energy absorption in the case of head-on collision [17]. From the optimization data, it is concentrated that, by increasing the length of the bumper beam in the lateral direction of the vehicle will increase energy absorption. The previous 9,912 Joules rose to 16,348 Joules for 10 mm optimization, 31,886 Joules for 15 mm optimization and 20,362 Joules for 20 mm optimization.

4. Conclusion

The kinetic energy absorbed during the crash test simulation of the complete structure (crash box and bumper beam) is 9,912 Joules. A smaller value when compared to the kinetic energy that the crash box can absorb, which is 14,066 Joules. Validation the error, when compared to the crash box simulation test alone, is 18,260 Joules, which is 22%. The deformation characteristic of the crash box is buckling and what occurs in the bumper beam is bending. Optimization of the crash box and bumper beam structural design can be done by increasing the lateral length of the bumper beam by 10 mm, 15 mm and 20 mm. The value of kinetic energy obtained after optimization increases from 9,912 Joules to 16,348 Joules for optimization of 10 mm length, 31,886 Joules for optimization of 15 mm length and 20,362 Joules for optimization of 20 mm length.

Acknowledgements

The authors are grateful to the Universiti Malaysia Pahang for funding this research under grant PGRS1903159. This research work is strongly supported by Mechanical Engineering Department, Engineering Faculty, Universitas Muhammadiyah Jakarta, which provide research material and equipment.

References

[1] POLRI, K., 2017. *Statistik laka*. Available: http://korlantas.polri.go.id/statistik-2/
[2] ASEAN N.C.A.P., 2017. *Honda cr-v* (Bangkok : International Motor Show).
[3] BPLJSKB 2017 *Resume Hasil Uji Honda CRV* (Indonesia : Bekasi).
[4] Kalshetti, A.S., and Sanjaysingh, V.P., 2015. *Energy absorption of varying thickness rectangular section crash box for quasi-static axial loading,* pp 1–5.
[5] Darmawan, A., 2016. *Aanalisis uji tabrak bodi mobil esemka dengan metode elemen hingga*, pp 1-14
[6] Jones, N., 2012. *Structural impact*. 2nd ed. (Liverpool: Cambridge University Press).
[7] Dassault System., 2014. *Abaqus Documentation.* Available : Latest Release | Abaqus - Dassault Systèmes® (3ds.com)
[8] Reynold, C.W. and Henry,C.P., 1991.*Termodinamika teknik.* 2nd ed.(Arizona: Erlangga).
[9] Kullgren, A. and Anders, Y., 2013. *Frontal impact with small partial overlap*. vol 3 (International Journal of Crashworthiness) pp. 335 - 346
[10] Otake, R.Y., 2015. RIKEN *accelerator-driven compact neutron source RANS.*
[11] Gere, J.M., 2004. *Mechanics of materials*. 6th ed. Thomson Learning.
[12] Logan, L.D., 2012. *First course in the finite element method learning.*
[13] Raymond, J, and Kamoji, M, A., 2015. *Crash analysis for energy absorption of frontal rails of a passenger car.* (International Research Journal of Engineering and Technology vol. 2) pp. 9 - 16.
[14] Redi, B, Andik, A, and Imam, K., 2016. *Analisis penyerapan energi crash box pola origami pada pengujian frontal impact posiss angular frontal.* (Jurnal Rekayasa Mesin vol. 8) pp.47-52.
[14] Salicone, and Simona., 2018. *Measuring Uncertainty within the Theory of Evidence*. (Milan: Springer) pp.17-36.
[15] Tawaf, N. and Asroni., 2015. *Analisa deformasi crash box dengan variasi diameter dengan simulasi sofware ansys14.5.* Jurnal Ummetro, 2, pp. 6 - 14.
[16] UN ECE 2013 *Uniform Provisions Concerning the Approval of Vehicles with Regard to the Protection of the Occupants in the Event of a Frontal Collision* (United State).
[17] Quanjin, M., Salim, M.S.A., Rejab, M.R.M., Bernhardi, O.E. and Nasution, A.Y., *2020. Quasi-static crushing response of square hybrid carbon/aramid tube for automotive crash box application.* Materials Today: Proceedings, 27, pp.683-690.

IOP Conf. Series: Materials Science and Engineering 1068 (2021) 012024 doi:10.1088/1757-899X/1068/1/012024

Solar PV Microgrids Implementation model: A case study of Local Self Governments in the Indian State of Kerala

Ajithgopi[1,2], K. Sudhakar[3,4, *], Ngui Wai Keng [3]

[1]Renewable Energy and Environmental Engineering Research Group, Universiti of Malaysia Pahang, Malaysia

[2]Agency for New and Renewable Energy Research and Technology (ANERT), Thiruvananthapuram, India

[3] Faculty of Mechanical and Automotive Engineering Technology, Universiti Malaysia Pahang, Pekan 26600, Malaysia

[4]Automotive Engineering Centre, Universiti Malaysia Pahang, Pekan 26600, Pahang Malaysia

*Email: sudhakar@ump.edu.my

Abstract. The State of Kerala in Southern part of India has significant potential for generation of power from renewable energy sources, especially solar energy. Most of the PV projects in the State are being implemented on roof-tops due to the unavailability of land area in the densely populated State for large utility scale PV power plants. The concept of implementing Solar PV projects by empowering Local Self Governments has been explained in this paper by illustrating the case study of the Indian State of Kerala. The respective Local Self Governments like Municipal Corporations are playing a key role in Kerala in this decision-making process to accomplish model carbon free solar communities by installing Solar Photovoltaic Projects. The requirement of local beneficiary will be analysed by local trained experts and feasibility study will be conducted for the beneficiary communities involving residential buildings, local industries and commercial institutions and educational institutions and the PV projects will be implemented utilising the local E.P.C players. Local Self Governments like Municipal Corporations were empowered to plan, formulate and implement their own Solar Photovoltaic Projects. This approach is being implemented in Kerala resulting in solar electrification of local communities/institutions through the decentralised approach. This created a new business model at the local level involving trained manpower and supply chain for meeting the Local Self Government targets for new PV projects in order to achieve the targets of carbon free communities.

1.Introduction

Kerala is a southern State in India mostly known as "God's Own Country" due to its scenic beauty. The State is also known for its specific model of development with incredible performance in health and education sectors even comparable with developed nations. After the 73rd and 74th amendments to the Constitution of India, the Local Self Government Institutions were strengthened in the State to implement projects in a decentralized fashion. These institutions include Municipal Corporations, District Panchayats, Block Panchayats and Grama Panchayats. All Local Self Governments are formulating their projects in all sectors based on the needs and requirement raised by the people's representatives. Recently renewable energy projects, especially PV projects take a major share in energy projects of Local Self Governments in Kerala because of the following main reasons:

1. All Local Self Governments are working for a self-sustainable model of generating power of their own.
2. Net metering policy is in place in the State which enables the utility to accept the generated power.
3. It makes sense for generating solar power since it has reached grid parity.

All Local Self Governments in Kerala are empowered to generate, transmit and distribute energy according to the above Constitution amendment. This paper describes the new strategy of implementing solar photovoltaic projects with local participation through Local Self Governments based on the experiences in Indian State of Kerala. Most of the projects are being implemented on shade-free roof-tops due to the unavailability of land area in the densely populated State for large utility scale PV power Plants.

2. Power Sector in Kerala

The State has been buying and importing power from outside the State to meet the increasing demand. More than 70% of power is being imported from the Central Grid. Electricity generation in Kerala is mostly from hydroelectric power stations (73.41%) as demonstrate [1]. The board has also set up several hydel power stations within the state. Because of deforestation and environmental concerns, further large hydroelectric projects could not be implemented. The State of Kerala has a total installed capacity of 2980.086MW, out of which 2089.31MW is from hydroelectric power plants, 710.67MW from thermal power plants, 110.831MW from solar PV power plants (both utility and roof-top), 59.275MW from wind and 10MW from co-generation. Pazheri et al. described the status of electricity generation in Kerala [2]. Electricity generation in Kerala from various resources is illustrated (Figure 1)

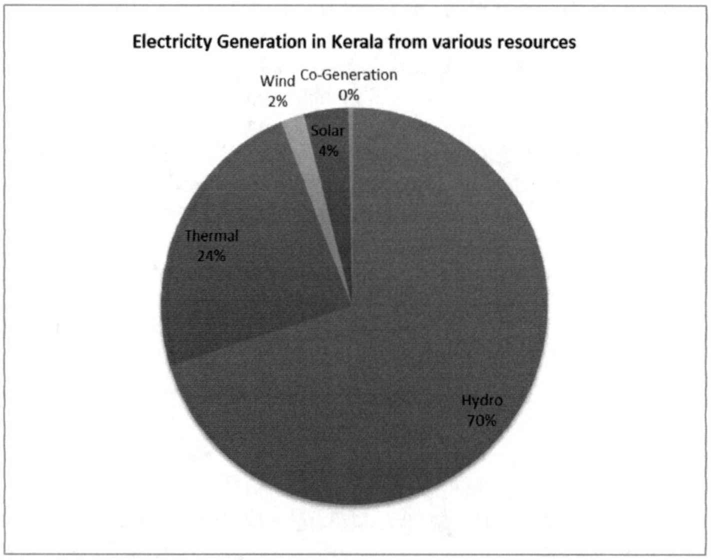

Figure 1: Electricity Generation in Kerala from various sources (Source – www.kseb.in)

Since only 30% of the power is being met from the internal generation, it is natural for the State to tap the abundant renewable energy sources to the maximum to meet it energy demands. There is a huge scope for avoiding various pollutants by utilising the significant potential of natural resources in Kerala, as described by Pazheri & Baby [2].

3. Solar Resources in Kerala

Kerala has significant potential for generation of power from renewable energy sources, especially solar energy. Except for June and July, the remaining months have significant number of clear sunny days. Interestingly, radiation analysis utilizing software (Meteonorm) also supports the view of local experts. Kerala is blessed with Solar Energy with 5 to 6 kWh/m^2 Solar Radiation per Day. Solar Radiation analysis utilizing software (Meteonorm) gives good results in the State of Kerala, which is a positive indication for achieving the energy security in the State using solar resources. The values of Global Horizontal Irradiance (GHI) and Direct Normal Irradiance (DNI) for each district of Kerala are reproduced below (Table 1)

Table. 1: Solar Radiation (DNI & GHI) in different Districts in the State (Source: The Energy Report Kerala, WWF India & WISE 2013)

Sl. No.	District	DNI (kWh/m^2/Day)	GHI (kWh/m^2/Day)
1	Alappuzha	4.29	5.52
2	Kannur	4.47	5.43
3	Ernakulum	4.36	5.44
4	Idukki	4.52	5.44
5	Kasargod	4.77	5.54
6	Kollam	4.27	5.50
7	Kottayam	4.22	5.45
8	Kozhikode	4.50	5.47
9	Malappuram	4.70	5.49
10	Palakkad	4.55	5.51
11	Pathanamthitta	4.52	5.62
12	Thrissur	4.60	5.53
13	Thiruvananthapuram	4.34	5.52
14	Wayanad	4.68	5.33
	Average	4.49	5.49

The Global Horizontal Irradiation (GHI) daily average value in the state is 5.49 kWh/m$_2$/day, or an annual value of 2,003 kWh/m2/year based on Meteonorm Data. As a general principle, any site with GHI more than 1,500 kWh/m2/year is suitable for solar PV technology. The State of Kerala is getting an average of 300 sunny days a year. Thus, a simple resource-based assessment indicates that Kerala is highly suitable for developing solar PV systems. The above fact is also evident from the Solar Resource Map of India prepared by NREL (Figure 2). Kerala gets an annual average solar insolation of 5.59 KWh/m^2 /day which indicates very good solar potential in the State. The districts wise average annual insolation is discussed by Ramachandra [3].

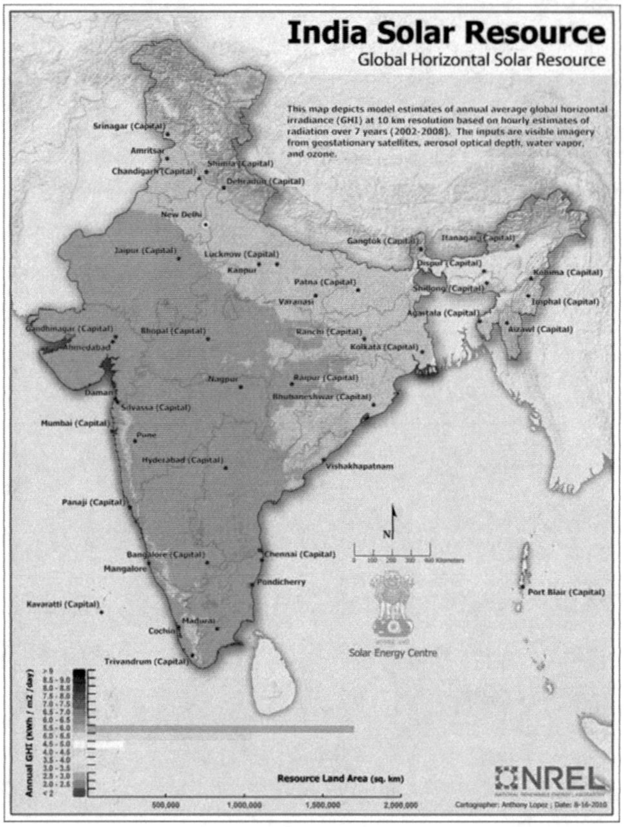

Figure 2: Solar Resource Map (Source – NREL)

4. Traditional Practices of PV Project Implementation

PV Projects were traditionally implemented based on the requirement from the community/ beneficiaries and the implementation was managed by ANERT, the State Nodal Agency of Renewable Energy under the Department of Power, through tender practices. There are various risks related with solar projects comes under the categories of political, financial, social and technical risks [3].

4.1 Role of ANERT

Agency for Non-Conventional Energy and Rural Technology (ANERT), is the autonomous body constituted under the Department of Power, Govt. of Kerala for the development and promotion of utilization of the renewable energy sources in the State. ANERT is also the State level Nodal Agency (SNA) of Central Ministry of New and Renewable Energy (MNRE). ANERT had been established in 1986 by the Government of Kerala, as an autonomous institution under the Science, Technology and Environment Committee (STEC) with the objectives of gathering and disseminating useful knowledge in various fields of Non–Conventional Energy, Energy Conservation and Rural Technology. ANERT, is currently under the Administrative Control of the Power Department of the Govt. of Kerala, implementing Renewable Energy Projects in the State (Figure 3).

Figure 3: A typical Remote Village Electrification Project

ANERT has implemented many successful solar projects in the last decades. The projects include Remote Village Electrification Programme (RVEP), 10000 solar roof-top programmes, etc. ANERT has successfully implemented programmes of the Central Ministry (MNRE – Ministry of New and Renewable Energy) like the dissemination of Solar Lanterns, Solar Home Systems and Solar Street Lighting Systems, etc. ANERT has recently come up with a Grid Connected Solar Programme for residential buildings in the State called "Solar Connect" with both Central and State financial assistance. The off-grid solar programme for the residential buildings in the State is called, "Solar Smart". ANERT has become the first solar Independent Power Producer (IPP) in the State by installing a 2MW Utility scale PV plant at Kuzhalmannam, Palakkad District in Kerala (Figure 4)

Figure 4: ANERT 2MW Kuzhalmannam PV power plant

ANERT is also positioned as a Technical Consultant in the State for helping the beneficiaries including various Govt and private institutions to install Solar PV Power Plants in their premises through its Deposit Work Scheme and Technical Consultancy Scheme. ANERT has already installed Grid Connected and Off-Grid PV Power Plants with a total capacity of more than 1MW under these schemes. 100kW Grid Connected PV Power Plant installed at the Kariavattom Campus of the Kerala University (Figure 5) and 45kW Off-Grid PV Power Plant installed at Cherthala Taluk Hospital are some examples

[4].

Figure 5: 100kWp Solar PV power plant at Kerala University Campus

4.2. Role of KSEB Ltd. and CIAL

Kerala State Electricity Board Limited (KSEBL), the major Power Utility in the State, has also taken up PV Power Plant projects in the State. KSEB has installed a 1MW Grid Connected PV Power Plant Project at Kanjikkode, Palakkad District. The biggest Solar farm installed in Kerala so far is the 30MW Plant at Cochin Airport premises by Cochin International Airport Limited (CIAL). This is the first Airport in the world to be completely powered on solar power (Figure 6) [5], [6]

Figure 6: 30MWp Solar PV power plant at Cochin Airport

5. Proposed New Strategy of Implementation

Solar Projects in communities can be implemented in a decentralized fashion by empowering Local Self Governments in planning and formulating the projects after feasibility studies. The Local Self Governments will include Solar PV projects in their Plan Document and implements projects with the help of suitable Consultants and E.P.C Companies. The warrantee and Annual Maintenance Contracts can be also arrived with the EPC companies for making the project sustainable. The local utility will accept the generated power based on the guidelines issued by the State regulatory commission. The

State Nodal Agency of the Central Ministry, will technically assess the PV Plant in communities during installation and commissioning.

5.1. Local Self Governments as the project developers

After the 73rd and 74th amendments to the Constitution of India, the Local Self Government Institutions were strengthened in the State to implement projects in a decentralized fashion as indicated (1). These institutions include Municipal Corporations, District Panchayats, Block panchayats and Grama Panchayats. In every year, all Local Self Governments in Kerala shall formulate their Plan Document after several brain storming sessions at the local level participating local people, experts, people's representatives, Government officials, NGOs etc. Based on their deliberations, a number of energy projects including renewable energy projects shall be formulated. Local trained experts will help the Local Self Governments to plan and formulate the PV projects based on the requirements based on the basic feasibility studies at the respective sites. These project reports will be evaluated at the district level after assessing the ground realities. One of the example for the above, is that the local Panchayat (local governance bodies) and citizen political dissent blocked the development of full-scale 200 MW Kasaragod solar park, which significantly reduced the overall land available for the initiative [7].

5.2. New approach

The decision will be made at the local level in this new approach. The respective Local Self Governments will play a key role in this decision-making process to accomplish model carbon free solar communities by installing solar photovoltaic projects. The requirement of local communities and beneficiaries will be analysed by local trained experts and feasibility study will be conducted for the beneficiary communities involving residential buildings, local industries, commercial institutions and educational institutions. Majorly, two major strategies are adopted for these feasibility studies:

- Identifying the remote communities where the conventional electricity could not be extended
- Detailed mapping of residential buildings, commercial establishments, educational institutions, religious places and industries within the boundary of Local Self Governments for finding the feasibility of installing either off-grid or grid connected PV power plants.

5.3. Solar Village Electrification of Remote Colonies

Based on the feasibility studies conducted on remote colonies, survey and detailed study has been conducted in remote villages of Kerala with the help of Local Self Governments in different districts. Mahapatra et al. discussed smart village models which has to be settled and this includes the availability of grid infrastructure, the location economic distance limit, the operating hours of the renewable energy systems, the life cycle cost of the energy generation equipment, customer engagement issues as well as the micropower system business strategy [8]

The remote villages have been identified and the following strategies were adopted for the solar electrification of remote colonies:

- Solar Home Systems are proposed for the households in remote colonies which are scattered in nature.
- Centralized Stand-alone PV plants or micro grids are proposed for the colonies which are clustered in nature

Beneficiary Societies have been constituted in all the colonies selected for solar electrification. All beneficiaries in each colony is a member of this society. The concerned Local Self Government President will be the Chairman of the beneficiary society. Selected members of the society will be

IOP Conf. Series: Materials Science and Engineering 1068 (2021) 012024 doi:10.1088/1757-899X/1068/1/012024

trained both by the EPC contractor and ANERT for topping up the batteries, cleaning the solar modules and timely reporting of faults. These members contribute a fixed amount per month to the society which is kept in a separate Bank account by the beneficiary society. This amount will be used for maintain the PV plant. Also, the concerned Local Self Governments are also financially helping these societies for maintenance and repair of the PV plants. Two representatives from the colony were trained for the operation maintenance of the SPV system by the EPC contractor [9].

5.4. Stand-alone or Micro Grids for remote tribal settlements

Remote tribal settlements are usually located in the midst of dense forests in Kerala and hence conventional grid lines cannot be extended due to the existing forest regulations. ANERT had identified such 11 numbers of remote tribal colonies in the State with the help of Local Self Governments. Local Self Governments had allocated funds for solar electrification of these tribal settlements in their budget based on the project reports of ANERT. Solar electrification of these tribal settlements had been conducted by installing PV micro grids. A typical PV micro grid will have a solar array in an open shade free area, a control room for placing the solar inverter/inverters and batteries. The list of PV Micro Grids installed in remote tribal colonies are given in Table 2.

Table. 2: PV Micro Grids installed in remote Tribal Colonies of Kerala as part of Kerala through Local Self Governments (Source: ANERT)

Sl. No	Name of Colony	Name of Grama Panchayat	Name of Block Panchayat	Name of District	No.of families	Capacity of the Micro Grid installed (kW_p)
1	Plamalakkudy	Adimali	Adimali	IDKI	**33**	3.36
2	Chambakkad	Marayoor	Devikulam	IDKI	**68**	5.60
3	Mavalikudi(Thayanamkudi)	Marayoor	Devikulam	IDKI	**20**	2.52
4	Kozhiyilakkudy	Munnar	Devikulam	IDKI	**48**	4.20
5	Mundakkadavu	Karulayi	Nilamboor	MPM	**30**	3.08
6	Nedumkayam	Karulayi	Nilamboor	MPM	**41**	3.92
7	Moolaganga	Sholayoor	Attappadi	PKD	**36**	3.64
8	Vellakulam	Sholayoor	Attappadi	PKD	**32**	3.36
9	Vechumaram	Athirappally	Thrissur	TCR	**33**	3.36
10	AnapandamST colony	Mattathoor	Kodakara	TCR	**50**	4.48
11	Kokkathode	Aruvappulam	Konny	PTA	**45**	2.52
			TOTAL		**436**	**40.04**

5.5. Solar Power Roof-top Power Plants for Residential Buildings and Institutions

Solar Roof-top Power Plants are becoming popular in the State since the cost of PV generation has almost reached grid parity. Though residential buildings started this trend initially, industrial &

IOP Conf. Series: Materials Science and Engineering 1068 (2021) 012024 doi:10.1088/1757-899X/1068/1/012024

commercial establishments, hospitals and educational institutions are adapting the technology in a faster pace presumably because of their higher tariffs of conventional energy use. Local Self Governments are allocating funds for establishing Solar Power Plants ANERT has initially implemented this programme through its Deposit Work and Consultancy schemes. All these projects are implemented with the financial assistance of Local Self Governments. Local Self Governments will allocate sufficient funding for solar electrification in their Plan Documents and the projects will be implemented with the help locally trained solar installers. The Local Self Governments has a pool of trained technicians as installers. The Operations and Maintenance (O&M) of the projects will be taken up by the EPC companies during the warrantee period. Once the warrantee period is completed, new Annual Maintenance Contract (AMC) for the upkeep of the PV power plants can be taken up with EPC companies. A list of Solar Power Plants (both Grid Connected and Off-Grid) implemented with the help of Local Self Government is given below in Table 3.

Table 3: List of Roof-top PV Power Plants installed with the help of Local Self Governments in the State of Kerala (Source: ANERT)

Solar Power Plants Commissioned - Deposit Work and Consultancy Scheme

I. Deposit Work Scheme

Sl No	District	Name of LSG/Dept/PSU	Capacity (kWp)	Offgrid/Grid connected
1	Kollam	Thrikkadavoor Grama Panchayath	3	offgrid
		Kulasekharapuram Grama Panchayath	2	offgrid
2	Pathanamthitta	Kalanjoor Grama Panchayath	2	offgrid
3	Idukki	Kumily Grama Panchayath	6	offgrid
4	Thrissur	Thrikkur Grama Panchayath	3	offgrid
		Govt. Mental Health Centre	5	offgrid
		Mathilakam Block Panchayath	5	offgrid
		Pig Production & Research Centre	4	offgrid
5	Kozhikode	Perambra Taluk Hospital	2	offgrid
		Chelannur Block Panchayath	4	offgrid
		CHC Narikuni	4	offgrid
		Govt. Homeopathic medical College	5	offgrid
6	Kannur	Kannur Block Panchayath	5	offgrid
		Aralam Wild Life Dormitory	2	offgrid
		Aralam Wild Life Division Office	2	offgrid

7	Wayanad	Govt. Polytechnic Meppady	3	offgrid
8	Kasaragod	Govt. College, Kasargod	10	offgrid
9	Eranakulam	Vengola Grama Panchayat	2	offgrid
10	Thiruvananthapuram	KSIHFW	15	offgrid
		CWRDM Neyyattinkara	4	offgrid
11	Alappuzha	Cherthala Taluk Hospital	45	offgrid
12	Thrissur	Irinjalakuda Municipality	25	offgrid
13	Kozhikode	Thamarassery Grama Panchayath	3	offgrid
14	Kannur	Kadannappally-Panappuzha Grama Panchayath	6	offgrid
15	Kasaragod	Pullur-Periya Grama Panchayath	4	offgrid
16	Thiruvananthapuram	University of Kerala	100	on grid
Total			**271**	
		2. Consultancy Scheme		
1	Kottayam	Assumption college, Changanassery	3	Offgrid
2	Thiruvananthapuram	Attingal Municipality	3	offgrid
3	Palakkad	Kozhinjanpara Grama Panchayath	4	Offgrid
4	Thrissur	Thrissur Corporation	200	ongrid
5	Thiruvananthapuram	Perumkadavila Block Panchayat-Anappara Community Health Centre	25	offgrid
6	Thiruvananthapuram	Technopark-Ganga&Yammuna Building	200	ongrid
7	Thrissur	Kerala Forest Research Institute (KFRI), Peechi	2	ongrid
Total			**437**	

6. Outcome of the Project in Kerala

This approach worked well in Kerala resulting in electrification of local communities / institutions through the decentralised approach. Local Self Governments like Municipal Corporations, District panchayats, Block panchayats and Grama Panchayats were empowered to formulate their own Solar Photovoltaic Projects according to the feasibility studies conducted utilising local trained technicians. It created a new business model at the local level involving trained manpower and supply chain for meeting the Local Self Government targets for new PV projects in order to achieve the targets of carbon free communities. A typical example is the 200kWp Grid Connected Solar Roof-top Project installed by Trissur Corporation as shown in Figure 7.

Figure 7: A 200kWp Solar Power Plant installed by Thrissur Corporation, a Local Self Government Institution in the State of Kerala

7. Conclusion

Based on a recent study of WWF and World Institute of Sustainable Energy (WISE), Pune, it is estimated that Kerala State has a domestic PV roof-top potential of 13079MW and industrial and commercial PV roof-top potential of 18066MW adding a total PV roof-top potential of 31145MW. Decentralization empowers the Local Self Governments and communities to take appropriate decisions at the right time with creativity and growth. All the Local Self Government Institutions will be trained for taking up this new challenge. The implementation of Solar PV Projects by through Local Self Governments is a unique model which can be implemented worldwide to make the projects sustainable. Local Governments like Municipal Corporations can thus play a crucial role in addressing global warming and climate change.

Acknowledgement

The authors are grateful to the financial support provided by the Universiti Malaysia Pahang (www.ump.edu.my) through the Post graduate Research Grant Scheme, PGRS 1903172. The authors acknowledge ANERT, State Renewable Energy Nodal Agency in Kerala for the information provided.

Conflict of interest statement

The authors declare that they have no known competing financial interests or personal relationships that could influence the work reported in the submitted manuscript. The opinion /findings/ discussions in the document are solely of the authors and do not necessarily reflect the opinion of any organization involved directly or indirectly. Hence the authors are not responsible for any consequences thereof with the use of information presented in this work.

Reference

1. Salim, M. (2016). Kerala-Gujarat Models : A Comparative Study With Respect To Socio-Economic Environment Assistant Professor of Economics. (7), 396–407.
2. Pazheri, F. R., & Baby, T. (2014). Rooftop Solar Projects Enhance the Environmental Friendly Power Dispatch in Kerala Saudi Aramco Chair in Electrical Power. 233–236.
3. Ramachandra, T. V. (2014). Scope for Solar Energy in Kerala and Karnataka. (November 2012).
4. B. Shiva Kumar and K. Sudhakar, "Performance evaluation of 10 MW grid connected solar photovoltaic power plant in India," Energy Reports, vol. 1, pp. 184–192, 2015, doi: 10.1016/j.egyr.2015.10.001
5. S. Sukumaran and K. Sudhakar, "Fully solar powered airport: A case study of Cochin International airport," J. Air Transp. Manag., vol. 62, pp. 176–188, 2017, doi: 10.1016/j.jairtraman.2017.04.004.
6. S. Sreenath, K. Sudhakar, and Y. Ahmad Fitri, "Airport-based photovoltaic applications," Prog. Photovoltaics Res. Appl., 2020, [Online]. Available: https://doi.org/10.1002/pip.3265
7. Bedi, H. P. (2019). " Lead the district into the light " : Solar energy infrastructure injustices in Kerala , India. Global Transitions, 1, 181–189. https://doi.org/10.1016/j.glt.2019.10.005
8. Mahapatra, S., & Dasappa, S. (2012). Energy for Sustainable Development Rural electri fi cation : Optimising the choice between decentralised renewable energy sources and grid extension. Energy for Sustainable Development, 16(2), 146–154. https://doi.org/10.1016/j.esd.2012.01.006

The Effect of Gain and Frequency on Eddy Current Testing for Copper Material Defect Inspection

W.Syahidah W.Mohd[1], Farraniza Attan[1], Damhuji Rifai[1], Kharudin Ali[1], M.Aizat M.Sulaiman[1], Nor Hana Mamat[1]

[1]Faculty of Electrical and Automation Engineering Technology, TATIUC, 24000 Terengganu, Malaysia. Tel./Fax +6-09-8601000/+6-09-8635863

*Email: syahidah@uctati.edu.my , farraniza@uctati.edu.my, damhuji@uctati.edu.my , kharudin@uctati.edu.my, aizatsulaiman@uctati.edu.my, norhana@uctati.edu.my

Abstract. Copper and the copper alloys are some of the most versatile materials available and used for applications in every type of industry with world consumption of exceeding 14 million tonnes per annum. In eddy current testing inspection (ECT), several forms of device calibration must be done before the assessment can be performed. The electronics equipment must be calibrated to ensure the accuracy of measurement. This paper is focused on investigate the optimum frequency and gain for copper material block. The material used to produce the sample test in this project was copper. They are many methods that involved in this project such as producing sample test and artificial defect, conductivity measure and perform calibration for copper material block. Two thickness of copper block were selected 8mm and 16 mm. Artificial defect with depth of defect 0.5 mm, 1mm and 2.0 mm are fabricate on the sample. Different frequency and gain of eddy current testing was used in order to identify the best setting for both parameters in ECT for copper material defect inspection. Experimental result showed that the thickness is directly proportional with the applied gain. From here the negative scanning will giving the good result of different slot of defect comparing the negative scanning. Besides the percentage of increment of signal amplitude being identify with increment 40% signal between 0.5mm and 1.0mm slot of defect. Lastly the depth of the slot is inversely proportional with applied gain.

1. Introduction

Various techniques are used in the assessment of defects in metal materials without damaging the metal structure and one of them is the Eddy Current Testing Technique (ECT). The use of ECT is not limited to identifying metal material damage only but it is also used to identify metal thickness, material and internal defect inspection and finally it is also used to measure electrical conductivity in heat treatment conditions for metal material [1,2]. This can be compared to other test methods where it will damage the physical and structure of the material while ECT on the other hand besides having high sensitivity for small defect measurement for rapid scan examination [3,4]. Additionally ECT can also be run without the need for contact between the surface of the metal material and the probe used and thus makes this technique suitable for monitoring disability issues[5]

A small change of the magnetic properties of the compositional material of steel or iron will to some extent will affect the eddy current testing. Therefore, low carbon steel substrates are widely used in this test by using a standard calibration and a set of standards used for the measurement of various different materials such as brackets or bolts, screws and washers [6].

Instrument calibration is an important step in eddy current inspection. The data collected from the inspection has to be calibrated for accurate defect analysis. Calibration acts as a form of compensation for variable instrument settings and probe variations[7], thus providing for a simpler comparison between measurements taken at different intervals. The process of calibration also provides for the development of simple calibration curves that map signal characteristics (such as phase angle or amplitude to flaw depth).

The physical properties of materials can be identified by using ECT in which it will identify the state of the material comprising the inner surface defects and subsurface defects, and the diversity of its thickness. Apart from that the conductivity and magnetic permeability properties can also be measured using ECT[8,9].

In this paper, test sample block of copper was produced with the depths of standard slot are 0.5mm, 1.0mm and 2.0 mm. Eddy current testing was performed within a frequency range of approximately 50Hz to 10MHz although most applications were performed well within the extremes of that range. Different gain parameters were tested in order to identify the best gain for copper ECT effect inspection.

2. Related Work

2.1. A Reference Standards

For each ECT test to be conducted, the calibration needs to be done using a reference standard (calibration block) and it was very important to obtain accurate values and readings from the instruments and probes used. It consists of a display of x and y (horizontal and vertical line) on the display screen where the clear observation that needs to be done to see changes in a signal testing. In the process of detecting defects, the standards reference consisting of materials, measurements and shapes similar to test equipment with the various types of artificial defects including drill holes, saw blades or milled walls to identify and simulate flaws[10].

To carry out the experimental process of thickness measurement, several samples as a reference standard were tested by knowing the actual thickness of the sample. The result of the signal obtained on the standard reference sample will be recorded and the recorded signal will be compared according to the thickness of the metal for the categorization process. To ensure that the inspection procedure during eddy is correct the appropriate reference standard is essential for proper calibration.

The same type of specimen material is essential during the standard calibration process. Multiple-dimensional defects including depth, width and length are used in measurement standards. It is important before the actual specimen is tested. This calibration method consists of scaling and rotating one or more to indicate the signal condition obtained. The application of the reference signal obtained on the calibration standard will be converted in the form of data and collected to compare the signal to the examination of the specimen tested [2]. Figure 1 depicts the rotation of a signal[11].

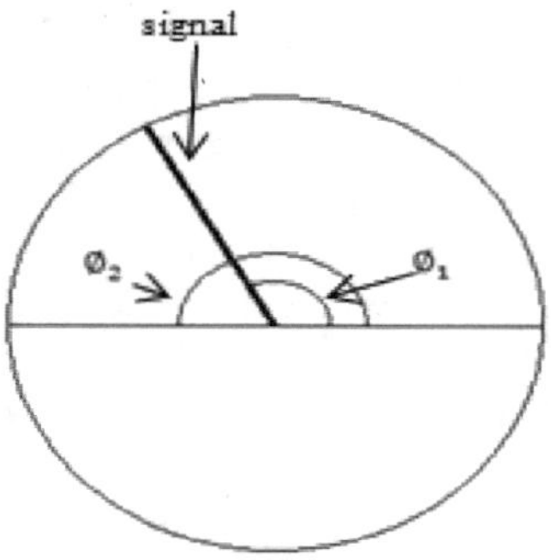

Figure.1 Rotation in Calibration

The rotation and scaling parameters are computed as follows. If $\emptyset_1$ is the angle of the signal and it has to be rotated to an angle $\emptyset_2$, then the rotation angle θ is given by:

$$\theta = \emptyset_2 - \emptyset_1 \quad (1)$$

The scale factor is determined as:

$$S = \frac{r_1}{r_2} \quad (2)$$

Where
S = scale factor
r_1 = desired peak to peak scaling of the signal
r_2 = original peak to peak value of the signal Eddy

2.2. Eddy Current Inspection Process

Eddy current inspection methods should follow these basic steps, namely:
• Selection of appropriate instruments and probes
• The selected frequency level according to the desired signal penetration depth
• Calibration standard according to the suitability of the instrument to obtain the response signal
• The surface of the probe should be placed on an empty plate surface without defects
• The movement pattern of the probe should be in the same direction over the area being inspected
• The probe orientation must be constant to avoid misinterpretation of the defect signal
• The resulting signal should be monitored to identify impedance changes resulting from defects on the plate surface. Eddy Current Testing Applications

Eddy current capability is used for various tests because it's widely used in industry. This can be seen based on the usual inspections made using ECT, namely:
• Welding inspection - ECT is used to scan for cracks on the surface of the test material for hot and covered areas by welding.
• Conduction test- Conductivity measurement can be made using Eddy current to identify alloy and non-ferrous iron to confirm heat treatment.
• Surface inspection - External cracks for metal stocks and machine parts can be detected using ECT. It is very important for critical parts including around the fasteners in the aircraft.
• Corrosion detection - Corrosion on metal parts can be measured using ECT. In addition, sensitive signals from ECT can be used for thin metals including aircraft skins that use aluminum. By using a low frequency, it will help in detecting corrosion on the inner metal layer which is layered in two layers where it cannot be achieved by ultrasonic examination.
• Bolt hole inspection - Cracks in the bolt hole can be identified by using a bolt hole probe with an automatic rotary scanner.
• Tube inspection - This inspection consists of tube field inspection such as heat exchanger and tube alignment inspection at the manufacturing stage for cracking and sample thickness.

3. Material and Method

3.1. Material

The design of the copper sample block is similar with the existing sample block for other material. The design specification for the copper sample test block is shown in Figure 2.

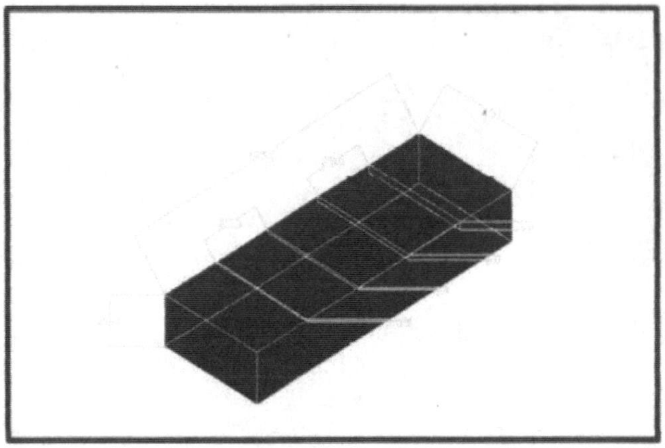

Figure 2. Design for the copper sample test block

The first step in producing standard sample blocks is the grinding surface. The grinding surface will produce a smooth finish on a flat surface. It uses a rough machining process in which a rotating wheel covered in a coarse particle (grinding wheel) of a metal or non-metallic material chip wound from a work-piece produces a flat and smooth face. There are three types of surface grinders such as Horizontal-spindle, Peripheral grinding and wheel-face grinding. However, this project only used the Horizontal-spindle (peripheral) grinder of the surface type, where the edge (flat edge) of the wheel is in contact with the work-piece, producing the surface. This is the important procedure to produce a smooth surface of the copper block as to perform the eddy current testing. Figure 3 shows the copper block surface before and after the surface grinding process.

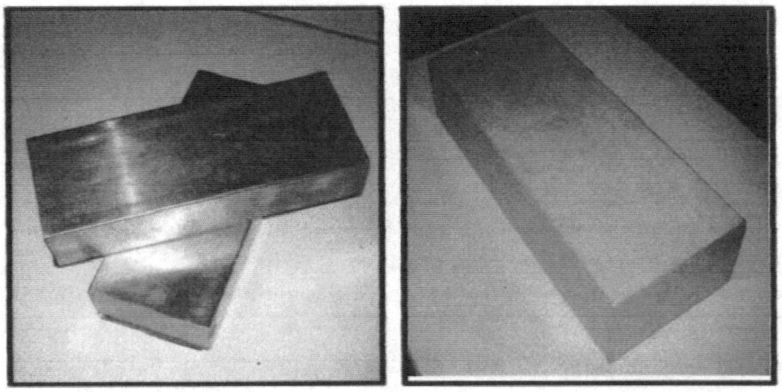

(a) (b)

Figure 3. Copper block; a) before surface grinding b) after surface grinding

The third step produces a standard sample block of manufacturing. the manufacturing process uses a rotary cutter to remove material from the work-piece advancing (or feeding) in that direction at an angle with the axis of the tool. It covers a variety of different operations and machines, on a scale from small individual parts to large, heavy-duty manufacturing operations. It is one of the most commonly used processes in the industry and machine shops today for machining parts for a variety of precise sizes and shapes.

The cut wire is the last step to produce a sample. This machine reduced the slot depth in 0.5 mm, 1.0 mm and 2.0 mm. Wire machined electric discharge (WEDM), is also known as wire-cut EDM and cutting wire it is a thin metal strand wire, usually copper, fed through the work piece, submerged in a dielectric liquid tank, the water is usually deionized.

3.2. Methods

The steps involved in this study are summarized in Figure.4.

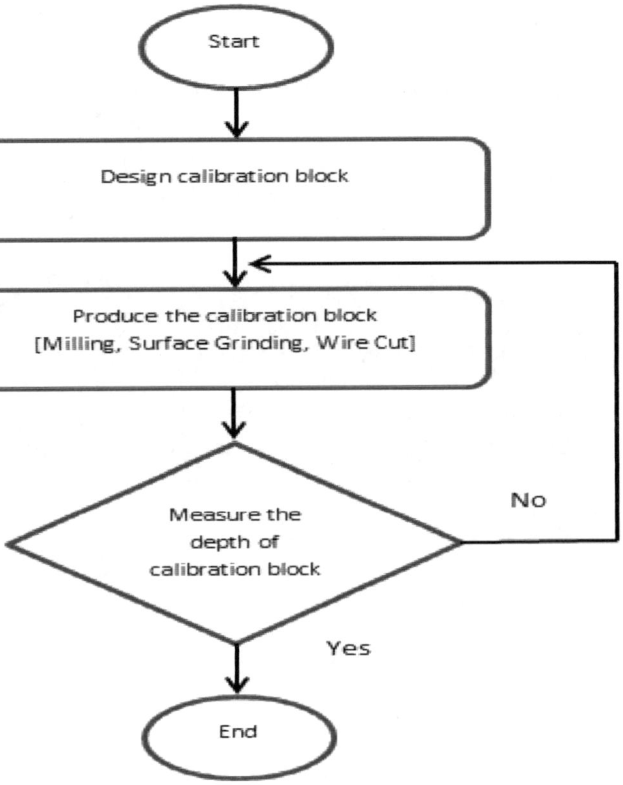

Figure 4. Process of fabricate sample block

The design of the sample block is perform using auto-CAD. Figure 5 shows two different block dimension.

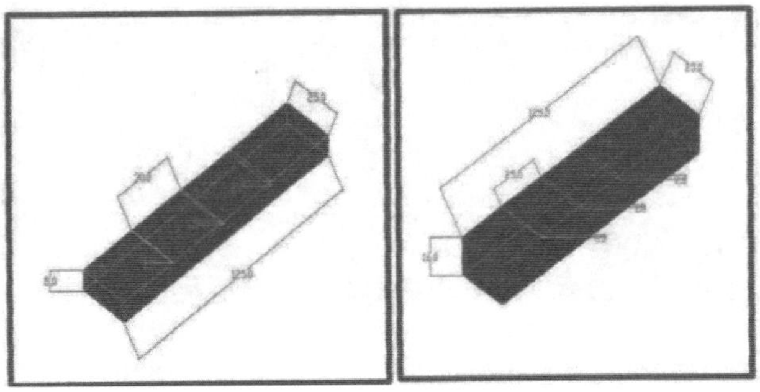

(a) (b)

Figure 5. Different of thickness; a)125mm x 25mm x 8mm b)125mm x 25mm x 16 mm

The first selected dimension for the sample block was used on the calculated frequency is higher than the eddy current set. The process of surface grinding is necessary to produce the smooth sample block. The copper was grinded about 200 mm, thus it did not affect the copper. After the grinding process, the copper block was cut into two pieces. The copper block is cut to two different thicknesses; the first dimension is 125mm x 25mm x 8mm and the second is 125mm x 25mmm x 16mm. The dimensions are selected for the best frequency when using weld probe. The copper block was cut using band sewing machine. Figure 6 show the process of copper cutting.

(a)

(b) (c)

Figure 6. Wire cutting of the sample block; (a) Copper cutting process (b) 16mm (c) 8 mm

3.3. Conductivity Measurement For Copper Block

In order to perform the conductivity testing, the conductivity probe was connected to eddy current flaw detector. The mode was automatically changed to conductivity. Figure 7 shows the configuration of eddy current flaw detector for conductivity testing and Table 1 show the configuration setting for the copper sample block.

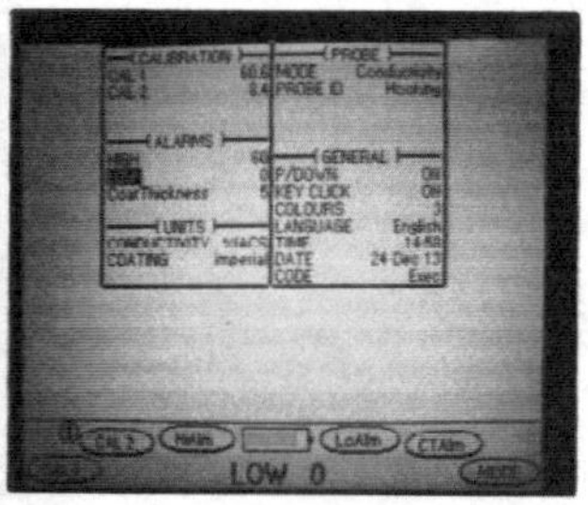

(a) (b)

Figure 7. Eddy current flaw detector set up; a) automatically change conductivity mode b) configuration for conductivity testing

Table 1. Configuration setting ECT instrument for Copper measurement

CAL 1	CAL 2
60.6 % IACS	8.45%IACS

The average calculation of the electrical conductivity for copper block was measured using (3).

$$Electrical\ conductivity = \frac{\sum A+B+C+D}{4} \quad (3)$$

The standard depth of penetration has been defined as the depth at which the Eddy current density is about 37 % of the surface density. In order to find the frequency of the copper block (δ), the standard depth of penetration must be 37% of the copper block thickness. The step of the finding the copper block frequency is summarized in Figure 8.

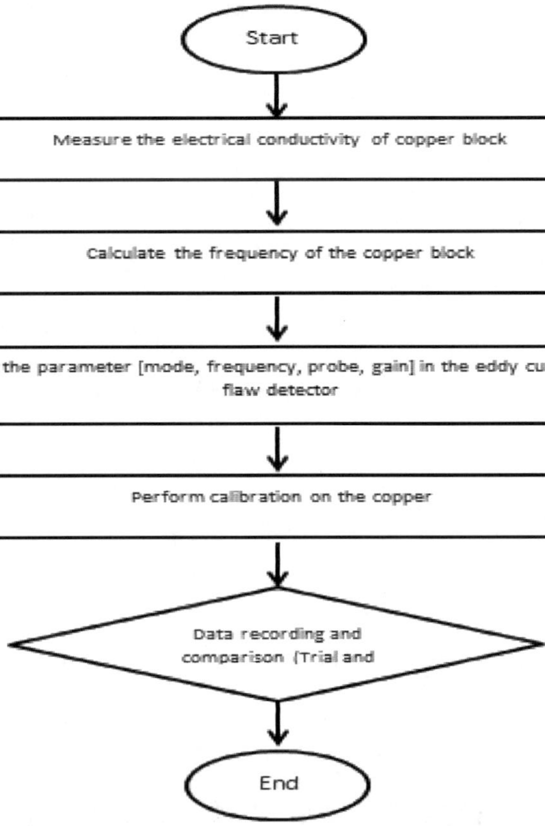

Figure 8. Step of finding the best frequency for the copper block

The thickness of the copper block are 8mm and 16 mm (π) is 3.14, (f) is test frequency (Hz) for two different thickness 16mm and 8mm,(μ) is magnetic permeability (1.25664 x 10-9H/mm), the electrical conductivity based on International Annealed Copper Standard (% IACS) (σ) is calculated from the average in Table 2.

Table 2. Conductivity of Copper block

Thickness Side	Conductivity IACS%		Conductivity Ms/m	
	8 mm	16 mm	8 mm	16 mm
A	102.67	101.64	58.83	58.92
B	102.58	101.69	59.10	59.01
C	102.52	101.54	58.98	59.08
D	102.44	101.60	59.99	58.95
Average	102.55	101.61	59.22	58.99

Formula to calculate of copper block frequency is given in (4).

$$\delta = \frac{1}{\sqrt{\pi f \mu \sigma}} \tag{4}$$

4. Result and Discussion

The measurements for the sample block were performed on the 8mm and 16mm thickness sample block. The positive scanning and negative scanning were inspected for all slots at different gain to find the most suitable gain for each slot. The sample block has three slots with depth of 0.5mm, 1.0mm and 2. 0mm.Each slot will be measured three times to find the best accuracy.

Frequency of 280 kHz was used for 8mm sample block according to the Table 3. Three different slots were measured with each slot measured three times to get the best accuracy. Three experiments supposed to produce the same reading. However, due to human error the readings were fluctuating each time the experiment was performed as in Table 3.

Table 3. Positive test scanning using 63.3 dB Gain

Slots	Experiment	Phase	Signal Amplitude
	1	110.2°	>93.7%
0.5 mm	2	110.9°	>92%
	3	111.0°	>97.5%
	1	112.2°	>100%
1.0 mm	2	110.9°	>100%
	3	109.8°	>100%
	1	110.2°	>100%
2.0 mm	2	109.8°	>100%
	3	109.8°	>100%

Figure 9 shows the signal amplitude of 8mm sample block based on gain 63.3 dB tests for three different slots. Based on Table 4 and Figure 9 the different of signal amplitude can be seen. The most suitable slots to be used with 63.3 dB is 0.5mm. It is because the signal amplitude of slot 0.5mm is 97.5%. Meanwhile, the FSH reading compared to slots 1.0mm and 2.0mm exceed 100% FSH.

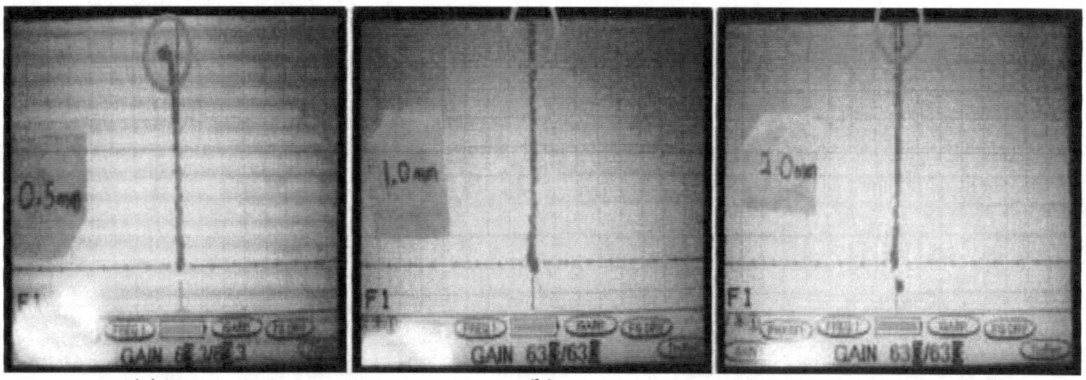

(a) (b) (c)

Figure 9. Signal amplitude of positive scanning for 63.3 dB; a) 0.5mm b) 1.0mm c) 2.0mm

Table 4 shows reading for the negative scanning with the same gain used in positive scanning.

Table 4. Negative test scanning using 63.3dB gain

Slots	Experiment.	Phase	Signal Amplitude
0.5 mm	1	110.2°	55%
	2	110.9°	52.50%
	3	111.0°	55%
1.0 mm	1	112.2°	97.50%
	2	110.9°	90%
	3	109.8°	97.50%
2.0 mm	1	110.2°	>100%
	2	109.8°	>100%
	3	109.8°	>100%

Figure 10 shows the signal amplitude of 8mm sample block in negative scanning based on gain 63.3 dB tests for three different slots. Based on Table 5 and Figure 10 the different of signal amplitude can be clearly seen. The negative scanning thus, the same slot (0.5mm) was confirmed as the best slot to be used with chosen gain.

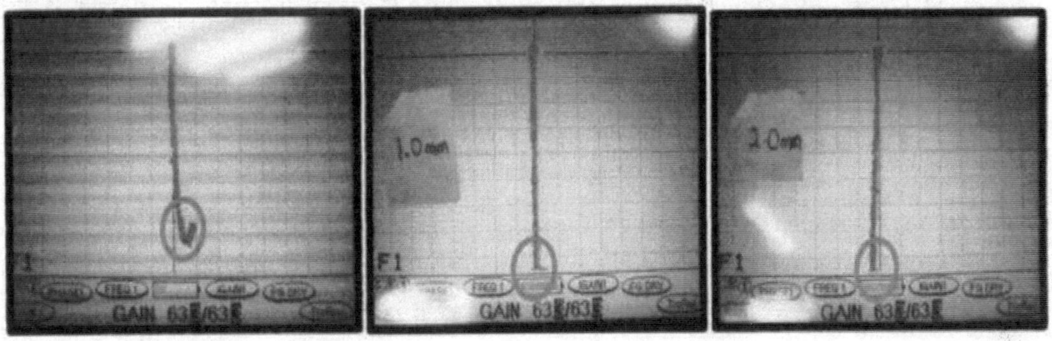

(a) (b) (c)

Figure10. Signal amplitude of negative scanning for 63.3 dB; a) 0.5mm b) 1 0mm c) 2.0mm

Table 5 summarizes the eddy current testing reading measured for three slots in sample block with 58.6 dB gain. The same frequency was maintained as the same thickness of sample block measured.

Table 5. Positive test scanning using 58.6 dB gain

Slots	Experiment	Phase	Signal Amplitude
0.5 mm	1	110.2°	>92.5%
	2	110.9°	>91.4%
	3	111.0°	>95%
1.0 mm	1	112.2°	>100%
	2	110.9°	>100%
	3	109.8°	>100%
2.0 mm	1	110.2°	>100%
	2	109.8°	>100%
	3	109.8°	>100%

Figure 11 shows the signal amplitude of 8mm sample block based on gain 58.6 dB test for three different slots. Based on TABLE.V and Fig.11 the different of signal amplitude can be seen. The most suitable slots to be used with 58.6 dB is 1.0mm. It is because the signal amplitude of slot 1.0mm is 97.5%. Meanwhile, the FSH reading compared to slots 0.5mm was not achieved the highest FSH and the signal amplitude for 2.0mm exceeded 100% FSH.

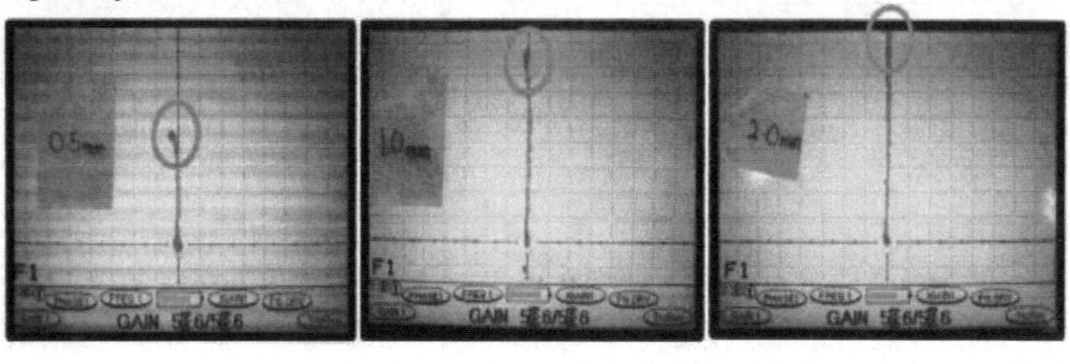

Figure 11. Signal amplitude of positive scanning for 58.6 dB a) 0.5mm b) 1.0mm c) 2.0mm

The same procedures were applied to measure the 16mm thickness copper sample block. Table 6 shows summary of the Eddy current testing reading taken for three slots in the sample block. The frequency applied was different as the thickness of the block is different. From the calculation in (4), Frequency of 71 kHz was used for 16mm thickness sample block.

Table 6 shows the summary of the suitable frequency and gain to be used with copper sample block for two different thicknesses.

Table 6. Copper sample block 8mm and 16mm thickness

Thickness		**8 mm**			**16 mm**	
Slots(mm)	**0.5**	**1.0**	**2.0**	**0.5**	**1.0**	**2.0**
Frequency (kHz)		280 kHz			71 kHz	
Gain (dB)	63.3	58.6	53.7	69.5	64.3	59.7

5. Conclusion

In this paper the copper calibration block has been designed and fabricated. These calibration blocks were meant to be used when testing defect on copper material. Two different thicknesses were produces which are 8mm and 16mm thickness. The appropriate frequency and gains best used with the block were identified and recorded. The results proved that the depth of slots is inversely proportional with the gain used. The dB gain gave the high impact on the signal amplitude. From here 63.3dB gain shows the good result especially on the defect slot 0.5mm and 1.0mm in the negative test scanning. When the width slots are increased then the signal amplitude achieve to 100%. An average of 0.5mm slot shows the 53.75% signal amplitude and 1.0mm slot is 93.75%. It shows the increasing of signal amplitude until 40%. The positive test scanning gave 100% signal amplitude start on 1.0mm slot and above. By comparing on positive and negative scanning the negative scanning shows the good result on amplitude signal because of diversity of signal amplitude based on different slot of defect.

6. References

[1] M. A. Faraj, A. N. Abdalla, F. Bin Samsuri, and D. Rifai, "Construct Coil Probe Using GMR Sensor for Eddy Current Testing," vol. **02**, pp. 1–8, 2018.

[2] M. A. Faraj, F. Samsuri, A. N. Abdalla, D. Rifai, and K. Ali, "Investigate the effect of lift-off on eddy current signal for carbon steel plate," vol. **06**, pp. 3–8, 2018.

[3] A. N. Abdalla, K. Ali, J. K. S. Paw, D. Rifai, and M. A. Faraj, "A novel eddy current testing error compensation technique based on mamdani-type fuzzy coupled differential and absolute probes," *Sensors (Switzerland)*, vol. **18**, no. **7**, 2018.

[4] M. A. Faraj, A. N. Abdalla, F. Bin Samsuri, D. Rifai, and K. Ali, "Investigate of the Effect of Width Defect on Eddy Current Testing Signals under Different Materials," *Indian J. Sci. Technol.*, vol. **10**, no. **2**, 2017.

[5] M. A. Faraj, F. Samsuri, A. N. Abdalla, D. Rifai, and K. Ali, "Adaptive neuro-fuzzy inference system model based on the width and depth of the defect in an eddy current signal," *Appl. Sci.*, vol. **7**, no. **7**, 2017.

[6] D. Rifai, A. N. Abdalla, Noraznafulsima, Kharudin, and Ruzlaini, "Defect signal analysis for nondestructive testing assessment," *ARPN J. Eng. Appl. Sci.*, vol. **11**, no. **4**, 2016.

[7] P. Gao, C. Wang, Y. Li, and Z. Cong, "Electromagnetic and eddy current NDT in weld inspection: A review," *Insight Non-Destructive Test. Cond. Monit.*, 2015.

[8] D. Rifai, A. N. Abdalla, N. Khamsah, and M. Aizat, "Subsurface Defects Evaluation using Eddy Current Testing," *Indian J. Sci. Technol.*, vol. **9**, no. March, 2016.

[9] K. B. Ali, A. N. Abdalla, D. Rifai, and M. A. Faraj, "Review on system development in eddy current testing and technique for defect classification and characterization," *IET Circuits, Devices Syst.*, vol. **11**, no. **4**, 2017.

[10] American Society for Testing and Materials, "Standard Practice for Electromagnetic (Eddy-Current) Examination of Nickel and," 2003.

[11] J. García-Martín, J. Gómez-Gil, and E. Vázquez-Sánchez, "Non-destructive techniques based on eddy current testing.," *Sensors (Basel).*, vol. **11**, no. **3**, pp. 25–65, Jan. 2011.

IOP Conf. Series: Materials Science and Engineering 1068 (2021) 012026 doi:10.1088/1757-899X/1068/1/012026

Thermal Characterization of Polyvinyl Alcohol Nanocomposites

M.Alif Faiz[1*], Nor Aisyah Mat Yahya[1], N.N.A.N. Yusoff[2], Norazlina Hashim[1], Lili Shakirah[1], Nabihah Abdullah[1*]
[1]Faculty of Engineering Technology, University College TATI, 24000 Kemaman, Terengganu , Malaysia.
[2] Faculty of Bioengineering and Technology, Universiti Malaysia Kelantan,17600 Jeli, Kelantan Malaysia.

[1*]Email: faizmansor07@gmail.com & nabihah@uctati.edu.my

Abstract. Nanocomposite films of PVA (polyvinyl alcohol) with GF(graphene flakes), GO (graphene oxide) and G(Graphite) as nanofillers have been synthesized. Nanocomposites were prepared to contain various weight amounts of nanofiller (0.3, 1, 1.5, 2 and 3%) within PVA matrix. The film is cast on a clean glass plate through the solution casting method and dried in the oven. Here a report of a simple and practical approach to synthesizing graphene-reinforced poly (vinyl alcohol) (PVA) composite films by incorporating other filler into PVA aqueous solution. Further, the investigation of thermal properties was used using thermal gravimetric analysis (TGA) and differential scanning calorimetry (DSC). Thermal stability shows the best stability with the nanofillers content of 1% for G, GO and GF. Nanocomposites with an additional of 1% GF gives the highest thermal stability compared to 1% of G and GO.

1. Introduction

Graphene, composed of a single layer of carbon atoms, was originally identified in a simple tabletop experiment by Geim, Novoselov and co-workers at Manchester University in 2004. Graphene has an ultra-high Young' 1 TPa module and an definitive strength of 130 GPa,[1] great thermal conductivity of 5000 W/(m·K)[2] and the highest known dielectric strength of up to 6000 S/cm[3] due to its exceptional two-dimensional (2D) nanostructure. In material science, all of these make graphene the wonder material and are considered a favourable candidate for polymer alteration[4].

Due to their peculiar mechanical, electrical, optical and thermal properties, the various carbon allotropes, such as diamond, graphite, graphene, and fullerene, with different geometric shapes (layered, single sheet, spherical, rods, and buds), have attracted considerable interest from both scientific and engineering points of view. Electronics, microcontrollers, photovoltaic panels and storage of data, optics and photonics, biology and medicine devices, tissue engineering and biomaterials, and functional nanocompositesare impressive potential applications[5].

Graphene has been experimentally studied for more than 40 years, and measurements of transport properties of graphene grown on (SiC) micromechanical exfoliated layers, large-area graphene grown on copper (Cu) substrates, as well as a range of studies involving the use of chemically modified graphene (CMG) to manufacture new materials, have contributed in part to an increase in the number of publications and in the number of studies.

Graphene oxide (GO) has been widely used as a nanofiller for fiber-reinforced polymer composites. To enhance the properties of interfacial bonding through the composites' surrounding polymer matrix, GO may be chemically painted on the surfaces of either carbon fibre or carbon fiber. GO has also been used recently for energy storage as a dielectric material. Between carbon fabric layers, which work on an electrode to form a structural capacitor, a thin layer of GO film is inserted. As demand for unmanned air vehicles (UAVs) increases, their fighting time is often restricted by the power storage system's power and the battery's weight[7,8].GO contains strongly oxygenated graphene sheets on the surface and edges with an epoxide, hydroxyl, carbonyl, and carboxylic functional groups. The graphene oxide (GO) layer is highly hydrophilic due to the high density of oxygen functionality[9].

Graphite consists of planer hexagonal units of covalently bonded sp^2 hybridized carbon atoms are referred to as graphene layers. Each carbon atom in the layer is joined by strong covalent bonds to only three other carbon atoms. The most popular crystal structure is the hexagonal graphite structure; the structure of the rhombohedral is not thermodynamically stable. As a result of its unique properties, natural graphite has numerous industrial applications, via electrical and thermal conductivity, lubricity thermal resistance and chemical inertness [10].Graphite has the thickest property compared to graphene flakes and graphene oxide. Because of this property, graphene flakes and graphene oxide disperse evenly in polymer matrix compared to graphite [9].

A synthetic polymer that is soluble in water is polyvinyl alcohol, also known as PVOH, PVA, or PVAL. It is productive in the formation, emulsification, and adhesive consistency of film. It is immune to grease, oils, and solvents and has no odor and is not harmful. It is ductile but solid, versatile and functions as a high barrier to oxygen and aroma. By annealing amorphous PVA films above 85°C, the glass transition temperature, semi-crystalline films of PVA were prepared [11]. PV The study and selected polyvinyl alcohol (PVA) in this research is a hydroxyl rich and soluble in water polymer that can establish and very easily develop new interfacial interaction with GO via the formation of hydrogen bond with GO. The strength and modulus of the PVA matrix can be significantly improved by adding small amounts of GO[4].In the manufacture of adhesives, fibers, paper, membranes, drug release media and pharmaceutical products, PVA is regularly used. It is tasteless, odorless and has excellent mechanical features. Unlike other polymers generally produced by polymerization, PVA is made from the hydrolysis of polyvinyl acetate [12]. The formation of PVA with the others carbon likes as GF and G. then, determined the thermal properties of PVA nanocomposites. The thermal properties of polyvinyl alcohol with three different nanofillers, graphite, graphene flakes and graphene oxides were determined in this research and has yet to determine by the previous researcher.

2. Materials & Methodology

2.1 Materials

Graphite(G) was purchased from ITO graphene industry. Graphene oxide(GO), graphene flakes(GF) and polyvinyl alcohol(PVA) were purchased from ACS material, Sigma Aldrich, Aldrich respectively. All other reagents were of analytical grade and were used without further purification.

2.2 Preparation of GF/PVA nanocomposites

2 g of PVA was dissolved under a hot plate in 30 ml of DI water at 95°C to form an aqueous solution. Graphene flake powder (0.3 wt. percent PVA) was gradually sonicated to form a black colloidal solution in 20 ml of DI water. For around 2 hours at 65 ° C, the as-obtained solutions were mixed under intense stirring to form a colloid. The vigorous stirring of the PVA colloid mixed with aqueous graphene flake is therefore essential to ensure full and uniform mixing. Finally, on a Teflon petri dish, 10 ml of this homogeneous GF mixed PVA colloid was poured and kept in a vacuum dryer for 48 hours at 40 C. For further characterizations, the dried films (hereinafter, PVA-GF) were peeled off the substrate.The same method was used to disperse graphite and graphene oxide in PVA. Table 1 shows the composition of fillers and PVA.

Table 1. The composition of the samples with different percentage of PVA and fillers

PVA(%)	Fillers(%)
99.7	0.3
99	1.0
98.5	1.5
98	2.0

2.3. Characterization

TGA was performed with a Mettler thermal analyzer at 30-850 °C under a nitrogen flow with a heating rate of 10 °Cmin-1. The thermal behavior was investigated on square-shaped coupons (±10 mg) over 30-350 °C with a heating rate of 5 °C min-1 using differential scanning calorimetry (DSC, Mettler-DSC822).

3. Results and Discussion

The thermal stability of the PVA and the filler materials was studied by TGA. In Figure 1, the TGA thermograms of PVA and G, GF and GO nanofillers with filler compositions ranging from 0.3 to 2.0 wt% are compared. The weight loss percentage depends on the composite material composition.

For figure 1(a) ,the pure PVA started to degrade at 150°C. Considering that the PVA decomposes completely within the composite material at 500° C. However, after the addition of graphite, the decomposition temperature of PVA nanocomposites was increasing from 250 to 400°C. The justification for the increase in polymer stability could be due to the restriction in mobility of the polymer chain and the drop in diffusivity inside the polymer matrix of attacking agents, both attributable to interactions between polymer and filler.In figure 1(b), the addition of 1% graphene flakes to the PVA matrix gives the highest thermal stability. Pure PVA gives the lowest thermal stability for figure 1(a-c), where it decomposed completely at 500 °C. no significant thermal stability effect for 0.3%, 1.5% and 2% of graphite and graphene flakes shown in figure 1(a) and (b). Improvement of PVA nanocomposites' thermal stability using graphene flakes has been reported in diverse studies [13,14]. Graphene flakes itself can resist burning up to 600 °C, but under oxygen-excluded condition, its resistivity reaches 2000 °C.

PVA with additional GO also gives the same trend as G and GF where 1% of GO shows the highest thermal stability compared to other percentages of GO (Figure 1c). Thermal stability determined using TGA (Figure 1 a-c), shows that PVA blend with 1 wt % of the different fillers has the highest temperature to decomposed which is at 300 – 450 °C compared to 0.3, 1.5 and 2 wt % of fillers. However, PVA nanocomposites with 1% GF gives the highest thermal stability compared to 1% of G and GO. At 800 °C(Figure 1b), PVA nanocomposites with 1% GF has 40% weight left whereas, PVA nanocomposites with 1% GO have only about 10% weight left. GF has the thinnest film compared to G and GO, and it will give good dispersion between the PVA matrix, thus enhance the thermal stability of PVA. GO has numbers of oxygen functional groups such as carboxyl and hydroxyl [15,16], thus the thermal stability of PVA with 1% GO is the lowest compared to 1% of G and GF. It is safe to say that all pure PVA in altogether figures (Figure 1a-1c) has the lowest thermal stability compared to the one that has nanofillers.

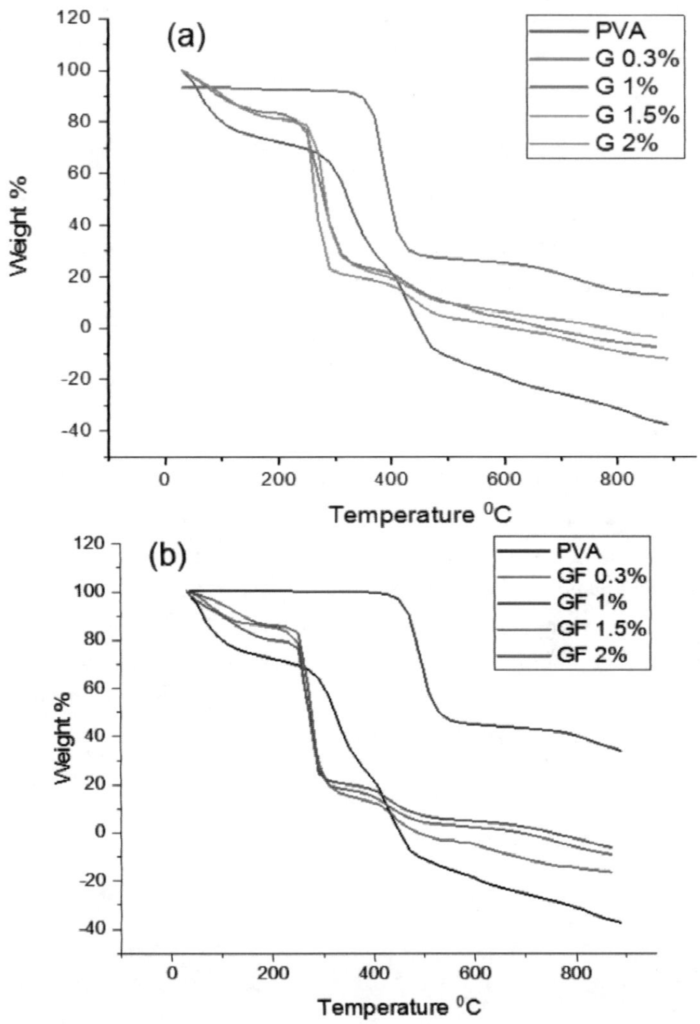

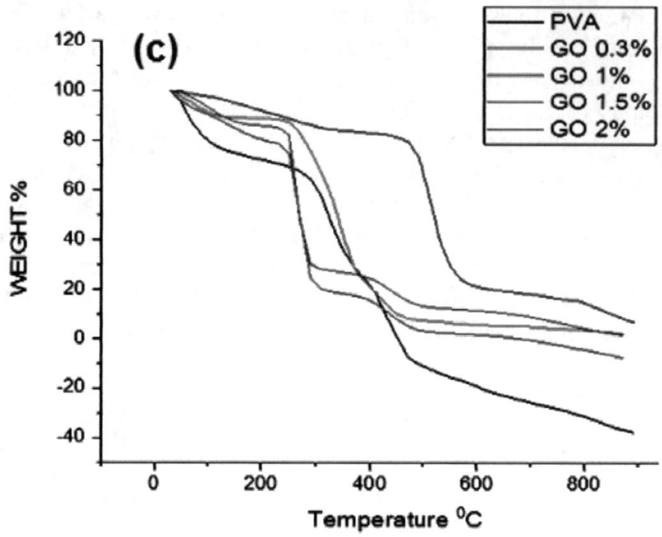

Figure 1. Thermal gravimetric analysis of (a) PVA/G, (b) PVA/GF (c) PVA/GO nanocomposites

Figure 2 showed PVA nanocomposites' DSC curves. Figure 2 (a) shown that T_g has happened at 150 °C while Figure 2 (b) happens at 135 °C and figure 2(c) happens at 170 °C. T_g means the glass transition temperature and this means at this time it gets more heat flow. This indicates that the polymer has also improved its heat capacity. This is because the polymers have a higher potential for heat just above the glass transition temperature than they do below it. It is also shown that different types of nanocomposites also affect the T_g happened.

Figure 2(a) showed that T_m occurred at 250 °C while figure 2(b) showed that T_m has happened at 240 °C and figure 2(c) showed that happened at 245 °C. This is also shown that different type of nanocomposite also affected Tm happen at the polymer. T_m is the melting temperature that shows the melting temperature is achieved by the polymers. Those polymer crystals, that is, they are melting, are starting to break apart. The chains come out of their organized configurations and openly start moving around.It also reveals the endothermic melting of PVA crystallites. In this experiment, the PVA sample used is basically atactic, but has a crystallinity level of about 40 percent. In addition, owing to hydrogen bonding between hydroxyl groups, PVA is a semi-crystalline polymer for which there are elevated physical interactions within polymer chains.

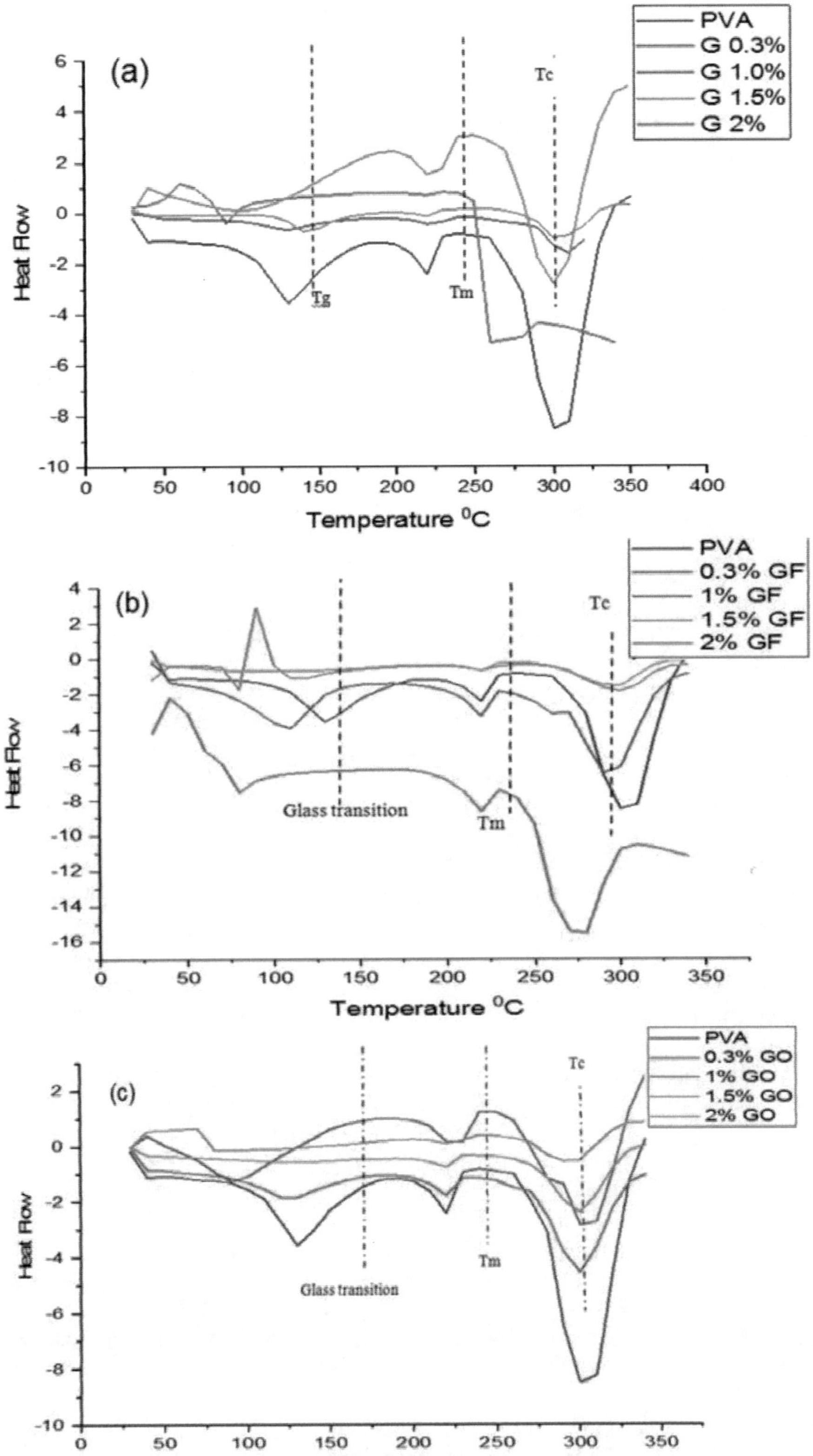

Figure 2. DSC curve of (a) PVA/G (b) PVA/GF (c) PVA/GO nanocomposites

4. Conclusion

Finally, PVA nanocomposite films have been successfully synthesized and characterized by integrating nanofillers such as GO, GF and G into the PVA matrix solution. 0.3, 1, 1.5 and 2 % of nanofillers were prepared for the sample. The 1 % film nanofillers demonstrates the improvement of thermal properties for G, GO and GF. The result showed that PVA has the highest thermal stability with 1 % GF compared with the other 1 % G and GO. This study opens up a new direction to optimize the interface structure and improve the comprehensive performance of nanocomposite polymer/nanofillers.

Acknowledgement

The study was supported by University College TATI under the Faculty of Engineering Technology.

References

[1] Zhang X, Zheng J and Ma Z, 2019 ,*IOP Conf. Ser.: Mater. Sci. Eng*, " Graphene-polymer composites for enhancing the mechanical properties," **538** 012001
[2] Du XU, Skachko I, Barker A, and Andrei EVAY, 2008 *Nat Nanotechnol.*, “Approaching ballistic transport in suspended graphene,” 491–495
[3] Abbasi H, Antunes M, and Velasco JI, 2019, *Prog. Mater. Sci.* “Recent advances in carbon-based polymer nanocomposites for electromagnetic interference shielding" **103**
[4] Wang J, Jin X, Li C, Wang W, Wu H, and Guo S, 2019, *Chem Eng J.* "Graphene and graphene derivatives toughening polymers: Toward high toughness and strength," **370** 831-854
[5] Deul B *et al.*, 2016, *Mater. Sci. Semicond. Process.* “Materials Science in Semiconductor Processing Effect of characteristic properties of graphene oxide on reduced graphene oxide / Si schottky diodes performance,” **44** 1–7
[6] Goumri M, Poilane C, Ruterana P, Ben Doudou B, Wery J, and Bakour A, 2017, *Chin J Phys.* “Synthesis and characterization of nanocomposites film with graphene oxide and reduced graphene oxide nanosheets,” **55** 412-422
[7] Balaji V, Lau K, Hui D, and Bhattacharyya D, 2018, *Compos. Part B*, “Graphene-based materials and their composites : A review on production , applications and product limitations,” **142** 200–220
[8] Mir A, Singh DK, and Shukla, 2018 *Mater. Chem. Phys.* “Size distribution of trilayer graphene flakes obtained by electrochemical exfoliation of graphite: Effect of the synthesis parameters,” **220** 87-97
[9] Kausar A,2019, *Polymer-Plastics Technology and Materials*” Poly(methyl methacrylate) nanocomposite reinforced with graphene, graphene oxide, and graphite: a review,’’58 821-842
[10] Kumawat, M. K., Thakur, M., Bahadur, R., Kaku, T., Prabhuraj, R. S., Ninawe, A., & Srivastava, R. (2019). *Materials Science & Engineering C*, Preparation of graphene oxide-graphene quantum dots hybrid and its application in cancer theranostics,**103** 109774.
[11] Elhajjar R, Law C, and Pegoretti A, 2018 *Prog. Mater. Sci.* “Magnetostrictive polymer composites: recent advances in materials, structures and properties,” **97** 204-229
[12] Aruldass S., Mathivanan V., Mohamed A. R., and Tye C. T., 2019 *J. Environ. Chem. Eng.* “Factors affecting hydrolysis of polyvinyl acetate to polyvinyl alcohol,” **7** 103238
[13] Kumar A., Sharma K., Dixit A.R,2019, *J Mater Sci*," A review of the mechanical and thermal properties of graphene and its hybrid polymer nanocomposites for structural applications",**54** 5992–6026
[14] Lawal A.T, 2020,*Cogent Chemistry*, "Recent progress in graphene based polymer nanocomposites",**6** 1833476.
[15] Yu W.,Sisi L, Haiyan Y.,Jie L., 2020, *RSC Advances*," Progress in the functional modification of graphene/graphene oxide: a review",**10** 15328

AUTHOR INDEX